Edited by
Vikas Mittal

**Surface Modification
of Nanotube Fillers**

Related Titles

Mittal, V. (ed.)
Optimization of Polymer Nanocomposite Properties

2010
ISBN: 978-3-527-32521-4

Mittal, V.
Polymer Nanotube Nanocomposites
Synthesis, Properties, and Applications

2010
ISBN: 978-0-470-62592-7

Mittal, V. (ed.)
Miniemulsion Polymerization Technology

2010
ISBN: 978-0-470-62596-5

Matyjaszewski, K., Müller, A. H. E. (eds.)
Controlled and Living Polymerizations
From Mechanisms to Applications

2009
ISBN: 978-3-527-32492-7

Leclerc, M., Morin, J.-F. (eds.)
Design and Synthesis of Conjugated Polymers

2010
ISBN: 978-3-527-32474-3

Cosnier, S., Karyakin, A. (eds.)
Electropolymerization
Concepts, Materials and Applications

2010
ISBN: 978-3-527-32414-9

Xanthos, M. (ed.)
Functional Fillers for Plastics

Second, Updated and Enlarged Edition
2010
ISBN: 978-3-527-32361-6

Pascault, J.-P., Williams, R. J. J. (eds.)
Epoxy Polymers
New Materials and Innovations

2010
ISBN: 978-3-527-32480-4

Edited by Vikas Mittal

Surface Modification of Nanotube Fillers

WILEY-VCH Verlag GmbH & Co. KGaA

The Editor

Dr. Vikas Mittal
Berner Weg 26
67069 Ludwigshafen
Germany

All books published by **Wiley-VCH** are carefully produced. Nevertheless, authors, editors, and publisher do not warrant the information contained in these books, including this book, to be free of errors. Readers are advised to keep in mind that statements, data, illustrations, procedural details or other items may inadvertently be inaccurate.

Library of Congress Card No.:
applied for

British Library Cataloguing-in-Publication Data
A catalogue record for this book is available from the British Library.

Bibliographic information published by the Deutsche Nationalbibliothek
The Deutsche Nationalbibliothek lists this publication in the Deutsche Nationalbibliografie; detailed bibliographic data are available on the Internet at <http://dnb.d-nb.de>.

© 2011 Wiley-VCH Verlag GmbH & Co. KGaA, Boschstr. 12, 69469 Weinheim, Germany

All rights reserved (including those of translation into other languages). No part of this book may be reproduced in any form – by photoprinting, microfilm, or any other means – nor transmitted or translated into a machine language without written permission from the publishers. Registered names, trademarks, etc. used in this book, even when not specifically marked as such, are not to be considered unprotected by law.

Typesetting Toppan Best-set Premedia Limited, Hong Kong
Printing and Binding Fabulous Printers Pte Ltd, Singapore
Cover Design Grafik-Design Schulz, Fußgönheim

Printed in Singapore
Printed on acid-free paper

ISBN: 978-3-527-32878-9

ISBN oBook: 978-3-527-63508-5
ISBN ePDF: 978-3-527-63510-8
ISBN ePub: 978-3-527-63509-2
ISBN Mobi: 978-3-527-63511-5

ISSN: 2191-0421

Contents

Preface *XI*
List of Contributors *XIII*

1 **Carbon Nanotubes Surface Modifications: An Overview** *1*
Vikas Mittal
1.1   Introduction *1*
1.2   Noncovalent Functionalization of Nanotubes *2*
1.3   Covalent Modifications of Carbon Nanotubes *12*
      References *20*

2 **Modification of Carbon Nanotubes by Layer-by-Layer Assembly Approach** *25*
Vaibhav Jain and Akshay Kokil
2.1   Introduction *25*
2.2   Layer-by-Layer Modification of the Carbon Nanotubes *28*
2.3   LbL Assembly on CNTs Using Click Chemistry *35*
2.4   LbL Assembly on Vertically Aligned (VA) MWNTs *37*
2.5   LbL on CNTs of Biological Molecules *38*
2.6   LbL on CNTs for Template Development *41*
2.7   Applications of LbL-Modified CNTs *43*
2.7.1 Donor Acceptor Assemblies *43*
2.7.2 Bioapplications *45*
2.8   Conclusions *47*
      References *48*

3 **Noncovalent Functionalization of Electrically Conductive Nanotubes by Multiple Ionic or π–π Stacking Interactions with Block Copolymers** *51*
Petar Petrov and Levon Terlemezyan
3.1   Noncovalent Functionalization of CNTs by π–π Stacking with Block Copolymers Bearing Pyrene Groups *52*

3.2	Noncovalent Functionalization of Electrically Conducting Nanotubes by Multiple Ionic Interactions with Double-Hydrophilic Block Copolymers 60
	Acknowledgment 64
	References 64

4	**Modification of Nanotubes with Conjugated Block Copolymers** 67
	Jianhua Zou, Jianhua Liu, Binh Tran, Qun Huo, and Lei Zhai
4.1	Introduction 67
4.2	Synthesis of P3HT Block Copolymer 69
4.2.1	Poly(3-hexylthiophene)-b-polystyrene (P3HT-b-PS) 70
4.2.2	Poly(3-hexylthiophene)-b-poly(methyl methacrylate) (P3HT-b-PMMA) 70
4.2.3	Poly(3-hexylthiophene)-b-poly(acrylic acid) (P3HT-b-PAA) 71
4.2.4	P3HT-b-poly(poly(ethylene glycol) methyl ether acrylate) (P3HT-b-PPEGA) 71
4.3	Dispersion of CNTs 71
4.3.1	Dispersing CNTs using P3HT-b-PS 71
4.3.2	Dispersing and Functionalizing CNTs by P3HT-b-PAA 83
4.3.3	Dispersing and Functionalizing CNTs by P3HT-b-PPEGA 85
4.4	Conclusions and Perspective 86
	Acknowledgments 87
	References 87

5	**Theoretical Analysis of Nanotube Functionalization and Polymer Grafting** 91
	Kausala Mylvaganam and Liang Chi Zhang
5.1	Introduction 91
5.2	Theoretical Techniques in Modeling Nanotube Functionalization 92
5.2.1	*Ab-initio* Quantum Mechanics Method 92
5.2.2	Molecular Dynamics Modeling 92
5.2.2.1	Scheme I 93
5.2.2.2	Scheme II 93
5.2.3	Hybrid QM/MM Approach 95
5.3	Functionalizing Carbon Nanotubes through Mechanical Deformation 95
5.3.1	Tension 96
5.3.2	Compression 96
5.3.3	Torsion 97
5.3.4	Bending 98
5.4	Functionalizing Carbon Nanotubes via Chemical Modification 99
5.4.1	Covalent Modification 99
5.4.1.1	Functionalization with Fluorine 100
5.4.1.2	Functionalization with Carboxylic Group, Amine Group, and Amide Group 100
5.4.1.3	Functionalization with Nitro Group 101

5.4.1.4	Functionalization with Transition Metal Complex	*101*
5.4.1.5	Functionalization via the Introduction of Free Radical	*102*
5.4.1.6	Functionalization via the Introduction of Anion	*102*
5.4.2	Noncovalent Modification	*103*
5.4.2.1	CHCl$_3$ Adsorption	*103*
5.4.2.2	Adsorption of Metalloporphyrin Complexes	*104*
5.4.2.3	Interaction of Aromatic Amino Acids	*104*
5.4.2.4	Encapsulation of Organic Molecules	*105*
5.5	Polymer Grafting	*105*
5.5.1	Grafting Polyethylene	*106*
5.5.1.1	"Grafting from" Technique	*106*
5.5.1.2	"Grafting to" Technique	*106*
5.5.2	Grafting Polypropylene Oxide	*108*
5.5.3	Grafting Oligo-N-Vinyl Carbazole	*109*
5.6	Summary	*109*
	References	*110*
6	**Covalent Binding of Nanoparticles on Carbon Nanotubes**	*113*
	Daxiang Cui	
6.1	Introduction	*113*
6.2	Covalent Binding of Quantum Dots on CNTs	*114*
6.3	Covalent Binding of Magnetic Nanoparticles on CNTs	*119*
6.4	Covalent Binding of Gold Nanoparticles on CNTs	*123*
6.4.1	Preparation of MWNT–COOH	*125*
6.4.2	Methyl Mercaptoacetate Modification of Gold Nanoparticles	*125*
6.4.2.1	Formation of Au–MWNT Conjugates	*125*
6.5	Growth of Poly(amidoamine) Dendrimers on Carbon Nanotubes	*126*
6.6	Application of Nanoparticles-Conjugated CNTs Composites	*129*
6.7	Concluding Remarks	*130*
	Acknowledgments	*131*
	References	*132*
7	**Amine-Functionalized Carbon Nanotubes**	*135*
	Ramaiyan Kannan and Vijayamohanan K Pillai	
7.1	Introduction	*135*
7.2	Functionalization Strategies for CNTs	*136*
7.3	Importance of Amine Functionalization	*138*
7.4	Methods of Amine Functionalization	*139*
7.4.1	Sidewall Functionalization	*140*
7.4.2	Tip and Defect Site Functionalization	*140*
7.4.3	Noncovalent Functionalization	*142*
7.5	Characterization Techniques	*142*
7.5.1	FT-IR	*142*
7.5.2	Raman Spectroscopy	*145*
7.5.3	AFM	*146*
7.5.4	XPS	*146*

7.5.5	Other Characterization Techniques	147
7.6	Degree of Amine Functionalization	147
7.7	Changes in the Band Structure	148
7.8	Applications of Amine-Functionalized CNTs	149
7.8.1	Solubility Manipulation	149
7.8.2	Selective Organic Reactions	150
7.8.3	Chemical and Biological Sensors	151
7.8.4	Molecular Electronics	152
7.8.5	Semiconductor Chemistry	153
7.8.6	Biocompatibility	153
7.8.7	Composite Electrolytes	153
7.8.8	Separation of Semiconducting Carbon Nanotubes	154
7.9	Limitations of Amine-Functionalized CNTs	154
7.10	Conclusions	155
	References	156

8 Functionalization of Nanotubes by Ring-Opening and Anionic Surface Initiated Polymerization 159

Georgios Sakellariou, Dimitrios Priftis, and Nikos Hadjichristidis

8.1	Introduction	159
8.2	Surface-Initiated Polymerization	160
8.2.1	Surface-Initiated Ring-Opening Polymerization	160
8.2.2	Surface-Initiated Anionic Polymerization	174
8.3	Conclusions	176
	References	176

9 Grafting of Polymers on Nanotubes by Atom Transfer Radical Polymerization 179

Chao Gao

9.1	Introduction	179
9.2	Grafting of Polymers on CNTs by ATRP	180
9.2.1	CNT-Based Macroinitiators	180
9.2.2	Linear Homopolymer-Grafted CNTs	183
9.2.2.1	CNT-Initiated ATRP of (Meth)acrylates	183
9.2.2.2	CNT-Initiated ATRP of Styrenes	194
9.2.2.3	CNT-Initiated ATRP of Acrylamides	195
9.2.2.4	Controllability of CNT-Initiated ATRP	197
9.2.3	Linear Block-Copolymer-Grafted CNTs	199
9.2.4	Hyperbranched Polymer-Grafted CNTs by a Combination of ATRP and SCVP	204
9.3	Functionalization of CNTs by Other CRPs	205
9.4	Grafting of Polymers on Other Nanotubes by ATRP	207
9.5	Conclusions	209
	Acknowledgments	210
	References	210

10	**Polymer Grafting onto Carbon Nanotubes via Cationic Ring-Opening Polymerization** *215*
	Ye Liu and Decheng Wu
10.1	Introduction *215*
10.2	Cationic Ring-Opening Polymerization of Cyclic Monomers in the Presence of Chain Transfer Reagents in One Pot *216*
10.2.1	Mechanism of Cationic Ring-Opening Polymerization of Cyclic Monomers in the Presence of Chain Transfer Agents in One Pot *217*
10.2.1.1	Cationic Ring-Opening Polymerization of 1,3-Dioxepane (DOP) in the Presence of Ethylene Glycol (EG) in One Pot *218*
10.2.1.2	Mechanism of Cationic Ring-Opening Polymerization of Glycidol *221*
10.2.2	Preparation of Functional Polyamine *224*
10.3	Preparation of Polymers-Grafted Carbon Nanotubes *226*
10.3.1	Preparation of PEI-grafted CNTs *226*
10.3.2	Preparation of Hyperbranched Polyether-Grafted CNTs *229*
10.4	Applications of Polymer-Grafted CNTs *233*
10.4.1	PEI-g-CNTs with Reduced Cytotoxicity and Good Capability to Enter Cells *233*
10.4.2	PEI-g-MWNTs Promising for Efficient DNA Delivery *235*
	References *237*
11	**Plasma Deposition of Polymer Film on Nanotubes** *239*
	Chuh-Yung Chen
11.1	Introduction *239*
11.2	Principle and Experiment *240*
11.2.1	Plasma Treatment of CNT *240*
11.2.2	Maleic Anhydride System *240*
11.2.2.1	Preparation of the mCNTs/Epoxy Nanocomposites *241*
11.2.2.2	Preparation of the mCNTs/Polyimide Nanocomposites *241*
11.2.3	AN System *242*
11.3	Results *243*
11.3.1	Maleic Anhydride System *243*
11.3.1.1	mCNTs/Epoxy Nanocomposites *243*
11.3.1.2	mCNTs/Polyimide Nanocomposites *247*
11.3.2	AN System *250*
11.3.2.1	CNT Identification *250*
11.3.2.2	Properties of CNT Nanocomposites *253*
11.4	Summary *254*
	References *255*
12	**Functionalization of Carbon Nanotubes by Polymers Using Grafting to Methods** *257*
	Jean-Michel Thomassin, Robert Jérôme, Christine Jérôme, and Christophe Detrembleur
12.1	Introduction *257*

12.2 Overview of the "Grafting to" Methods 258
12.2.1 The Two-Step Method 259
12.2.1.1 "Grafting-onto" CNTs Pre-functionalized by Carboxylic Acid Groups (CNTs-COOH) 261
12.2.1.2 Grafting onto CNTs Pre-functionalized by Acyl Chloride Groups (CNTs-C(=O)Cl) 263
12.2.1.3 Grafting onto CNTs Pre-functionalized by Hydroxyl Groups (CNTs-OH) 264
12.2.1.4 Grafting onto CNTs Pre-functionalized by Primary Amines (CNTs-NH$_2$) 266
12.2.1.5 Grafting onto CNTs Pre-functionalized by Alkynes (CNTs-alkyne) 267
12.2.1.6 Other Functionalized CNTs (-NCO, -S-S-Py, -anhydride, -anion) 271
12.2.2 The One-Step Method 272
12.2.2.1 Radical Grafting 273
12.2.2.2 Anionic Grafting 279
12.2.2.3 Azide Grafting 281
12.3 Conclusions 282
Acknowledgments 283
References 283

13 Organic Functionalization of Nanotubes by Dipolar Cycloaddition 289
Vassilios Georgakilas and Dimitrios Gournis
13.1 Introduction 289
13.2 The Case of Azomethine Ylide 290
13.3 The Case of Pyridinium Ylides 296
13.4 The Case of Nitrile Oxide 300
13.5 The Case of Nitrone 302
13.6 The Case of Nitrile Imines 307
13.7 Conclusions 307
References 307

Index 309

Preface

Nanotubes have emerged as a very high potential inorganic reinforcement for the polymer materials to enhance mechanical, electrical, and transport properties of the pristine polymers. Though they are relatively new class of polymer fillers, but the factors like matrix compatibility, uniform distribution in the polymer matrix, etc., are same as conventional fillers. In order to achieve uniform dispersion of nanotubes in the polymer matrices thus to achieve their full potential, it is very important to ensure their surface compatibility with the polymer matrix. As polymer matrices may have different polarity and functional groups present in their structure, therefore, the surface of nanotubes is required to be altered accordingly. There are numerous ways of surface functionalization reported in the literature, the compilation of which is presented in this book. Physical and chemical modes of functionalization generate different surface properties of nanotubes and hence different ways of interaction with the polymer matrices. Different modes of surface functionalization of nanotubes have their own advantages and limitations; the choice of the functionalization method is based on the application at hand or based on the required properties.

Chapter 1 presents an overview of the different nanotube surface modification techniques. Advantages and limitations of these modification techniques have been pointed. The use of modified fillers with a variety of polymer matrices has been demonstrated. Chapter 2 details the layer-by-layer assembly approach for the noncovalent modification of nanotubes. Thus, the surface properties of nanotubes can be controlled by the nature of the layer associated with the surface. Chapter 3 describes the noncovalent functionalization of electrically conductive nanotubes by multiple ionic or π–π stacking interactions using block copolymers. Use of conjugated block copolymers for the surface modification of nanotubes has been reported in Chapter 4. Chapter 5 presents in-depth analysis of the theoretical analysis of nanotube modifications and polymer grafting on the surface of nanotubes. Chapter 6 describes the interesting concept of covalent binding of nanoparticles on the surface of nanotubes. Thus, by controlling the surface properties of nanoparticles, the similar behavior can also be induced on the nanotube surface. Chapter 7 presents the amine-functionalized carbon nanotubes. Such nanotubes can be used for subsequent chemical reactions to generate a variety of functional groups on the surface of nanotubes. Functionalization of nanotubes by ring-opening and anionic-surface-initiated polymerization has been described in Chapter 8. The use of living atom transfer radical polymerization for the grafting of polymers on nanotube surfaces has been reported in Chapter 9. Cationic ring-opening polymerization has been demonstrated in Chapter 10 for the polymer grafting on the nanotube surface. Use of plasma for the deposition of polymer films on the surface of nanotubes has been discussed in Chapter 11. Grafting to method has been described for the functionalization of carbon nanotubes in Chapter 12. Dipolar cycloaddition processes for the organic functionalization of nanotube surfaces have been the focus of Chapter 13.

It gives me immense pleasure to thank Wiley-VCH for kind acceptance to publish the book. I dedicate this book to my mother for being constant source of inspiration. I express heartfelt thanks to my wife Preeti for her continuous help in co-editing the book as well as for her ideas to improve the manuscript.

Vikas Mittal

List of Contributors

Chuh-Yung Chen
National Cheng Kung University
Department of Chemical Engineering
Room 93B15, 70101, Tainan
Taiwan

Daxiang Cui
Department of Bio-Nano Science
and Engineering
National Key Laboratory of Micro/
Nano Science and Technology
Institute of Micro/Nano Science
and Technology
800 Dongchuan Road
Shanghai, 200240
China

Christophe Detrembleur
University of Liège
Center for Education and Research
on Macromolecules (CERM)
Sart-Tilman B6a
4000 Liège
Belgium

Chao Gao
Zhejiang University
Department of Polymer Science and
Engineering
MOE Key Laboratory of
Macromolecular Synthesis and
Functionalization
38 Zheda Road
Hangzhou, 310027
China

Vassilios Georgakilas
NCSR "Demokritos"
Institute of Materials Science
15310 Ag. Paraskevi, Athens
Greece

Dimitrios Gournis
University of Ioannina
Department of Materials Science and
Engineering
45110 Ioannina
Greece

Nikos Hadjichristidis
University of Athens
Department of Chemistry
Panepistimiopolis Zografou
Athens 15771
Greece

Qun Huo
University of Central Florida
Nanoscience Technology Center and
the Department of Chemistry
12424 Research Parkway Suite 400
Orlando, FL 32826
USA

Vaibhav Jain
Naval Research Laboratory
Optical Sciences Division
Washington, DC
USA

Christine Jérôme
University of Liège
Center for Education and Research
on Macromolecules (CERM)
Sart-Tilman B6a
4000 Liège
Belgium

Robert Jérôme
University of Liège
Center for Education and Research
on Macromolecules (CERM)
Sart-Tilman B6a
4000 Liège
Belgium

Ramaiyan Kannan
National Chemical Laboratory
Physical and Materials Chemistry
Division
Pune 411008
India

Akshay Kokil
University of Massachusetts Lowell
Department of Chemistry
Center for Advanced Materials
Lowell, MA
USA

Jianhua Liu
University of Central Florida
Nanoscience Technology Center and
the Department of Chemistry
12424 Research Parkway Suite 400
Orlando, FL 32826
USA

Ye Liu
Institute of Materials Research and
Engineering
3 Research Link
Singapore 117602

Vikas Mittal
BASF SE, Polymer Research
67056 Ludwigshafen
Germany
and
Chemical Engineering Program
The Petroleum Institute
2533, Abu Dhabi
UAE

Kausala Mylvaganam
The University of New South Wales
School of Mechanical and
Manufacturing Engineering
J17, NSW 2052
Australia

Petar Petrov
Bulgarian Academy of Sciences
Institute of Polymers
Akad G Bonchev Str
Block 103A, BG-1113 Sofia
Bulgaria

Vijayamohanan K Pillai
National Chemical Laboratory
Physical and Materials Chemistry
Division
Pune 411008
India

Dimitrios Priftis
University of Athens
Department of Chemistry
Panepistimiopolis Zografou
Athens 15771
Greece

Georgios Sakellariou
University of Athens
Department of Chemistry
Panepistimiopolis Zografou
Athens 15771
Greece

Levon Terlemezyan
Bulgarian Academy of Sciences
Institute of Polymers
Akad G Bonchev Str
Block 103A, BG-1113 Sofia
Bulgaria

Jean-Michel Thomassin
University of Liège
Center for Education and Research
on Macromolecules (CERM)
Sart-Tilman B6a
4000 Liège
Belgium

Binh Tran
University of Central Florida
Nanoscience Technology Center
and the Department of Chemistry
12424 Research Parkway Suite 400
Orlando, FL 32826
USA

Decheng Wu
Institute of Materials Research and
Engineering
3 Research Link
Singapore 117602

Lei Zhai
University of Central Florida
Nanoscience Technology Center and
the Department of Chemistry
12424 Research Parkway Suite 400
Orlando, FL 32826
USA

Liang Chi Zhang
The University of New South Wales
School of Mechanical and
Manufacturing Engineering
J17, NSW 2052
Australia

Jianhua Zou
University of Central Florida
Nanoscience Technology Center and
the Department of Chemistry
12424 Research Parkway Suite 400
Orlando, FL 32826
USA

1
Carbon Nanotubes Surface Modifications: An Overview
Vikas Mittal

1.1
Introduction

Carbon nanotubes are allotropes of carbon and are regarded as the ultimate carbon fibers [1–3]. The credit for realizing the nanotubes in an arc discharge apparatus is given generally to Iijima who successfully proved the existence of first multiwalled carbon nanotubes (MWCNTs) mixed with other forms of carbon [4], though the existence of these materials was realized earlier also, for example, by Endo in 1976 [5]. Subsequently, single-walled carbon nanocomposites (SWCNTs) were discovered and a significant research effort followed thereafter [6–9]. Carbon nanotubes have unique mechanical, electrical, magnetic, optical, and thermal properties [10]. Although the organic–inorganic nanocomposites can conventionally contain inorganic fillers which differ by the virtue of their primary particle dimensions [11–16], nanotubes containing composites generate much high-end application potential thus signifying the importance of nanotubes. The synthesis methods for the generation of nanotubes include high-temperature evaporation using arc-discharge [17–19], laser ablation [20], chemical vapor deposition, etc. [21–26].

Owing to their inert nature, the nanotubes tend to form bundles with each other and thus do not disperse well in the organic matrices in their pristine state. Suitable enhancement of the surface of the nanotubes is thus required in order to optimize their dispersion in the organic matrices. Out of various possible ways to achieve the surface functionalization, noncovalent means of surface modification are quite common. In this methodology, polymer chains are wrapped around the nanotubes or various surfactant molecules are adsorbed on the surface of nanotubes. In this case, the modification molecules are only physically bound to the surface. In the covalent means of surface modification, polymer chains can be grafted to or from the surface of the nanotubes. Controlled living polymerization methods have also been used to control the architecture of these grafts. These methods generate surface modifications which are chemically bound to the surface of the tubes, but such chemical means of modifications also disturb the structural homogeneity of the tubes, thereby decreasing the mechanical properties. Thus different surface functionalization methods have their own benefits

Surface Modification of Nanotube Fillers, First Edition. Edited by Vikas Mittal.
© 2011 Wiley-VCH Verlag GmbH & Co. KGaA. Published 2011 by Wiley-VCH Verlag GmbH & Co. KGaA.

and limitations, and the choice of these methods is dictated by the applications required from the generated nanocomposite materials. Pi–pi stacking is also one of the means to organophilize the nanotubes. The surface modification of the nanotubes forms a critical phase in the nanocomposite synthesis as the interactions of the surface modification molecules with the polymer chains dictate if the nanotubes can be homogenously dispersed in the organic phase or not. Therefore, various methods of surface modification of nanotubes are required to be reviewed in detail in order to tune the nanocomposite morphology and subsequent properties.

1.2
Noncovalent Functionalization of Nanotubes

Noncovalent mode of nanotube functionalization has received a lot of academic interest owing to the noninvasive mode of surface functionalization, which keeps the original nanotube properties intact. However, for some load-bearing applications, the presence of physically bound modifier molecules on the surface can also be a concern. The following studies detail the noncovalent functionalization of nanotubes for eventual dispersion in polymer matrices.

Zhu *et al.* reported the noncovalent functionalization of nanotubes based on the colloid stabilization principles [27]. In this approach, charged inorganic ZrO_2 nanoparticles were used as stabilizing media as shown in Figure 1.1. HiPco SWNTs were first purified following a two-step procedure in which the nanotubes were heated in an O_2 atmosphere at 300 °C and subsequent removal of metal catalysts in HCl at 60 °C for 2 h. The purified SWCNTs were then mixed with ZrO_2 nanoparticle aqueous solutions and were sonicated. The suspensions were allowed to stand for few hours to few days to remove the unstable large bundles of nanotubes. After this gravity-driven sedimentation of the uncoated nanotubes, the suspensions of nanotubes with nanoparticles were observed to be very stable for long periods of time and the suspensions were transparent. The microscopic investigations by AFM also confirmed the existence of single nanotubes in the dispersed mode. It was suggested that charge repulsion originating from ZrO_2 particles can be the most plausible phenomena and there was no direct evidence of nanoparticle haloing of ZrO_2 around SWNTs. The external stimuli were observed to self-assemble the dispersed nanotubes into macroscopic materials in solution indicating that the functionalization of the nanotubes with the nanoparticles was very efficient in tuning the surface properties.

Proteins (like streptavidin) adsorb spontaneously on the surface of the nanotubes [28]. Nonspecific binding of the proteins has also been proven microscopically on nanotubes after exposure of the nanotubes with solution of streptavidin. A number of polymer systems have been used for the prevention of nonspecific binding of proteins on the surfaces by forming coatings and self-assembled monolayers. In an interesting study on the interactions of proteins with nanotubes (Figure 1.2), it was demonstrated that by noncovalent functionalization of nano-

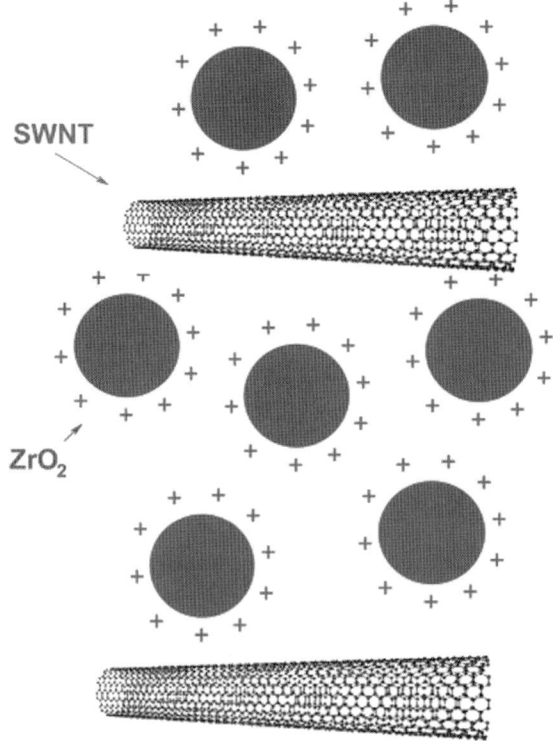

Figure 1.1 Schematic representation of dispersion of nanotubes in water by using inorganic nanoparticles. Reproduced from Ref. [27] with permission from American Chemical Society.

tubes, nonspecific binding of the proteins on the nanotube surface could be eliminated and specific binding of proteins after surface functionalization could be achieved [29]. Poly(ethylene glycol) is the most commonly used polymer to eliminate the nonspecific binding of the proteins on the surface [30–32]. However, it was observed by the authors that the poly(ethylene glycol) though irreversibly adsorbed on the nanotube surface, but appreciable adsorption of streptavidin was still observed indicating that the coverage of nanotube with the polymer was not optimum. The authors suggested that surfactant can be used for improving the coverage of the surface with poly(ethylene glycol). Triton-X surfactant containing an aliphatic chain and a short hydrophilic poly(ethylene glycol) unit was used to bind to the nanotube surface by hydrophobic interactions. Its adsorption alone on the surface was observed to resist the nonspecific binding of the proteins on the surface and after the adsorption of poly(ethylene glycol) on the surface, no streptavidin adsorption was observed. The authors also suggested means to specifically bind proteins on the surface by following the similar noncovalent functionalization of nanotubes. Biotin moiety was attached to the poly(ethylene glycol) chains by using amine-functionalized poly(ethylene glycol) and covalently linking it with

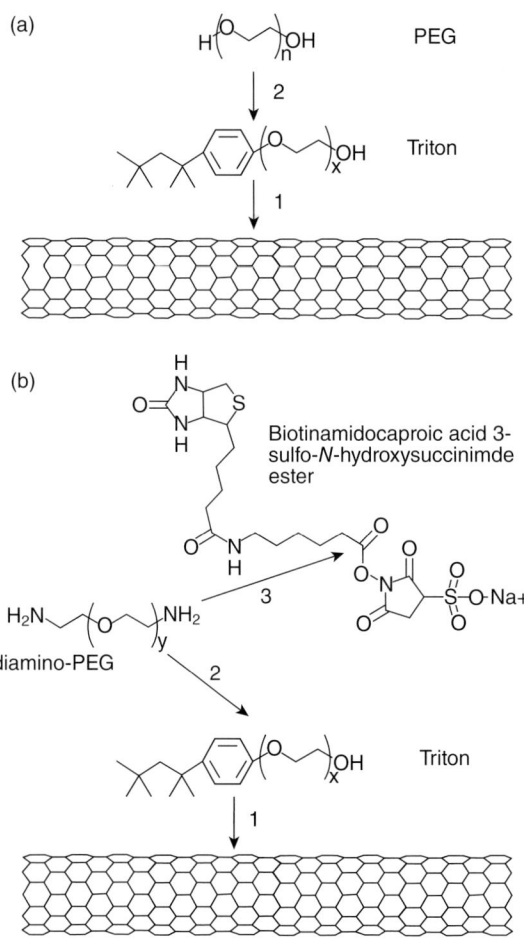

Figure 1.2 (a) Functionalization of SWNTs for preventing nonspecific binding of protein and (b) strategy for introducing selective binding of streptavidin with prevention of nonspecific binding. Reproduced from Ref. [29] with permission from American Chemical Society.

an amine reactive biotin reagent. As streptavidin demonstrates high affinity to biotin, the triton-PEG-biotin functionalized nanotubes were ideal substrates for the very specific adsorption of streptavidin on them.

Polymer wrapping method was reported by O'Connell et al. for the noncovalent functionalization of polymer nanotubes [33]. The SWCNTs were dispersed in 1% sodium dodecyl sulfate (SDS) in water by using shear mixer and ultrasonication followed by the addition of aqueous solution of poly(vinyl pyrrolidone) (PVP). Excess amount of the polymer was used for wrapping process, and the excess of SDS and PVP were subsequently removed by ultracentrifugation. The AFM images shown in Figure 1.3 confirm that most of the nanotubes existed as single

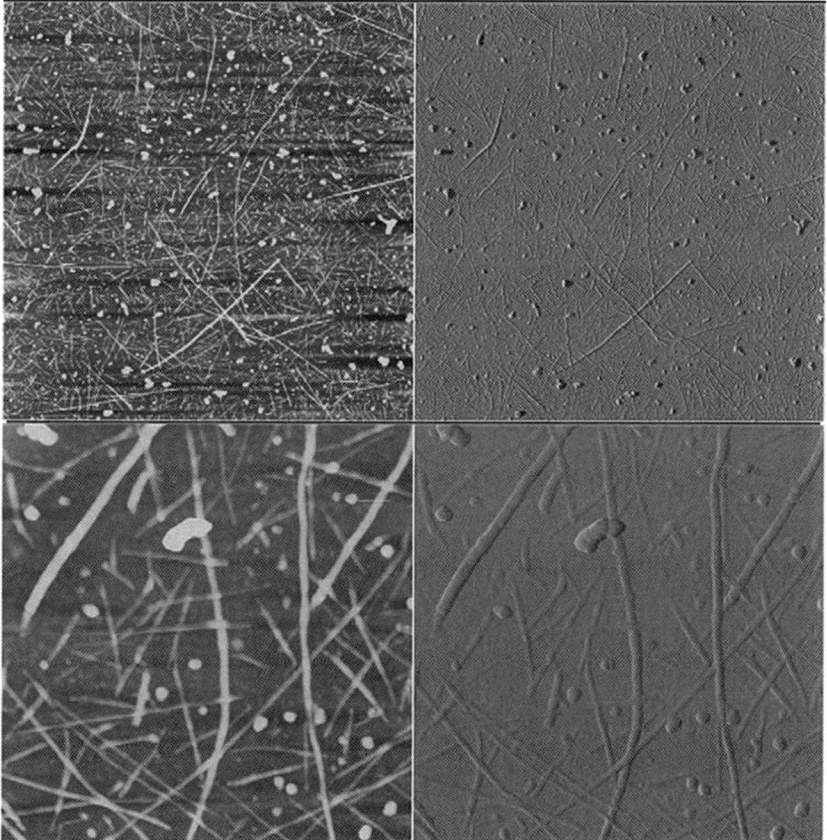

Figure 1.3 AFM images of PVP–SWNT system adsorbed on a functionalized substrate. Top two images correspond to 5 μm height and amplitude images, respectively, whereas bottom two images correspond to 1 μm counterparts. Reproduced from Ref. [33] with permission from Elsevier.

tubes associated with at most a single layer of polymer and only a smaller number of aggregates consisting of more than one nanotube. The suspensions were observed to be stable for months and once dried, the nanotubes were reported to disperse in water with minimal ultrasonication. Nonwrapped nanotubes were observed to assemble into mats of tangled and seemingly endless ropes, whereas it was not the case in the polymer-wrapped nanotubes. The modified nanotubes were also tested for the strength of bonding between the polymer and the nanotubes. The modified tubes were subjected to a cross flow in the channel of field-flow fractionation device and there was no effect on the nature of the nanotubes. The authors suggested that the nanotube wrapping is driven largely by the thermodynamic drive to eliminate the hydrophobic interface between the tubes and

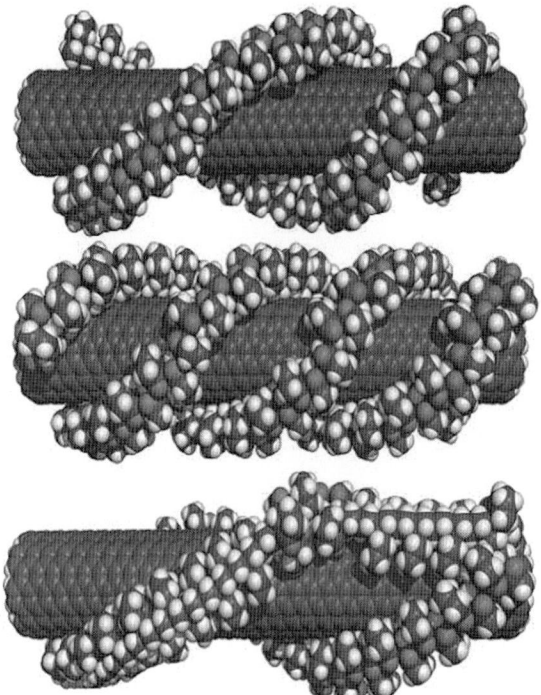

Figure 1.4 Possible arrangements of the polymer wrapping around the nanotubes. Top image corresponds to a double helix, the middle scheme is a triple helix, whereas the bottom arrangement shows switch backs, which allow multiple parallel wrapping strands to come from the same polymer chain owing to the backbone bond rotations. Reproduced from Ref. [33] with permission from Elsevier.

the dispersion medium. The molecular level image of the nanotubes wrapped with polymer chains is also demonstrated in Figure 1.4. The authors suggested that the helical wrapping of the chains around the nanotubes takes place due to the free rotation about the backbone bonds. However, single tight coil was suggested to generate significant bond angle strain in the polymer backbone. Therefore polymer coverage by multihelical wrapping was suggested. This mode of wrapping provides high extents of surface coverage with low backbone strain and locally multiple strands of polymer coil around the nanotube close to their nascent backbone curvature.

Apart from polymer layer medications, the coverage of the nanotube surfaces with an inorganic silica layer has also been reported and the generated silica layer was bound noncovalently to the nanotubes [34]. Functionalizing the sidewalls of carbon nanotubes with a layer of SiO_2 also opens the possibilities of further modification of the nanotube surfaces by utilizing the surface reactions available for silica surfaces. The process of nanotube functionalization is demonstrated in

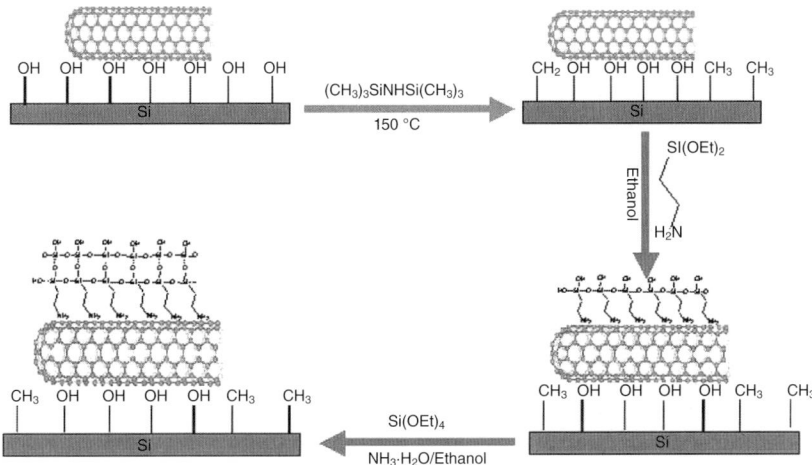

Figure 1.5 Growth of thin SiO$_2$ films on the walls of single-walled carbon nanotubes by using 3-APTES as promoter. Reproduced from Ref. [34] with permission from American Chemical Society.

Figure 1.5. A promoter, 3-aminopropyltriethanoxysilane (APTES), was absorbed on the sidewalls of the nanotubes by the interaction of sidewalls of nanotubes with the amine groups followed by the polymerization of the molecule by heat treatment. This, thus, generates an irreversible coating on the nanotube surfaces. As a next step, thin shell of SiO$_2$ of thickness roughly 1 nm was generated around the nanotubes by using tetraethyl orthosilicate as the precursor. The thickness of the coating can be adjusted by controlling the reaction time and the concentration of tetraethyl orthosilicate. This, thus, generates two advantages: first, the thin silica layer is bound noncovalently to the surface; and second, this can then further allow using the SiO$_2$ functionalization chemistries to generate different functionalities on the SiO$_2$ surface. The nanotubes were grown by chemical vapor deposition method in this study, and the authors also opined that the similar inorganic surface modifications can also be extended to other oxide materials. Similarly, norbornene polymerization was achieved on the nanotube surface via noncovalently bound ring-opening metathesis initiator [35].

A graft copolymer polystyrene-g-(glycidyl methacrylate-co-styrene) (PS-g-(GMA-co-St)) was reported to noncovalently modify the surface of nanotubes [36]. The graft copolymer was synthesized by free radical melt grafting of GMA on PS chains. The PS chains owing to their affinity of the surface of the MWNTs led to the noncovalent modification of the nanotubes. The MWNTs, which were produced by catalytic pyrolysis of propylene in this study, were dispersed in THF by mixing with excess of copolymer. The surface modification allowed the nanotubes to be dispersed in a variety of polar and nonpolar solvents. Also, when the acid-treated MWNTs were used, the presence of GMA in the copolymer provided the

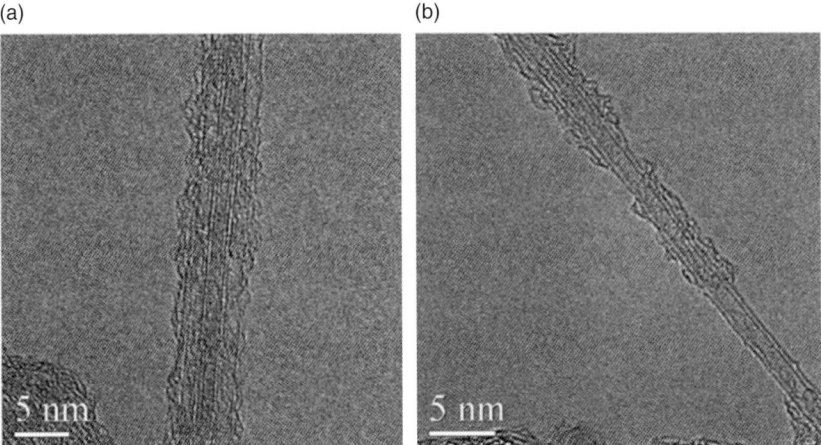

Figure 1.6 High-resolution TEM images of nanotubes modified by the polymer (a) noncovalently and (b) covalently. Reproduced from Ref. [36] with permission from Elsevier.

epoxy functional groups to which the acidic groups present on the surface of nanotubes could covalently attach to form esters. Figure 1.6 shows the nanotubes functionalized by the copolymer by following both the above-mentioned functionalization approaches. The nanotubes, which had only noncovalent functionalization, were observed to be uniformly modified on the surface by the polymer layer. However, in the nanotubes, which were acid-treated and hence had a covalent binding of the polymer chains, the polymer was bound to the surface only at sites where the carboxyl groups were present. Different modification mechanisms were also observed to significantly affect the solubility characteristics of the nanotubes. The noncovalently modified nanotubes were observed to swell well in different polar and nonpolar solvents, whereas the covalently functionalized nanotubes were swollen in solvents in which polystyrene is soluble. In the case of noncovalently functionalized nanotubes, the polystyrene component of the copolymer was attached to the nanotubes surface, thus leaving the chains containing gylcidyl methacrylate dangling away from the nanotube surface. As glycidyl methacrylate is readily soluble in many solvents, the swelling of noncovalently functionalized nanotubes was improved significantly. However, in the case of covalently functionalized nanotubes, the polystyrene chains formed the external "capsules" owing to the chemical reaction of the GMA block of the copolymer with the surface of nanotubes. Thus, the swellability of nanotubes was dictated by the swelling or solubility characteristics of polystyrene chains.

In another study, the authors stressed the need of noncovalent functionalization of nanotubes to retain their attractive electronic and mechanical properties. To achieve this, the authors used an approach for the functionalization of nanotubes by using polymer multilayers [37]. The formation of polymer electrolyte layers has been reported in the literature as a promising strategy to functionalizing the surfaces [38–45]. The thickness of the layers depends on the nature of the polymer used for adsorption as well as pH tuning and the ionic strength of the solution used for adsorption. In this study, hydrolyzed-poly(styrene-*alt*-maleic anhydride) (h-PSMA) was adsorbed noncovalently onto nanotubes surface via hydrophobic interactions. As the copolymer contained carboxylic acid groups, these groups could be used to further attach a second layer of polyethyleneimine covalently forming a cross-linked polymer bilayer. The same process was followed to attach second layer of h-PSMA and subsequently a second layer of polyethyleneimine. In the end, a layer of polyacrylic acid was generated on the surface of the nanotubes. The functionalization of the surfaces by the polymer multilayers was confirmed by using polarized infrared (IR) internal reflectance spectroscopy. Gold nanoparticles could also be immobilized on the surface by generating a third layer of polyethyleneimine. As gold particles were negatively charged, they could bind to the positive charges in the polyethyleneimine layer by electrostatic interactions. A control sample which was untreated substrates was also immersed in aqueous suspension of gold nanoparticles. A negligible amount of gold nanoparticles was observed to attach to this sample, thus confirming the tuning of the surface properties by using polymer multilayers approach. Figure 1.7 shows the functionalization of nanotubes by the polymer multilayers approach. Thus, this noncovalent mode of surface functionalization retains the properties of nanotubes as well as generates reactive functional groups on the surface to tune the properties of nanotubes accordingly.

Chen *et al.* reported the pi-stacking approach to immobilize biomolecules on carbon nanotubes [46]. Noncovalent functionalization of the sidewalls of SWCNTs was achieved, subsequent immobilization of various biological molecules was observed, and the process was reported to be controllable and site-specific.

For the noncovalent functionalization of nanotubes, a bifunctional molecule, 1-pyrenebutanoic acid, succinimidyl ester was irreversibly adsorbed onto the inherently hydrophobic surfaces of SWNTs in dimethylformamide or methanol. The pyrenyl group is highly aromatic in nature and, therefore, is known to interact strongly with the basal plane of graphite via pi-stacking [46, 47]. The authors also observed the similar interaction of the pyrenyl group with the nanotubes surface. It was also observed that the ester molecule was irreversibly adsorbed on the surface and resisted desorption in aqueous solutions.

Succinimidyl ester groups are highly reactive to nucleophilic substitution by primary and secondary amines present in excess on the surface of most proteins; therefore the functionalization of nanotubes with these groups opens the possibilities for nanotubes to generate specific interactions with the biomolecules. Figure 1.8 demonstrates these functionalization processes. The authors suggested that the proteins immobilization on nanotubes involves the nucleophilic substitution of *N*-hydroxysuccinimide by an amine group on the protein, resulting in the formation of an amide bond. Successful immobilization of both ferritin and streptavidin proteins could be achieved by this method, thus indicating the versatility of this functionalization approach to make the surfaces of nanotubes more usable. Apart from proteins, any other chemical molecules can also be reacted on the

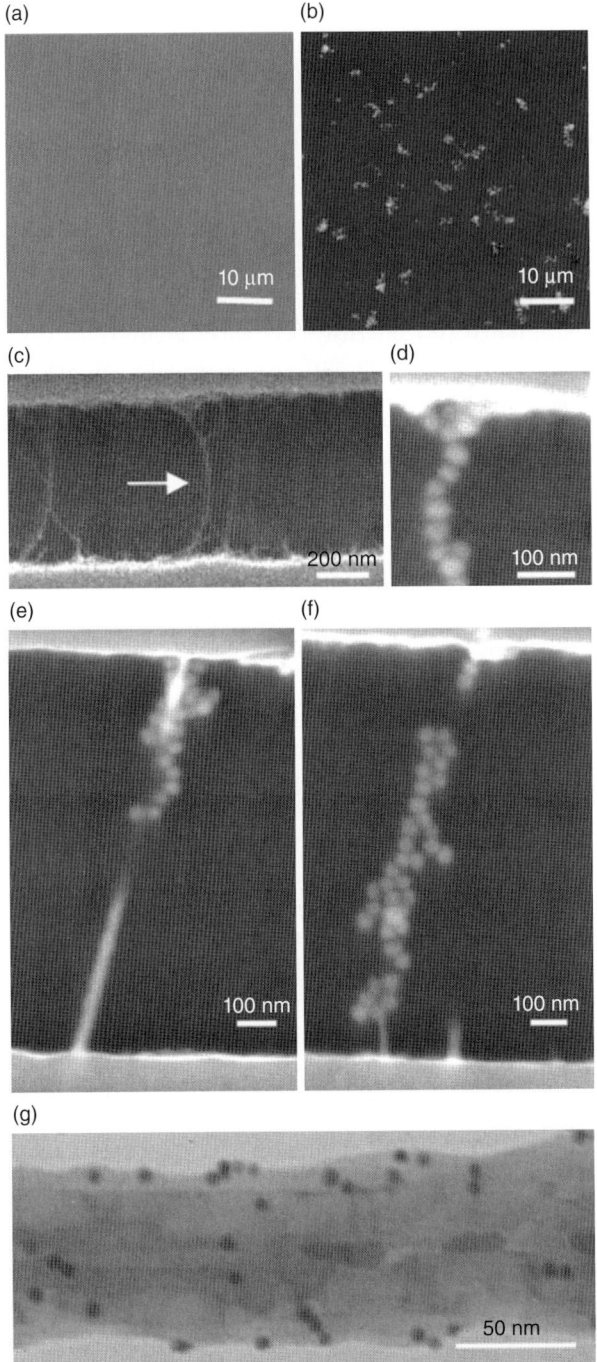

Figure 1.7 Functionalization of graphite and nanotube surfaces: (a) fluorescence micrograph of polymer-coated graphite, (b) SEM image of gold nanoparticles immobilized on polymer-coated graphite, (c) SEM image of substrate-grown SWNTs, (d)–(f) SEM images of gold nanoparticles immobilized on polymer-coated SWNTs, and (g) TEM image of gold nanoparticles immobilized on polymer-coated MWNTs. Reproduced from Ref. [37] with permission from American Chemical Society.

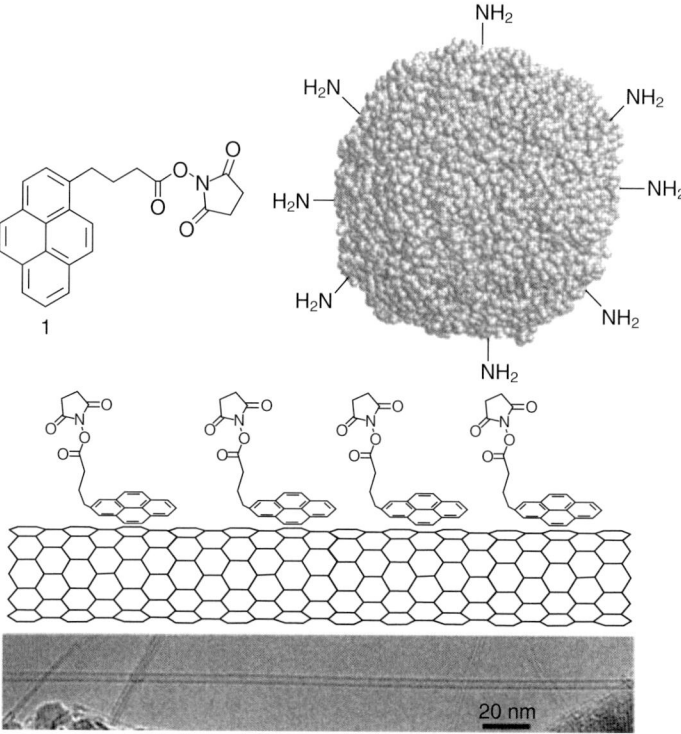

Figure 1.8 Irreversible adsorption of 1-pyrenebutanoic acid, succinimidyl ester on the side walls of nanotubes. Reproduced from Ref. [46] with permission from American Chemical Society.

surfaces treated with ester and thus the surface properties can be tuned or controlled according to the requirement.

Kang et al. [48] reported an interesting approach of encapsulating polymer nanotubes within cross-linked, amphiphilic copolymer micelles, similarly reported by other researchers [49, 50]. The modified nanotubes could be dispersed in a variety of polar and nonpolar solvents and polymer matrices. The authors used an amphiphilic poly(styrene)-block-poly(acrylic acid) copolymer, which was dissolved in dimethylformamide (DMF), a solvent in which both the blocks of the copolymer are soluble and do not form micelles. The nanotubes were subsequently suspended in the copolymer solution by ultrasonication. Micellization of the copolymer was achieved by the gradual addition of water to the suspension, leading to the encapsulated nanotubes inside these micelles. The poly(acrylic acid) blocks were subsequently cross-linked by the addition of diamine linker and carbodiimide activator. The nanotubes were observed to remain dispersed throughout the cross-linking process. It was also observed that full cross-linking of poly(acrylic acid) blocks was not necessary to achieve stabilization; conversion of 25% acid groups to amides was enough to bring about the stability of nanotube suspension. The

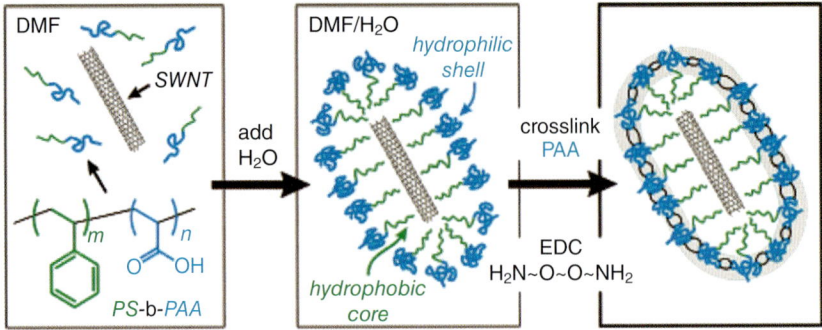

Figure 1.9 Encapsulation of nanotubes by using poly(styrene)-block-poly(acrylic acid) copolymer. Reproduced from Ref. [48] with permission from American Chemical Society.

treated nanotubes were washed off the excess reagents and empty micelles, and the encapsulated nanotubes could be successfully redispersed in different solvents. Apart from that, these suspensions were stable over long period of time as neither the visible absorbance nor the scattered light intensity of the suspensions changed as a function of time. It was hypothesized that high solubility of these encapsulated nanotubes resulted because of solvation of at least one block of the encapsulating copolymer in the solvents used. The authors reported in some cases the solvation of even both the blocks of the copolymer in the solvent. The solvents were also analyzed and it was confirmed that no polymer was desorbed from the surface of nanotubes. Figure 1.9 shows the schematic of the encapsulation process.

Many other studies have been reported for the noncovalent functionalization of nanotubes. Conjugated luminescent polymer poly(metaphenylenevinylene) (PmPv) could be coated on to the nanotubes surface [51]. Water-soluble polyvinyl alcohol was coated around the nanotubes, thus allowing the dispersion of nanotubes in water [52]. Carbon nanotubes were also wrapped with starch to generate starched nanotubes [53]. Long-chain polymers like Gum Arabic were also physically adsorbed to disperse the nanotubes individually in solvents [54]. Aromatic polyimides were also used as functionalizing polymer to achieve noncovalent functionalization of carbon nanotubes [55].

1.3
Covalent Modifications of Carbon Nanotubes

Covalent functionalization of nanotubes though may produce defects in the wall structure of the nanotubes, this mode of surface functionalization is required when the load-transfer properties are concerned. Thus, the choice of the modifications on the nanotubes surface may be a direct result of the properties required from them. Following paragraphs explain the various methodologies reported in the literature in order to achieve covalent functionalization of nanotubes.

Velasco-Santos et al. reported the advantages of chemical functionalization of nanotubes in improving the composite properties [56]. In the study, *in situ* polymerization of methyl methacrylate (MMA) was achieved both in the presence of unfunctionalized and acid-functionalized nanotubes. Interesting phenomena regarding the mode of polymerization and resulting composite properties were reported. Some studies have earlier reported that the unfunctionalized nanotubes can have chemical interactions with the polymer through opening of the pi bonds in the nanotube structure and subsequently nanotube can take part in polymerization reaction, thus grafting the polymer chains covalently bound to the clay surface [57, 58]. The authors also observed in this study that the position of vibration bands in the IR spectra was observed to increase, indicating that the pi bonds in the nanotubes could have been opened. However, it was suggested that the covalent functionalization provides much more attractive route as the generated reactive groups on the tips and surface of nanotubes can be much more useful in binding the nanotubes to the polymer chains during the course of polymerization. Composites generated using functionalized nanotubes were confirmed through IR spectroscopic characterization to have a chemical connection between the formed polymer and nanotubes. It was concluded that covalent surface functionalization in the presence of *in situ* polymerization to attach the polymer chains covalently to the surface of nanotubes is the best method for generating nanotube nanocomposites.

As one-dimensional materials have strong potential as building blocks in electronics, nanohybrid materials from CNTs and nanoparticles were reported in which the nanoparticles were covalently immobilized on the nanotubes surface [59–61]. To achieve theses functionalizations, the nanotubes were treated with nitric acid followed by thionyl chloride to generate COCl groups on the surface. Magnetite colloid solution was prepared in ethanol solution, to which 3-aminopropyltrimethoxysilane (APTS) was added with rapid stirring. APTS-coated magnetite nanoparticles were dried under vacuum at room temperature, which resulted in amine-terminated magnetite nanoparticles. These particles were reacted with COCl-treated nanotubes at room temperature using ultrasonification as explained in Figure 1.10. Transmission electron microscopy observations also revealed that the magnetite nanoparticles were deposited and homogenously distributed on the surface of the carbon nanotubes. It also demonstrated the effectiveness of the pretreatment of nanotubes in generating active sites on the carbon nanotubes. The resulting nanohybrids were also easily dispersible in aqueous solvents and were observed to be stable for more than 2 months; no precipitation of the inorganic material was observed. It should be noted that noncovalent immobilization of nanoparticles on the nanotubes surface has also been reported, but the covalent functionalization was reported to be more efficient and stronger in interactions between the components.

Covalent functionalization method was reported for the modification of SWCNTs with enzymes and amines [62]. In a typical experiment, SWNTs were refluxed in 4 M HNO_3 for 24 h and were exposed to 1 M HCl. The carboxylated SWNTs were then filtered and washed with water followed by drying in air.

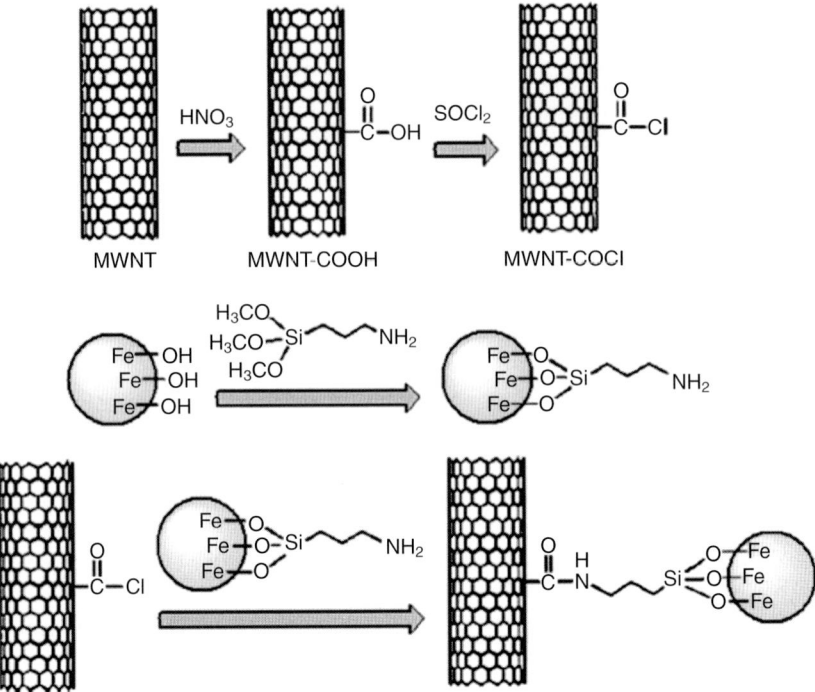

Figure 1.10 Covalent attachment of magnetite nanoparticles on the carbon nanotube surface. Reproduced from Ref. [61] with permission from Elsevier.

Acylation reactions with the acid-treated nanotubes were carried out by stirring the nanotubes in a 20:1 mixture of thionyl chloride and DMF at 70 °C for 24 h. The nanotubes were then filtered, washed with anhydrous THF, and dried under vacuum at room temperature. The so-obtained nanotubes were then reacted with desired amines using DMF as solvent, at 110 °C. The amines were used in excess, which after the reaction was washed first with DMF, followed by anhydrous THF. In the case of the enzymes, the reactions were carried out at low temperatures. Successful reactions of the following molecules with the surface groups on nanotubes could be achieved: porcine pancrease lipase (PPL), amino lipase (AK), cis-myrtanylamine, 2,4-dinitroaniline, 2,6-dinitroaniline, N-decyl-2,4,6-trinitroaniline, and N-(3-morpholinopropyl)-2,4,6-trinitroaniline). It was observed that the generated defects and covalent binding of amines on the surface after functionalization caused the reduction in electrical conductivity in the functionalized nanotubes. It was also observed that both the single nanotubes as well as bundles could be grafted, indicating that initial homogenous dispersion of nanotubes during functionalization is immensely important. Successful immobilization of a variety of amine as well as biological entities on the surface confirmed the potential as well as robustness of the functionalization

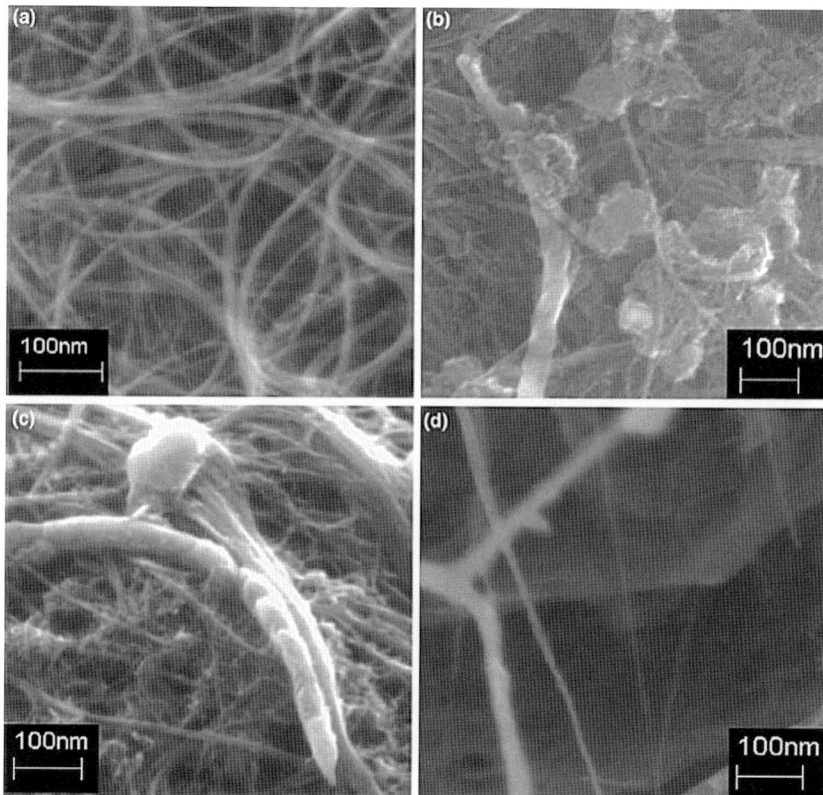

Figure 1.11 SEM images of (a) pristine SWNTs, (b) *cis*-myrtanylamine functionalized SWNTs, (c) AK functionalized SWNTs, and (d) PPL functionalized SWNTs. Reproduced from Ref. [62] with permission from Elsevier.

method. Figure 1.11 shows the SEM images of the untreated and treated nanotubes [62].

Grafting from the surface methods has been studied widely in detail and can also be directly applied on the nanotubes [63–68]. Qin *et al.* reported the polymerization of *n*-butyl methacrylate from the surface of nanotubes by using controlled living polymerization method [69]. To achieve such surface grafting, nanotubes were first refluxed in nitric acid, which generates acid functional groups on the sidewalls of the tubes. These acid groups were subsequently reacted with thionyl chloride to convert them into acyl chloride groups. An atom transfer radical polymerization agent 2-hydroxyethyl-2-bromopropionate (HEBP) was synthesized by the reaction of ethylene glycol and 2-bromopropionlybromide. The acyl chloride groups on the surface of nanotubes were then reacted with the ATRP initiator. After the ATRP ignition moiety was bound to the surface of nanotubes, these nanotubes could then be used to graft brushes of poly(*n*-butyl methacrylate) in the presence of CuCl and BiPy. Figure 1.12 shows the schematic of the functionalization process. The grafted brushes had significant effect on the solubility properties of nanotubes. This process also offers many other advantages: the amount of ATRP initiator on the surface can be controlled, the chain length of the grafts can also be tuned according to requirement, and a large number of monomers can be polymerized to graft brushes of various functionalities from the surface of nanotubes. Gao et al. also followed the similar grafting from the surface approach and polymerized MMA on the surface by using atom transfer radical polymerization [70]. Figure 1.13 shows the images of the pristine nanotubes as well as treated nanotubes where reaction conditions have been altered to change the grafting characteristics. Polymer thicknesses from 3.8 to 14 nm were grafted by changing the feed ratios, and the nanotubes after functionalization with poly(methyl methacrylate) could also be further reacted to form another layer of hydroxyethyl methacrylate (HEMA).

Ge et al. reported the chemical grafting of polyetherimides on the surface of nanotubes [71]. A polyetherimide (BisADA-DCB) was synthesized via the condensation polymerization of 4,4′-bis(4,4′-isopropylidene diphenoxy)-bis(phthalic anhydride) (BisADA) and 2,2′-dichloro-4,4′-biphenyldiamine (DCB) in m-cresol with isoquinonline at elevated temperatures via one-step imidization. The resulting BisADA-DCB containing diamine endgroups was observed to be soluble in many solvents such as chloroform, methylene chloride, THF, and cyclopentanone. The carboxylic acid functionalized MWNTs were reacted with polyetherimide and new characteristic bands were detected in the attenuated total reflection Fourier trans-

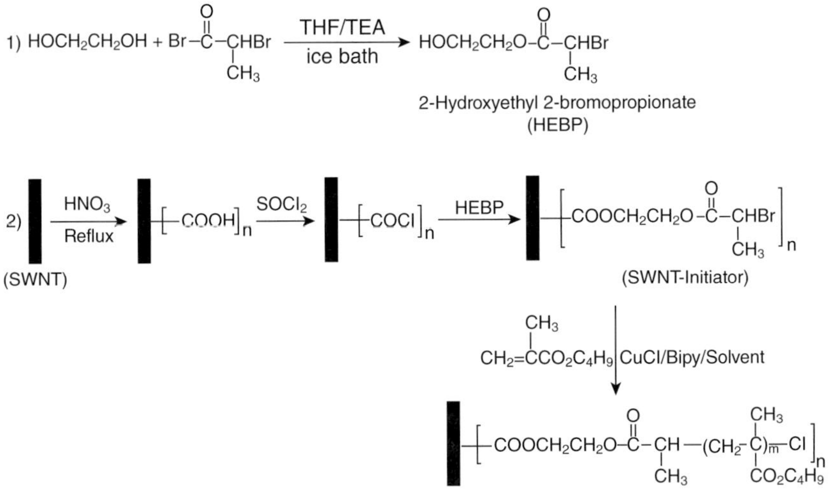

Figure 1.12 Atom transfer radical polymerization method from the grafting of poly(n-butyl acrylate) chains from the surface of nanotubes. Reproduced from Ref. [69] with permission from American Chemical Society.

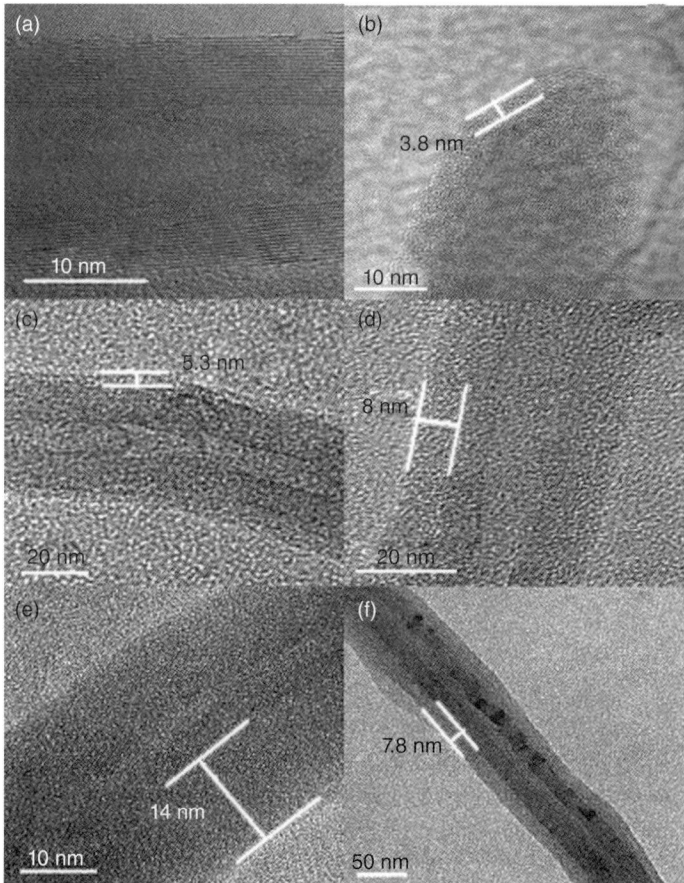

Figure 1.13 (a) Pristine nanotubes and (b–f) nanotubes modified with varying thicknesses of polymer film grafted from the surface. Reproduced from Ref. [70] with permission from American Chemical Society.

form IR spectroscopy. Transmission electron microscopy was also used to investigate the MWNTs before and after chemical grafting. A large population of MWNTs was coated by BisADA-DCB. A thin layer of BisADA-DCB with a thickness of 5–20 nm was observed. Figure 1.14 shows the schematic of functionalization process.

Attachment of diglycidyl ether of bisphenol A on the nanotubes surfaces was studied [72] to covalently functionalize the nanotubes. MWCNTs with acidic groups on the surface were reacted with epoxide-terminated molecule as shown in Figure 1.15. As the molecule had epoxy groups in both the ends, both the groups

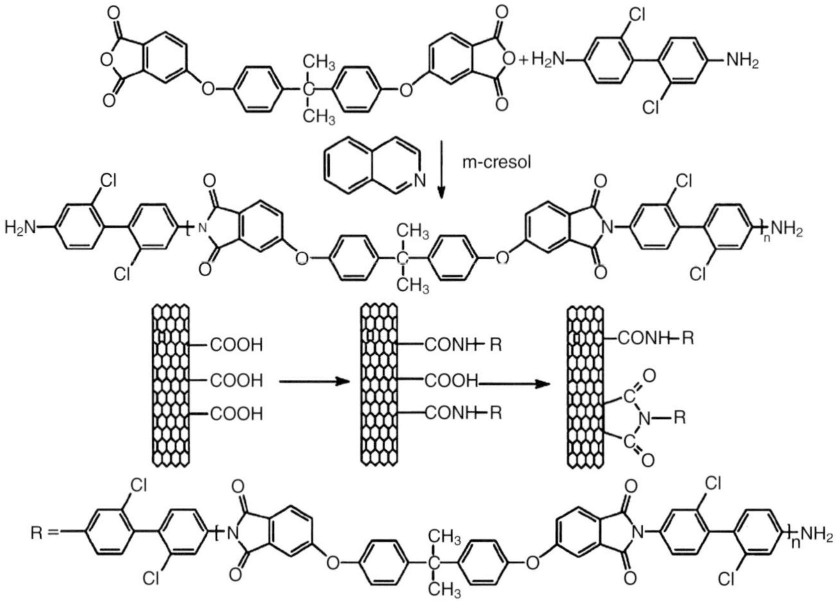

Figure 1.14 Functionalization of MWNT with polyetherimide. Reproduced from Ref. [71] with permission from American Chemical Society.

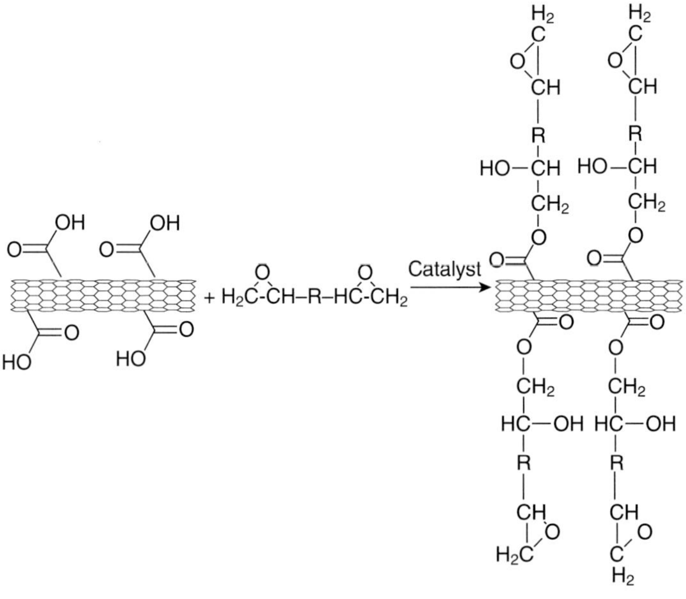

Figure 1.15 Epoxy functionalization of the nanotubes. Reproduced from Ref. [72] with permission from American Chemical Society.

could take part in the functionalization reactions, which was also confirmed with IR spectrometry. It was observed that the epoxide groups reacted either by the epoxide–epoxide reactions followed by the reaction of the epoxide ring with the carboxylic acid group on the nanotube, or both the epoxide groups can react with the acid groups present on the nanotube surface.

A novel approach to *in situ* surface functionalization of the nanotubes by the attachment of polystyrene chains to SWCNTs based on anionic polymerization scheme was reported by Viswanathan et al. [73] as shown in Figure 1.16. As produced nanotubes without any purification or surface treatment were used for the functionalization process, these procedures were suspected to introduce functionalities that hinder carbanion formation. To achieve the grafting, SWCNTs were dispersed by sonication in purified cyclohexane to which a slight excess of butyl lithium was added. This introduced carbanions on the SWCNT surface, which help to exfoliate the nanotube bundles because of the repulsion between the negatively charged nanotubes and also provided initiating sites for the polymerization of styrene. The small diameters of nanotubes offer high reactivities for the carbanion addition reactions owing to their high curvatures. Raman spectroscopy was used to confirm the evidence of formation of carbanions and subsequent attachment of polymer chains. On addition of styrene, both free butyl lithium and the nanotube carbanions were observed to initiate polymerization, resulting in an intimately mixed composite system. The polymerization could efficiently be terminated by the addition of degassed *n*-butanol.

Low-molecular-weight chitosan, a natural and green polymer, was covalently bound to the sidewalls of MWCNTs by a nucleophilic substitution reaction on the surface of nanotubes [74]. Amino and primary hydroxyl groups of chitosan were reported to contribute mainly to the formation of MWNT–chitosan structures. These groups reacted with COCl groups generated on the nanotube surfaces by

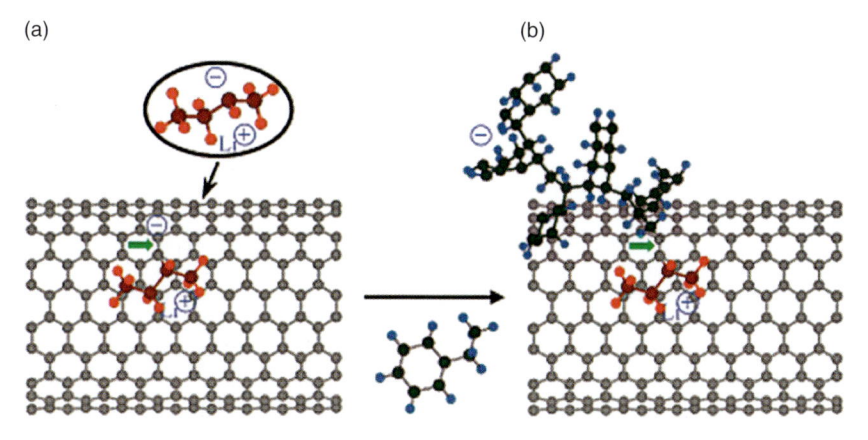

Figure 1.16 Covalent functionalization of nanotubes by anionic polymerization approach. Reproduced from Ref. [73] with permission from American Chemical Society.

(a) (b) (c)

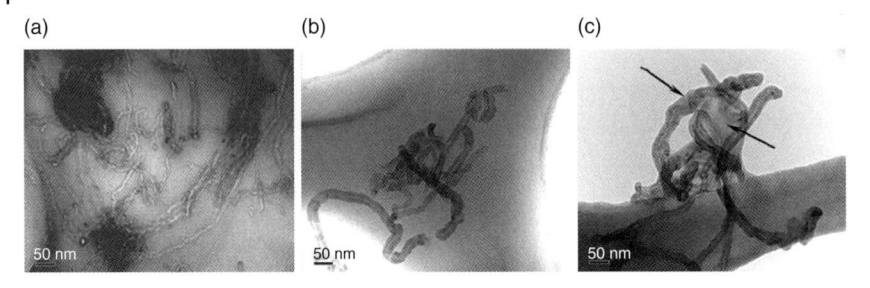

Figure 1.17 TEM images of (a) raw MWNTs, (b) cut and purified MWNTs, and (c) the MWNT-modified with chitosan. Arrows indicate the typical features of the attached polymer. Reproduced from Ref. [74] with permission from American Chemical Society.

acid treatment. By thermogravimerty, the extent of chitosan in the organically modified nanotubes was approximately 58 wt%, indicating a significant amount of chitosan grafting to the surface. It was observed from XPS data that a nitrogen-to-carbon ratio of 0.091 was present in the modified nanotubes, which was used to estimate the average degree of functionalization in the nanotubes. On the basis of this information, it is calculated that approximately four molecular chains of the LMCS were attached to 1000 carbon atoms of the nanotube sidewalls. The functionalized nanotubes were soluble in a variety of solvents like dimethylformamide, dimethyl acetamide, dimethylsulfoxide, and acetic acid aqueous solution. Figure 1.17 also shows the TEM images of the nanotubes before and after the modification of the tubes with chitosan.

Microwave irradiation for the acid functionalization of both single and MWCNTs was also reported, which caused the generation of significant amount of hydrophilic functional groups on the surface [75]. Immobilization of DNA on the nanotubes was reported by covalent functionalization of nanotubes with polyethylenimine (PEI) [76]. Plasma deposition of polymers as thin films on the surface of nanotubes has also been reported [77, 78]. Fluorinated SWCNTs have also been reported to demonstrate higher potential for further derivatization due to higher reactivity than the unfunctionalized nanotubes [79–81].

References

1 Ajayan, P.M. and Ebbesen, T.W. (1997) *Rep. Prog. Phys.*, **60**, 1025.
2 Dresselhaus, M.S., Dresselhaus, G., and Eklund, P.C. (1996) *Science of Fullerenes and Carbon Nanotube*, Academic Press, New York.
3 Dai, H. (2002) *Acc. Chem. Res.*, **35**, 1035.
4 Iijima, S. (1991) *Nature*, **354**, 56.
5 Endo, M. (1978) (In Japanese), PhD Thesis, Nagoya University, Japan.
6 Harris, P.J.F. (2001) *Carbon Nanotubes and Related Structures, New Materials for the Twenty-First Century*, Cambridge University Press, Cambridge.
7 O'Connell, M.J. (2006) *Carbon Nanotubes Properties and Applications*, CRC Taylor & Francis, Boca Raton, FL.
8 Harris, P.J.F. (2001) *Carbon Nanotubes and Related Structures*, Cambridge University Press, Cambridge.

9 Ajayan, P.M. (1999) *Chem. Rev.*, **99**, 1787.
10 Xie, X.-L., Mai, U.-W., and Zhou, X.-P. (2005) *Mater. Sci. Eng. R*, **49**, 89.
11 Pavlidoua, S. and Papaspyrides, C. (2008) *Prog. Polym. Sci.*, **33**, 1119.
12 Mark, J.E. (1996) *Polym. Eng. Sci.*, **36**, 2905.
13 Reynaud, E., Gauthier, C., and Perez, J. (1999) *Rev. Metall./Cah. Inf. Tech.*, **96**, 169.
14 Calvert, P. (1997) *Carbon Nanotubes* (ed. T.W. Ebbesen), CRC Press, Boca Raton, FL.
15 Favier, V., Canova, G.R., Shrivastava, S.C., and Cavaille, J.Y. (1997) *Polym. Eng. Sci.*, **37**, 1732.
16 Mittal, V. (2006) Organic modifications of clay and polymer surfaces for specialty applications, PhD Thesis, ETH Zurich.
17 Journet, C., Master, W.K., Bernier, P., Loiseau, A., Lamy de la Chapelle, M., Lefrant, S., Deniard, P., Lee, R., and Fischer, J.E. (1997) *Nature*, **388**, 756.
18 Ebbesen, T.W. and Ajayan, P.M. (1992) *Nature*, **358**, 220.
19 Colbert, D.T., Zhang, J., McClure, S.M., Nikolaev, P., Chen, Z., Hafner, J.H., Owens, D.W., Kotula, P.G., Carter, C.B., Weaver, J.H., Rinzler, A.G., and Smalley, R.E. (1994) *Science*, **266**, 1218.
20 Thess, A., Lee, R., Nikolaev, P., Dai, H., Petit, P., Robert, J., Xu, C., Lee, Y.H., Kim, S.G., Rinzler, A.G., Colbert, D.T., Scuseria, G.E., Tomanek, D., Fischer, J.E., and Smalley, R.E. (1996) *Science*, **273**, 483.
21 Cheng, H.M., Li, F., Su, G., Pan, H.Y., He, L.L., Sun, X., and Dresselhaus, M. (1998) *Appl. Phys. Lett.*, **72**, 3282.
22 Rao, C.N.R., Govindaraj, A., Sen, R., and Satishkumar, B. (1998) *Mater. Res. Innov.*, **2**, 128.
23 Andrews, R., Jacques, D., Rao, A.M., Derbyshire, F., Qian, D., Fan, X., Dickey, E.C., and Chen, J. (1999) *Chem. Phys. Lett.*, **303**, 467.
24 Ren, Z.F., Huang, Z.P., Xu, J.W., Wang, J.H., Bush, P., Siegel, M.P., and Provenico, P.N. (1998) *Science*, **282**, 1105.
25 Singh, C., Shaffer, M.S.P., Kinloch, I.A., and Windle, A.H. (2002) *Physica B*, **323**, 339.
26 Kuzmany, H., Kukovecz, A., Simon, F., Holzweber, M., Kramberger, C., and Pichler, T. (2004) *Synth. Metals*, **141**, 113.
27 Zhu, J., Yudasaka, M., Zhang, M., and Ijima, S. (2004) *J. Phys. Chem. B*, **108**, 11317.
28 Balvlavoine, F., Schultz, P., Richard, C., Mallouh, V., Ebbeson, T.W., and Mioskowski, C. (1999) *Angew. Chem. Int. Eng. Ed.*, **38**, 1912.
29 Shim, M., Kam, N.W.S., Chen, R.J., Li, Y., and Dai, H. (2002) *Nano Lett.*, **2**, 285.
30 Harris, J.M., and Zalipsky, S. (1997) *Poly(Ethylene Glycol): Chemistry and Biological Application*, American Chemical Society, Washington, DC.
31 Szycher, M. (ed.) (1983) *Biocompatible Polymers, Metals and Composites*, Technomic, Lancaster, PA.
32 Ostuni, E., Chapman, R.G., Holmlin, R.E., Takayama, S., and Whitesides, G.M. (2001) *Langmuir*, **17**, 5605.
33 O'Connell, M.J., Boul, P., Ericson, L.M., Huffman, C., Wang, Y., Haroz, E., Kuper, C., Tour, J., Ausman, K.D., and Smalley, R.E. (2001) *Chem. Phys. Lett.*, **342**, 265.
34 Fu, Q., Lu, C., and Liu, J. (2002) *Nano Lett.*, **2**, 329.
35 Gomez, F.J., Chen, R.J., Wang, D., Waymouth, R.M., and Dai, H. (2003) *Chem. Commun.*, 190.
36 Liu, Y.-T., Zhao, W., Huang, Z.-Y., Gao, Y.-F., Xie, X.-M., Wang, X.-H., and Ye, X.-Y. (2006) *Carbon*, **44**, 1613.
37 Carillo, A., Swartz, J.A., Gambe, J.M., Kane, R.S., Chakrapani, N., Wei, B., and Ajayan, P. (2003) *Nano Lett.*, **3**, 1437.
38 Decher, G., Lvov, Y., and Schmitt, J. (1994) *Thin Solid Films*, **244**, 772.
39 Stroock, A.D., Kane, R.S., Weck, M., Metallo, S.J., and Whitesides, G.M. (2003) *Langmuir*, **10**, 2466.
40 Korneev, D., Lvov, Y., Decher, G., Schmitt, J., and Yaradaikin, S. (2005) *Physica B*, **213-214**, 954.
41 Hammond, P.T. (1999) *Curr. Opin. Colloid Interface Sci.*, **4**, 430.
42 Schmitt, J., Grünewald, T., Decher, G., Pershan, P.S., Kjaer, K., and Lösche, M. (1993) *Macromolecules*, **26**, 7058.
43 Wang, T.C., Chen, B., Rubner, M.F., and Cohen, R.E. (2001) *Langmuir*, **17**, 6610.

44 Caruso, F., Caruso, R.A., and Mohwald, H. (1998) *Science*, **282**, 1111.
45 Chapman, R.G., Ostuni, E., Liang, M.N., Meluleni, G., Kim, E., Yan, L., Pier, G., Warren, H.S., and Whitesides, G.M. (2001) *Langmuir*, **17**, 1225.
46 Chen, R.J., Zhang, Y., Wang, D., and Dai, H. (2001) *J. Am. Chem. Soc.*, **123**, 3838.
47 Katz, E. (1994) *J. Electroanal. Chem.*, **365**, 157.
48 Kang, Y., and Taton, T.A. (2003) *J. Am. Chem. Soc.*, **125**, 5650.
49 Huang, H.Y., Kowalewski, T., Remsen, E.E., Gertzmann, R., and Wooley, K. (1997) *J. Am. Chem. Soc.*, **119**, 11653.
50 Zhang, L., Shen, H., and Eisenberg, A. (1997) *Macromolecules*, **30**, 1001.
51 Star, A., Stoddart, J.F., Steuerman, D., Diehl, M., Boukai, A., Wong, E.W., Yang, X., Chung, S.-W., Choi, H., and Heath, J. (2001) *Angew. Chem. Int. Ed.*, **40**, 1721.
52 Vedala, H., Choi, Y.C., Kim, G., Lee, E., and Choi, W.B. (2004). Proceedings of 2nd International Latin American and Caribbean Conference for Engineering and Technology, Florida, USA.
53 Star, A., Steurmann, D.W., Heath, J.R., and Stoddart, J.F. (2002) *Angew. Chem. Int. Ed.*, **41**, 2508.
54 Bandyopadhyaya, R., Nativ-Roth, E., Regev, O., and Yerushalmi-Rozen, R. (2002) *Nano Lett.*, **2**, 25.
55 Yang, Z., Chen, X., Chen, C., Li, W., Zhang, H., Xu, L., and Yi, B. (2007) *Polym. Compos.*, doi: 10.1002/pc.20254.
56 Velasco-Santos, C., Martinez-Hernandez, A.L., Fischer, F.T., Ruoff, R., and Castano, V. (2003) *Chem. Mater.*, **15**, 4470.
57 Jia, Z., Wang, Z., Xu, C., Liang, J., Wei, B., Wu, D., and Zhu, S. (1999) *Mater. Sci. Eng.*, **A271**, 395.
58 Geng, H.Z., Rosen, R., Zheng, B., Shimoda, H., Fleming, L., Liu, J., and Zhou, O. (2002) *Adv. Mater.*, **14**, 1387.
59 Pan, B., Gao, F., and Gu, H. (2005) *J. Colloid Interface Sci.*, **284**, 1.
60 Xu, Y., Gao, C., Kong, H., Yan, D., Jin, Y.Z., and Watts, P.C.P. (2004) *Macromolecules*, **37**, 8846.
61 Xu, P., Cui, D., Pan, B., Gao, F., He, R., Li, Q., Huang, T., Bao, C., and Yang, H. (2008) *Appl. Surf. Sci.*, **254**, 5236.
62 Wang, Y., Iqbal, Z., and Malhotra, S. (2005) *Chem. Phys. Lett.*, **402**, 96.
63 Matyjaszewski, K., Miller, P.J., Shukla, N., Immaraporn, B., Gelman, A., Luokala, B.B., Siclovan, T.M., Kickelbick, G., Vallant, T., Hoffmann, H., and Pakula, T. (1999) *Macromolecules*, **32**, 8716.
64 Vestal, C.R. and Zhang, Z.J.J. (2002) *J. Am. Chem. Soc.*, **124**, 14312.
65 Jordan, R. and Ulman, A. (1998) *J. Am. Chem. Soc.*, **120**, 243.
66 Jordan, R., Ulman, A., Kang, J.F., Rafailovich, M.H., and Sokolov, J. (1999) *J. Am. Chem. Soc.*, **121**, 1016.
67 Weck, M., Jackiw, J.J., Rossi, R.R., Weiss, P.S., and Grubbs, R.H. (1999) *J. Am. Chem. Soc.*, **121**, 4088.
68 Mittal, V. (2007) *J. Colloid Interface Sci.*, **314**, 141.
69 Qin, S., Qin, D., Ford, W.T., Resasco, D.E., and Herrera, J.E. (2004) *J. Am. Chem. Soc.*, **126**, 170.
70 Kong, H., Gao, C., and Yan, D. (2004) *J. Am. Chem. Soc.*, **126**, 412.
71 Ge, J.J., Zhang, D., Li, Q., Hou, H., Graham, M.J., Dai, L., Harris, F.W., and Cheng, S.Z.D. (2005) *J. Am. Chem. Soc.*, **127**, 9984.
72 Eitan, A., Jiang, K., Dukes, D., Andrews, R., and Schadler, L. (2003) *Chem. Mater.*, **15**, 3198.
73 Viswanathan, G., Chakrapani, N., Yang, H., Wei, B., Chung, H., Cho, K., Ryu, C.Y., and Ajayan, P.M. (2003) *J. Am. Chem. Soc.*, **125**, 9258.
74 Ke, G., Guan, W., Tang, C., Guan, W., Zeng, D., and Deng, F. (2007) *Biomacromolecules*, **8**, 322.
75 Kakade, B.A. and Pillai, V.K. (2008) *Appl. Surf. Sci.*, **254**, 4936.
76 Liu, Y., Wu, D.-C., Zhang, Y.-D., Jiang, X., He, C.-B., Chung, T.S., Goh, S.H., and Leong, K.W. (2005) *Angew. Chem. Int. Ed.*, **44**, 4782.
77 Shi, D., Lian, J., He, P., Wang, L.M., van Ooij, W.J., Schulz, M., Liu, Y., and Mast, D. (2002) *Appl. Phys. Lett.*, **81**, 5216.
78 Tseng, C.-H., Wang, C.-C., and Chen, C.-Y. (2007) *Chem. Mater.*, **19**, 308.

79 Khabashesku, V.N., Billups, W.E., and Margrave, J.L. (2002) *Acc. Chem. Res.*, **35**, 1087.
80 Khabashesku, V.N. and Margrave, J.L. (2004) *Encyclopedia of Nanoscience and Nanotechnology* (ed. H.S. Nalwa), American Scientific Publishers, Stevenson Ranch, CA.
81 Pulikkathara, M.X., Kuznetsov, O.V., and Khabashesku, V.N. (2008) *Chem. Mater.*, **20**, 2685.

2
Modification of Carbon Nanotubes by Layer-by-Layer Assembly Approach
Vaibhav Jain and Akshay Kokil

2.1
Introduction

The discovery of carbon nanotubes (CNTs) in 1991 by Iijima [1] offered an excellent material that possesses exceptional mechanical and electrical properties [2, 3]. As opposed to the spherical nature of fullerenes [4], CNTs can be visualized as seamless rolled-up hollow cylinders of graphite sheets. They are obtained as single-walled tubes (SWCNTs), represented by a hollow single cylinder or as multiwalled tubes (MWCNTs), which are a collection of multiple graphene concentric hollow cylinders. CNTs are mainly synthesized by three methods: electric arc discharge, laser ablation, or chemical vapor deposition (CVD). From the time of their discovery, CNTs have had a significant impact on research both in academia and in industry. Due to their outstanding mechanical, thermal, and electronic properties, CNTs offer wide-ranging applications in materials science, molecular electronics, and in the biomedical sciences [2, 5]. Around the world, significant strides have been made in CNT research and this area is ever-evolving. Newer applications have been envisioned for CNTs owing to their superior properties. Although great strides have been achieved in the development of CNTs, their covalent functionalization still requires at times intricate protocols, which can limit their applicability. Covalent functionalization of CNTs can also have a degradative effect on their electronic properties. Noncovalent functionalization of CNTs aids in preserving the sp^2 hybridized structure of the nanotube and thus maintains their novel electronic properties. Deposition of polymers and small molecules on CNTs in layers introduces a relatively facile technique for CNT functionalization.

Layer-by-layer (LbL) deposition of polymeric thin films (also known as ionic self-assembled multilayers (ISAMs) of polyelectrolytes) is a technique developed by Decher *et al.*, [6, 7] which uses the attractive forces between the molecules of opposite charges to form the films [8, 9]. Figure 2.1 shows the steps of the fabrication process along with its brief description. In this technique, a high concentration of polyelectrolytes (PE) is deposited on the substrate that was carrying a negative charge. These highly concentrated solutions lead to excess adsorption,

---

Surface Modification of Nanotube Fillers, First Edition. Edited by Vikas Mittal.
© 2011 Wiley-VCH Verlag GmbH & Co. KGaA. Published 2011 by Wiley-VCH Verlag GmbH & Co. KGaA.

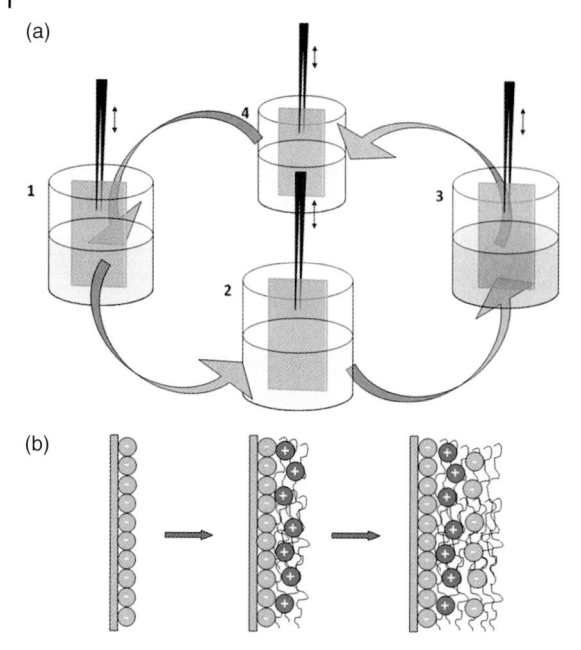

Figure 2.1 (a) Schematic of the film deposition process using slides and beakers. Steps 1 and 3 represent the adsorption of a polyanion and polycation, respectively, and steps 2 and 4 are washing steps. The four steps are the basic buildup sequence for the simplest film architecture, (A/B)n. The construction of more complex film architectures requires only additional beakers and a different deposition sequence. (b) Simplified molecular picture of the first two adsorption steps, depicting film deposition starting with a positively charged substrate. Counterions are omitted for clarity. The polyion conformation and layer interpenetration are an idealization of the surface charge reversal with each adsorption step.

which means that there is complete charge neutralization and resaturation bringing charge reversal [10]. Brent et al. proved this phenomenon by adopting a surface-force measurement technique [11]. Alternate deposition of these highly concentrated PE of positive and negative charge increases the thickness of the film and also provides flexibility in choosing the number of bilayers while still maintaining nanoscale control of the thickness. One of the important advantages of the LbL film assembly is that it is very easy to use without the need of any kind of expensive or special apparatus. Various combinations of PE with other materials such as gold colloids [12], silica [13, 14], clay minerals [15], proteins [16, 17], viruses [18], dendrimers [19], etc., have also proven that the LbL technique is very flexible and easy to use.

Soon after Decher's work, Rubner et al. demonstrated the ability to deposit thin uniform LbL films of conductive materials (polymers like polyaniline, polypyrrole, etc.) on substrates of different shapes and sizes [20–22]. Thin-film formation by molecular self-assembly of p-type-doped electrically conductive polymers with either a conjugated or nonconjugated polyanion was demonstrated. Partially doped conjugated polymers have delocalized defect sites along the polymer backbone that carry positive charges (polarons and bipolarons) and that charge has been used with the negative charge of the polyanion to deposit the multilayers [23]. The range of the applications of LbLs is very broad, but a few of the most common ones that have been demonstrated in the last 15 years are thin films for nonlinear optics [23, 24], selective area patterning [25], sensors [26], electrochromic films [27, 28], electrocatalysis [29], thin films for light-emitting diodes [30], antireflection coatings [31], and active enzymes [32].

A change in pH, ionic strength, and salt concentration of the PE solution results in changing the thickness of the LbL film bilayers. Rubner *et al.* have explained the influence of changing pH of the solution on the film morphology, physical properties, and composition [33, 34]. Variation in the charge density of a weak PE (e.g., poly(allylamine hydrochloride) and poly(acrylic acid)) can result in a change in the adsorption behavior, that finally results in a change in the behavior of the films resulting in very thick to very thin films over a narrow pH range [35]. At low pHs, positively charged PE (polycations) are highly ionized, but lose protons as the pH goes higher than a characteristic pH known as the pK_a, hence resulting in more deposition of material to neutralize the same amount of charge from the prior layer. This results in thicker film deposition at higher pH for polycations with globular (loops and tails) morphology. The opposite happens in the case of a negatively charged PE (polyanion), which forms thicker films at lower pH and thinner at higher. In brief, the film thickness of an adsorbed layer is dependent on the charge density (or pH) of the adsorbing PE and remains relatively independent of the morphology of the previously deposited PE or bilayers. However, in some cases at neutral pH for one of the PE, it was observed that the thickness also depends on whether the layers deposited before are above or below the value of that PE's pK_a [36]. Lvov *et al.* [36] first introduced the idea of changing the ionic strength of the PE solution to adjust the desired layer thickness; later this idea was further developed by many scientists [35, 37]. Screening-enhanced adsorption [38] explains the increase in the thickness of adsorbed PE layer by the addition of a small amount of the salt, but needs to be carefully used because a higher salt concentration can result in delamination of films from the substrates and also surface defects by increased surface roughness [39]. Schlenoff *et al.* have explained, in detail, the effects of surface charge and mass balance in the PE multilayers and how the salt content within the multilayers can be controlled easily by using electrochemically active multilayers [39]. There are also secondary interactions (H bonding) that occur between opposite charged PE layers, but they are weak interactions and excluded from the focus of this review. Rubner *et al.* [40] and Garnick *et al.* [41] have discussed the presence of secondary interactions in weak PE, which can be used to fabricate films by inkjet printing water of controlled pH.

Significant research has been performed around the world on processing of CNTs using LbL into thin films for varying applications and has been covered in

2.2
Layer-by-Layer Modification of the Carbon Nanotubes

The demonstration of depositing CNTs in the LbL technique for various applications has been extensively studied in the last decade [43, 44]. In contrast to this, the technique of assembling LbL films of polyions and nanoparticles around CNTs were introduced within the last few years [45, 46]. Self-assembly of PE nanoshells on individual CNTs with water-soluble PE, was first reported by Noy *et al.* (Figure 2.2) [47]. The authors first synthesized the CNTs as "bridges" between the two sides of the openings of a copper transmission electron microscopy (TEM) grid via CVD method utilizing pyrolyzed ethylene as a carbon source.

It has been observed that the surface tension of water aids in the desorption of the CNTs from the substrate. To overcome this issue, the authors immersed the TEM grid with the CNTs in water through a hexane layer, which replaced the water–air interface. Since pyrene irreversibly adsorbs onto CNTs through strong π-stacking interactions, the authors utilized charged pyrene molecules (namely

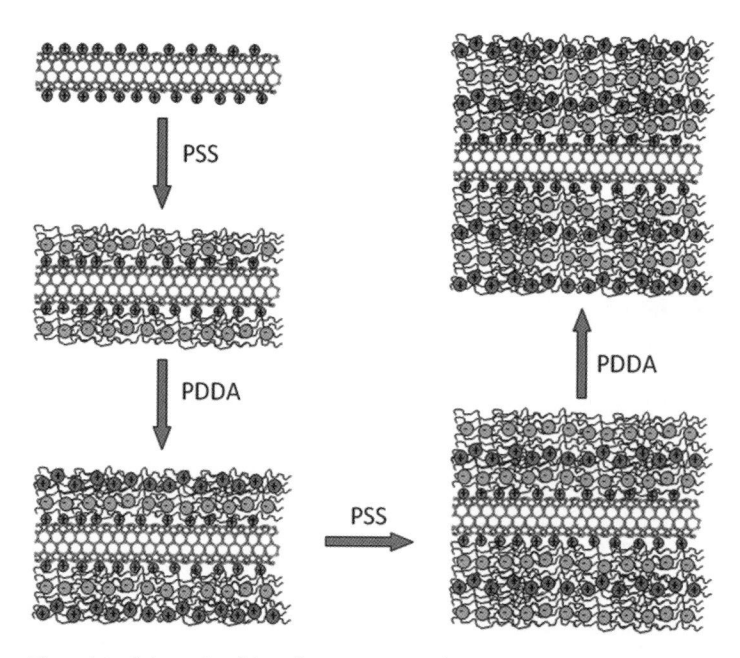

Figure 2.2 Schematic of the adsorption steps, depicting deposition starting with a positively charged CNT. Counterions are omitted for clarity. The polyion conformation and layer interpenetration are an idealization of the surface charge reversal with each adsorption step.

1-pyrenepropylamine hydrochloride and 1,3,6,8-pyrenetetrasulfonic acid tetrasodium salt) adsorbed onto the nanotube as an ionic layer for subsequent PE deposition; the details of this protocol were same as adopted by Dai et al. [48] Polystyrene sulfonate sodium salt (PSS) and poly(diallyldimethylammonium chloride) (PDDA) were then deposited on the CNTs as alternating layers. TEM was utilized for determining the adsorption of the bilayers on the CNTs (Figure 2.3) and energy-filtered TEM was employed to confirm the alternating nature and presence of both the PE on the CNT surface. This technique served as a good starting point to confirm the proof-of-concept of LbL deposition on CNTs.

For charged spherical substrates, the adsorption of PE depends on a variety of parameters, such as the persistence length of the PE, particle diameter, etc. [49]. The persistence length of the PE is in turn dependent on the ionic strength of the solution. As the particle size decreases, the stiffness of the PE chain should also decrease to obtain good surface coverage of the deposited layer. An increase in the ionic strength of the PE solution aids in screening the charges present on the polymer backbone resulting in a more flexible chain. Noy et al. reported the effect of the persistence length (directly related to the ionic strength of the PE solution) on the LbL assembly of PE on CNTs utilizing a similar system as stated above [50]. The authors reported that the thickness of the deposited layer depended on the total ionic concentration of the solutions (Figure 2.4). As compared to floating multilayers of the same polymers, the thickness of the deposited bilayers on CNTs was lower at low ionic strengths and comparable at high ionic strengths. The authors discussed that after a threshold ionic strength, the persistence length of the PE was smaller than the curvature of the nanotube and the deposition occurred

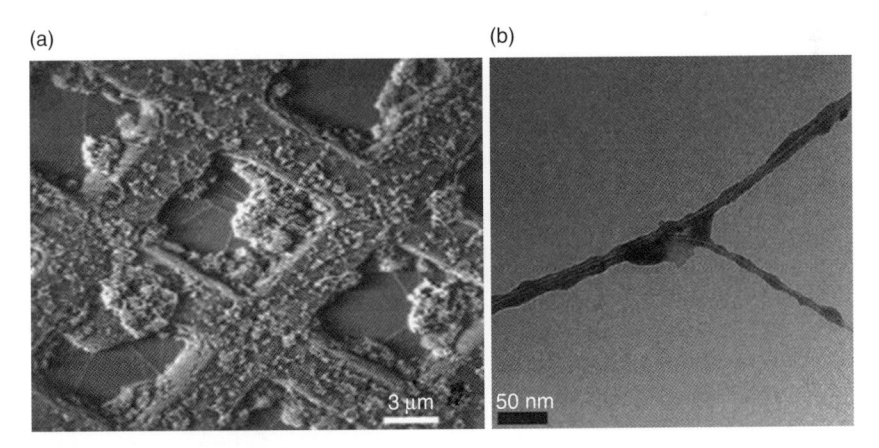

Figure 2.3 (a) Differential SEM images of CNTS grown over the TEM grid holes after dipping in aqueous solution through a hexane layer. (b) TEM image of as-grown CNTs coating with four bilayers of PDDA/ PSS after surface modification with pyrene derivatives. Reprinted with permission from A. B. Artyukhin, O. Bakajin, P. Strove, A. Noy, Langmuir, **2004**, 20, 1442. Copyright 2004 American Chemical Society.

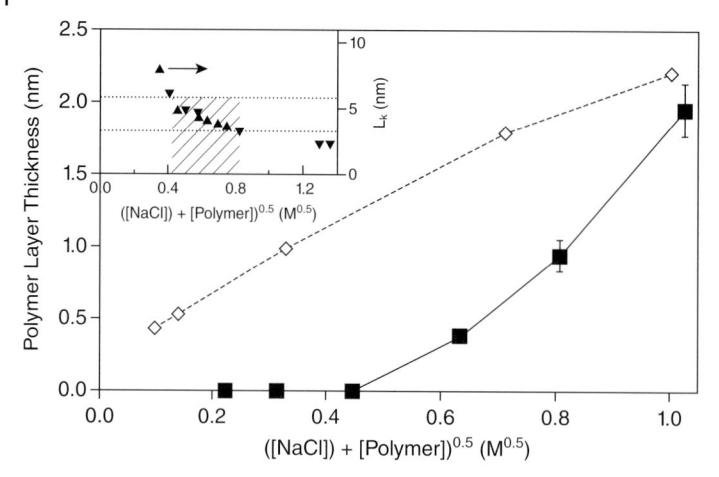

Figure 2.4 Average measured polymer thickness per layer of polyelectrolyte (■) on CNT surfaces as a function of electrolyte concentration. Experimental data for floating multilayers of the same polymer pair (PAH/PSS) (◇) are shown for comparison [51]. (Inset) Experimental data for PSS persistence length (L_k) as a function of electrolyte concentration (▲ [52], ▼ [53]). Dashed lines indicate the upper and lower boundary for the half circumference of CNTs used in our experiments. Shaded area corresponds to the electrolyte concentration region, where persistence length crosses these boundaries. Reprinted with permission from S. C. J. Huang, A. B. Artyukhin, Y. Wang, J. W. Ju, P. Stroeve, A. Noy, *J. Am. Chem. Soc.* **2005**, *127*, 14176. Copyright 2005 American Chemical Society.

(Figure 2.4 inset). For ionic strengths lower than the threshold value, the PE chains were stiff and deposition did not proceed. It was also reported that after deposition of the first few bilayers, the dependence on curvature reduced and deposition was possible even at lower ionic strengths. However, since the effective curvature of the CNT bundles is larger than that of an individual CNT, such curvature effects were not observed. This suggests that the polymers might wrap around the CNTs during the deposition process.

Utilizing the PAH/PSS LbL assembly on CNTs, Noy and coauthors were also successful in functionalizing the coated nanotubes with a 1D lipid layer [54]. Since CNTs are inherently hydrophobic, the presence of hydrophilic PE at the surface of the CNTs was critical in the formation of the lipid bilayer. A luminescent lipid probe was incorporated for bilayer visualizations. The authors observed that the fluorescence of a bleached spot on the nanotube recovered over time, which indicated that the lipid layer on the surface was mobile.

Gao and coworkers reported an interesting protocol for functionalization of the CNTs. They performed covalent functionalization of MWNTs with polyacrylic acid (PAA) and PSS [55]. The polymer chains were grown on MWNT-Br macroinitiators using atom transfer radical polymerization (ATRP). They reported that higher than 55 wt.% of the polymer was grafted on the surface of the MWNTs. Utilizing the

MWNTs grafted with PE, the authors obtained better surface coverage and smooth coating on the nanotubes after LbL assembly. The authors utilized as-prepared MWNT-COOH, MWNT-PAA, and MWNT-PSS as the substrate for LbL assembly. Poly(2-(N,N-dimethylaminoethyl) methacrylate (PDMAEMA) or hyperbranched poly(sulfone amine) (HPSA) were used as linear and branched polycations, respectively. PSS was employed as the polyanions to investigate the LbL self-assembly behavior on the nanosurfaces. The amount of PE deposition was studied using TGA. It was reported that in the case of oxidized CNTs (MWNT-COOH), lower loading of the PE was obtained as compared to the PAA- or PSS-grafted CNTs. The authors also reported that deposition of the linear polyelelctrolytes was more efficient as compared to the hyperbranched PE. It is interesting to note that according to the "wrapping" mechanism, the deposition of hyperbranched polymers should be hindered as compared to that of the linear ones; however, in the case of the hyperbranched PE the increased charge density on the polymers might be playing a role in anchoring the polymers on the CNTs. Utilizing HRTEM, the authors reported a uniform coating of the PE on the oxidized nanotubes. Even in the case of the hyperbranched polycation, which has a globular structure, the authors observed uniform surface coverage (Figure 2.5). Smooth and uniform coating of the PE was also reported on the functionalized nanotubes with thickness greater than 10 nm. Using scanning electron microscopy (SEM), the authors observed that after deposition of the PE on the tube surfaces, a composite was obtained in which the tubes were uniformly dispersed, suggesting that CNT/polymer composites can be obtained by the LbL self-assembly. Because the dendritic polymers can play the role of molecular vessels to load guest compounds, this protocol for adsorbing hyperbranched PE opens new avenues for applications in the areas of nano-, bio-, and supramolecular chemistry.

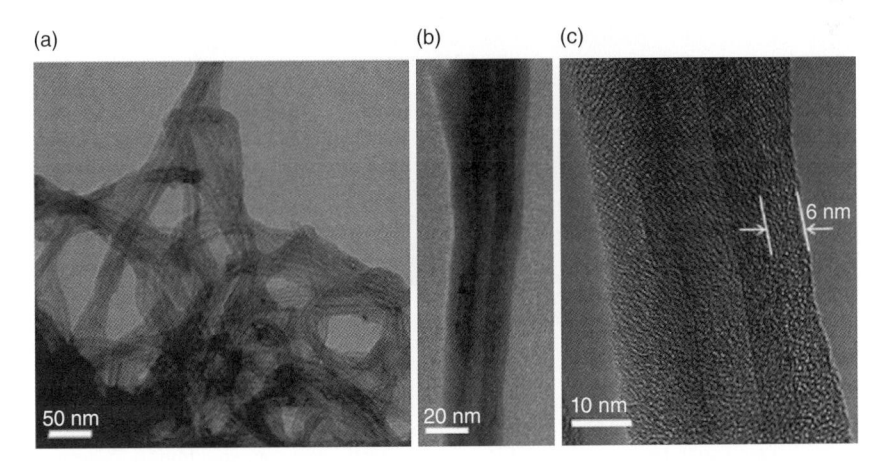

Figure 2.5 TEM images of HPSA hyperbranched polycation assembled on a CNT using LbL technique. Reprinted with permission from H. Kong, P. Luo, C. Gao, D. Yan, *Polymer* **2005**, *46*, 2472. Copyright 2005 Elsevier B.V.

In an interesting study by Liu and coauthors, solution ^{1}H NMR was utilized for the determination of the nature of the top layer of water-soluble poly(ethyleneglycol)-graft-MWCNT coated with LbL assembly of PSS and poly(allylaminehydrochloride) (PAH). The authors reported that due to high dispersibility of the coated CNTs, recording NMR in solution was possible. To obtain clear signals in solution, NMR mobility of the solute is important. After coating the CNT with PAH, the authors observed discrete peaks that were assigned to the protons in PAH. The CNTs were then coated with the PSS layer and only signals ascribable to PSS were observed. Upon adsorption of the next PAH layer, only the PAH proton signals were observed. The authors discuss that this behavior was a result of mobile polymer chains of the top layer of the LbL assembly. Upon adsorption of the polymer with counterchange since only the signals of the top layer were observed, the authors concluded that the polymer chains of the bottom layers lost their mobility due to the formation of PE complexes.

Zykwinska et al. were successful in depositing commonly used PE for LbL on CNTs. Alternative layers of PDDA and PSS were deposited on standard 30–40% pure MWNTs of 10–25 nm diameter and 1–5 µm length [45]. TEM analysis of the number of bilayers suggested that the thickness of the adsorbed PE around the MWNTs increases linearly with the number of bilayers, but plateaus at around 6 nm or so. A nice surface characterization of the coated MWNTs was done to investigate the surface homogeneity of the coated layer. Equal parameters comparison (Figure 2.6) was done between a coated (two bilayers of PDDA/PSS) and a noncoated MWNT lying on a silicon substrate. No clear contrast difference in the AFM phase image was observed between the uncoated MWNT and silicon

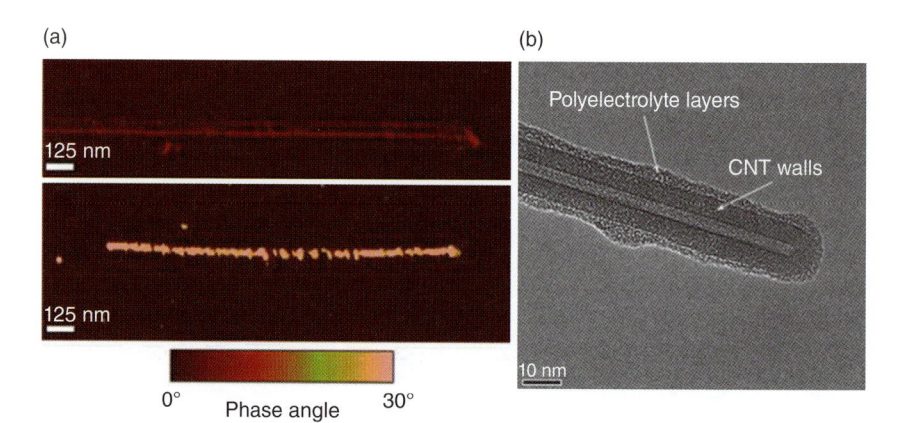

Figure 2.6 (A) AFM phase contrast images (2 µm width) of an uncoated CNT (a) and a functionalized nanotube by two PDDA/PSS bilayers (b) deposited onto the silicon substrate. The same experimental conditions and phase angle scale were used. (B) High-resolution TEM image of MWCNT after coating with four bilayers of PDDA/PSS. Reprinted with permission from A. Zykwinska, S. Radji-Taleb, S. Cuenot, *Langmuir* **2010**, *26*, 2779. Copyright 2010 American Chemical Society.

substrate, conforming high hardness of both materials. Coated MWNT revealed a higher contrast phase shift for the softer polymer region around the nanotube, confirming the uniformity of the wrapped LbL films.

The approach adopted by Rivadulla et al. was primarily targeted toward controlling the conductivity of the nanotubes and adjusting the polymer coating for the application of gate dielectric material for the fabrication of thin film transistors [46]. Figure 2.7 explains the experimental procedure of depositing LbL films of positively charged PAH and negatively charged PSS. To confirm an increase in the number of polymer layers deposited on the CNTs, high-resolution transmission electron microscopy (HR-TEM) was performed. Figure 2.8 showed a gradual increase in the coating thickness in 1.8 ± 0.2 nm steps per layer [46]. LbL film thickness increases as high as 12 nm till seven bilayers, which is higher than maximum thickness of 6 nm observed by Zykwinska et al. As explained in Figure 2.7b, films formed by the LbL-coated CNTs on cellular filter substrates were tested for the temperature dependence of the tunneling resistance by the number of bilayers.

As explained by O'Connell et al., the LbL of CNTs is according to the ionic (noncovalent) bonding based on the thermodynamic preference of CNT–polymer interactions over CNT–water [56]. Because of this, the intrinsic nanotube sp^2 structure conjugation stays undisturbed and hence the original CNT electronic structure kept intact.

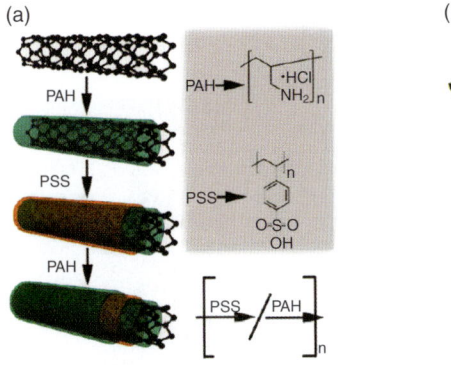

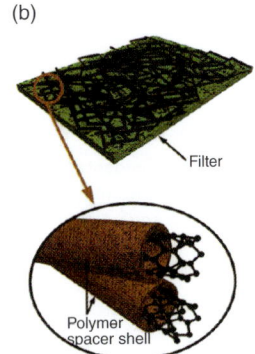

[a]The tube-to-tube distance in the film is determined by the number of layers previously deposited onto the CNT.

Figure 2.7 (a) Illustration of the synthetic process comprising the individual CNT functionalization using PAH as a polymer wrapping agent followed by a sequential deposition of the oppositely charged polyelectrolytes (PSS/PAH) through the LbL self-assembly technique and (b) the latter solution after filtration and formation of the film of CNT on a cellulose filter substrate.[a] Reprinted with permission from F. Rivadulla, C. Mateo-Mateo, M. A. Correa – Duarte, J. Am. Chem. Soc. **2010**, *132*, 3751. Copyright 2010 American Chemical Society.

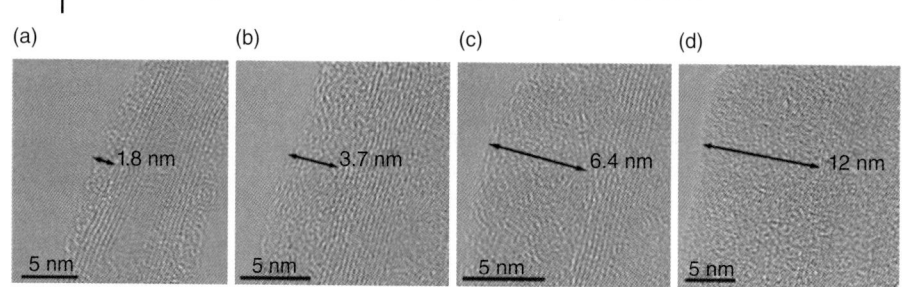

Figure 2.8 HRTEM images of the CNT coated with (a) one, (b) two, (c) three, (d) seven polyelctrolyte layers, corresponding to polymer thickness of 1.8, 3.7, 6.4, and 12 nm, respectively. Reprinted with permission from F. Rivadulla, C. Mateo-Mateo, M. A. Correa – Duarte, *J. Am. Chem. Soc.* **2010**, *132*, 3751. Copyright 2010 American Chemical Society.

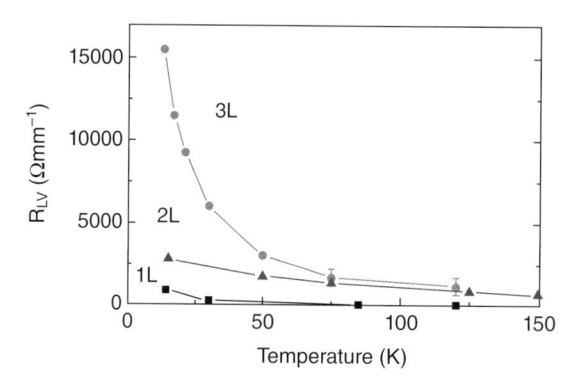

Figure 2.9 Temperature dependence of the low-voltage tunneling resistance in films of CNT wrapped by 1L (black squares), 2L (red triangles), and 3L (green dots) of polymer. Reprinted with permission from F. Rivadulla, C. Mateo-Mateo, M. A. Correa – Duarte, *J. Am. Chem. Soc.* **2010**, *132*, 3751. Copyright 2010 American Chemical Society.

Figure 2.9 shows the dependency of the resistance of CNTs with one, two, and three bilayers of PAH/PSS coating with increase in temperature. An exponential increase in the nonlinear resistance with an increase in the number of polymer layers was observed, which confirms that the electronic tunneling resistance dominates the conductivity in the polymer layer at lower temperatures because of the energy barrier by the polymer layer during the intertube charge transfer [46]. However, thermal energy overcomes this barrier easily at higher temperatures bringing the linear regime.

2.3
LbL Assembly on CNTs Using Click Chemistry

In a novel approach, Gao and coworkers recently reported covalent functionalization of MWNTs by employing click chemistry during the LbL deposition process [57]. In the past decade, click chemistry has received enormous interest since it is a relatively facile reaction with high efficiency and tolerance of the presence of many other functional groups [58, 59]. The combination of LbL technique and click chemistry offers many advantages, such as the use of aqueous as well as nonaqueous (using common organic solvents viz. ethanol, methanol, tetrahydrofuran) conditions. Such conditions for the LbL assembly process permit the use of polymers that are not PE. The triazole linkages resulting from the click reaction between azides and alkynes are also highly stable to hydrolysis, oxidation, and reduction [60]. The residual clickable moieties on the top layer of the LbL assembly can also be utilized for postfunctionalization with a variety of molecules, including dyes, drugs, biomolecules, and polymers [61–63]. In the report, the authors utilized Cu(I)-catalyzed azide/alkyne click reaction, a variation of the Huisgen 1,3-dipolar cycloaddition reaction between azides and alkynes.

The employed clickable polymers poly(2-azidoethyl methacrylate) (poly(AzEMA)) and poly(propargyl methacrylate) (poly(PgMA)) were synthesized utilizing ATRP of 2-azidoethyl methacrylate and reverse addition fragmentation chain transfer (RAFT) polymerization of propargyl methacrylate, respectively. Alkyne-modified MWNTs (MWNT-Alk) [64] were utilized as substrates for the LbL "click" chemistry (LbL-CC) process (Figure 2.10). For the typical LbL-CC process, MWNT-Alk were suspended in DMF and the suspension was deoxygenated by bubbling with nitrogen. PolyAzEMA and polyPgMA were then added successively for clicking the alternating layers on the MWNTs. CuBr, and N,N,N',N',N''-pentamethyldiethylenetriamine (PMDETA) were added under the protection of nitrogen flow. The mixture was stirred at room temperature for 24 h, at the end of which the solid was separated from the mixture by centrifugation. The collected solid was redispersed in DMF and separated by centrifugation. This purification cycle was repeated thrice. After purification, the resulting solid was dried overnight in vacuum, obtaining clickable polymer-functionalized CNTs. The authors assembled three layers using this technique and the layer deposition was monitored using TGA, Fourier transform infrared spectroscopy (FTIR), Raman, X-ray photoelectron spectroscopy (XPS), and TEM. An exponential increase in the content of the clicked polymer layer was reported using the logarithm of the relative mass increment (Inc%) of the functionalized MWNTs as a function of the clicked polymer layer. The authors observed that the thickness of the polymer shell increased considerably with increasing layers of polymer clicked on MWNTs. An average layer thickness of approximately 2.0, 5.0, and 7.5 nm was reported after the deposition of the first, second, and third layer, respectively (Figure 2.11).

Since the increase in mass of the adsorbed polymer is a function of the polymer volume, the linear relationship between the thickness and LbL times is in accordance with the exponential increase in grafted polymer amounts. To study the utility

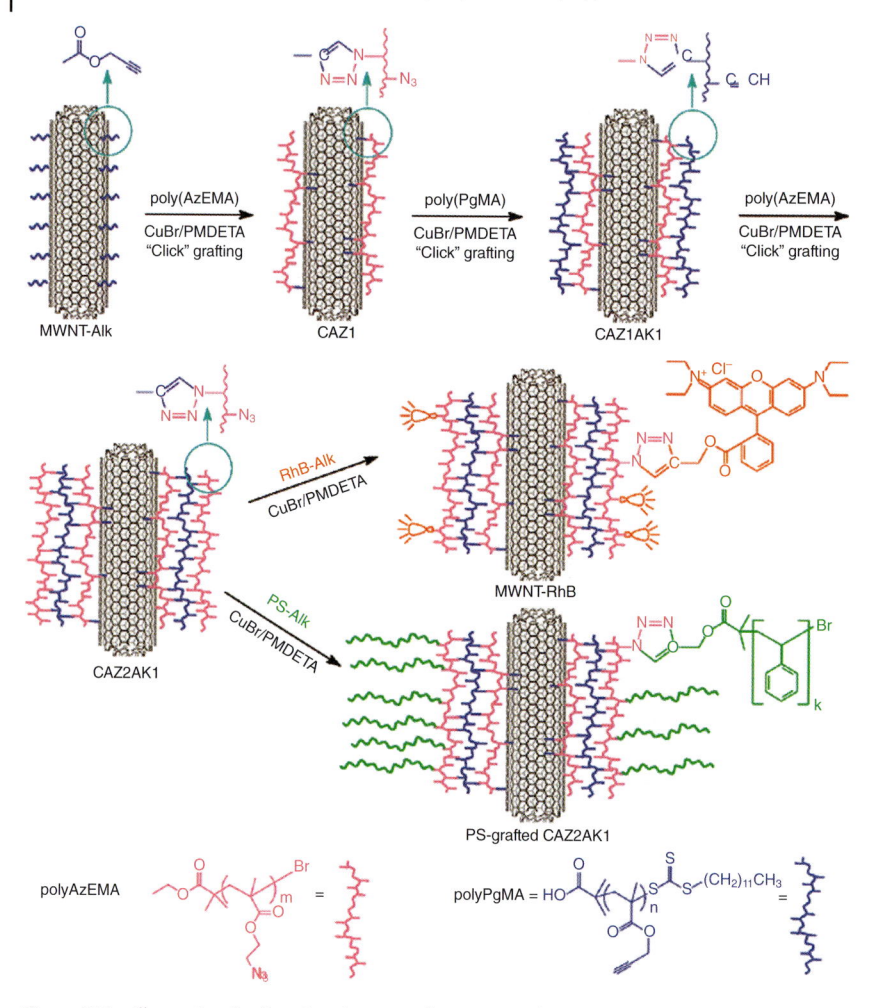

Figure 2.10 Illustration for functionalization of MWNTs utilizing LbL-CC approach. Reprinted with permission from Y. Zhang, H. He, C. Gao, J. Wu, *Langmuir* **2009**, *25*, 5814. Copyright 2009 American Chemical Society.

of the azide groups at the surface of the top layer for further functionalization of the CNT LbL assembly, the authors clicked alkyne-modified polymer (polystyrene) and dye (rhodamine B, RhB-Alk) onto the top layers. Confocal laser scanning microscopy performed on a single tube affirmed the presence of the dye clicked onto the surface of the CNTs (Figure 2.12). This novel approach of LbL-CC assembly not only provides freedom for the choice of the polymers used for LbL assembly, but also introduces a means for further functionalization of the MWNT-LbL assembly via the excess "clickable" moieties present on the top layer of the assembly.

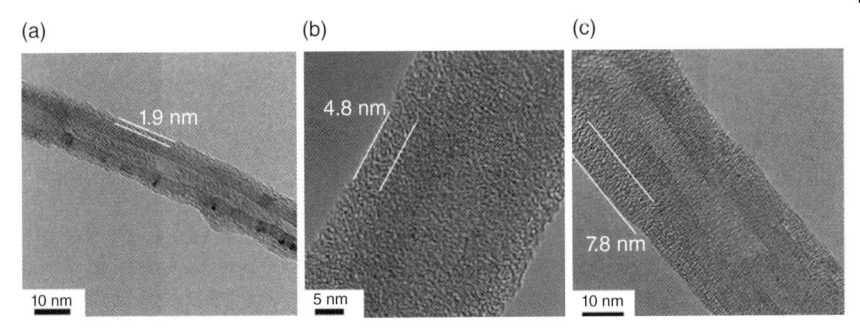

Figure 2.11 TEM images of MWNTs functionalized utilizing LbL-CC approach after the deposition of first layer (a), second layer (b), and third layer (c). Reprinted with permission from Y. Zhang, H. He, C. Gao, J. Wu, *Langmuir* **2009**, *25*, 5814. Copyright 2009 American Chemical Society.

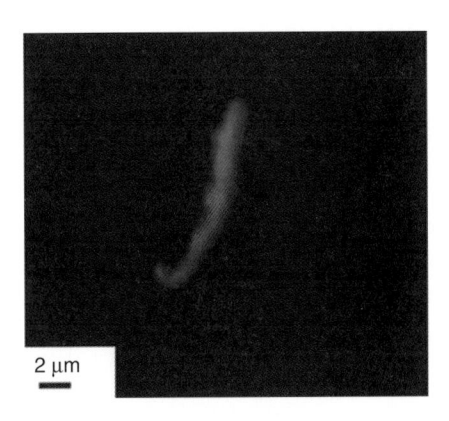

Figure 2.12 Confocal laser scanning image of a single MWNT-LbL assembly functionalized with rhodamine B utilizing LbL-CC approach. Reprinted with permission from Y. Zhang, H. He, C. Gao, J. Wu, *Langmuir* **2009**, *25*, 5814. Copyright 2009 American Chemical Society.

2.4
LbL Assembly on Vertically Aligned (VA) MWNTs

Cui et al. were the first to coat vertically aligned CNTs with LbL assembly of poly(vinylimidazole) (PVI) and nafion to improve the performance in electrochemical sensing of dopamine (DA) levels [65]. MWNTs were vertically grown by a method previously explained in detail by the same group. In brief, Co-coated Ta plates were used and combined with CVD technique to grow high-density vertically aligned MWCNTs with ethylenediamine as a precursor at 880 °C for 5–10 min [66, 67].

Then, the Co-coated Ta plate containing vertically aligned MWNTs is connected to the glassy carbon (GC) electrode. Following it, as shown in Figure 2.13, a derivative of PVI–poly(vinylimidazole) complexes of [Os(4,4-dimethylbpy)2Cl](PVI-

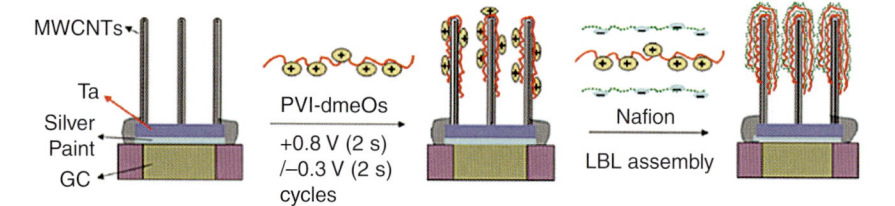

Figure 2.13 The schematic construction of MWCNT electrode and the electrodeposition of PVI-dmeOs and LbL assembly of Nafion/PVI-dmeOs films on MWCNT electrode. Reprinted with permission from H. Cui, Y. Cui, Y. Sun, K. Zhang, W.D. Zhang, *Nanotechnology* **2010**, *21*, 215601. Copyright 2010 IOP Publishing Ltd.

dmeOs) was electrodeposited to provide positive charge and also to establish a good template for LbL deposition. Subsequent deposition of LbL films of nafion (negatively charged polyeletrolyte) and PVI-dmeOs were done and confirmed by FESEM (not shown here), which shows a clear increase in the thickness around the individual MWCNT. The authors used linear sweep voltammetry (LSV) technique to study the DA responses of the bare CNT electrode and the LbL modified electrode. CNTs are supposed to possess strong electrocatalytic activities to DA, but its detection is difficult due to a significant presence of ascorbic acid (AA) in the physiological samples and hence coating the CNTs with nafion under neutral conditions helps as it blocks the response to AA (negatively charged), while still increasing the sensitivity for DA (positively charged) [68, 69].

This is exactly what the authors have expected and observed in the LSV experiments (Figure 2.14) that an increase in the number of LbL bilayers of nafion/PVI-dmeOs would increase the oxidation peak current of DA, hence increasing the sensitivity.

2.5
LbL on CNTs of Biological Molecules

In an interesting study, Guldi and coworkers deposited enzymes on CNTs using electrostatic assembly [70]. The authors first functionalized SWNTs with water-soluble pyrene derivatives with a negatively charged head group. A series of water-soluble metalloporphyrins (MP^{8+} where M = Zn, H_2, Fe, and Co) were then deposited on the SWNTs through a combination of associative van der Waals and electrostatic interactions. Exfoliation of the CNTs from their aggregated forms is critical for the electrostatic assembly process, and the authors reported that deposition of the charged pyrene moieties aided in breaking the CNT bundles. The authors also reported that upon photoexcitation of the nanohybrid systems with visible light, a rapid intrahybrid charge separation causes the reduction in the electron-accepting SWNT and, simultaneously, the oxidation of the electron-donating MP^{8+}. In another approach, Guldi and coworkers first grafted

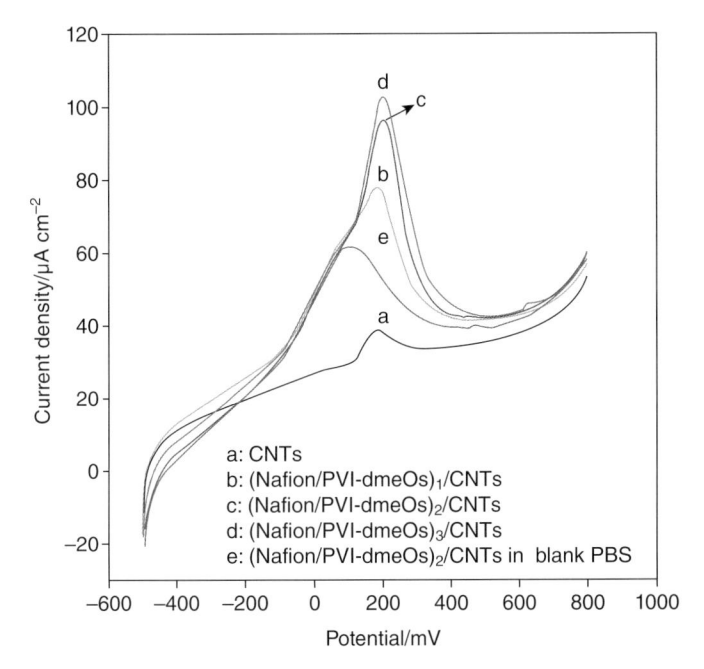

Figure 2.14 LSVs in phosphate buffer solution (PBS)-containing 10 μM DA at CNT without modification (a), or modified with growing (Nafion/PVI-dmeOs)$_n$ films: (b) $n = 1$; (c) $n = 2$; (d) $n = 3$. Curve (e) is the LSV response of blank PBS at the CNT electrode modified with a (Nafion/PVI-dmeOs)$_2$ film. Scan rate: 50 mV/s, scan range: −0.5 to +0.8 V. Reprinted with permission from H. Cui, Y. Cui, Y. Sun, K. Zhang, W.D. Zhang, *Nanotechnology* **2010**, *21*, 215601. Copyright 2010 IOP Publishing Ltd.

poly(vinyl pyridine) onto SWNT (SWNT-PVP), which were then utilized for the deposition of zinc tetraphenylporphyrin (ZnP) via coordination between ZnP and PVP [71]. The authors reported that upon photoexcitation of the nanohybrid, static electron-transfer quenching occurs. This leads to the formation of a radical-cation (single-electron oxidized ZnP), radical-anion (single-electron reduced SWNT) pair.

Using an alternative approach, Mann and coworkers reported an LbL deposition based on programmed biomolecular assembly to produce a multicomponent, multilayered CNT-based conjugate [72]. The conjugate consisted of a MWCNT core coated with four functionalized layers that successively comprised protein-encapsulated iron oxide nanoparticles (biotinylated ferritin (*b*Fn)), the tetravalent biotin-binding protein, streptavidin (SA), 24-base three-stranded biotin-terminated oligonucleotide duplexes, and oligonucleotide-coupled Au nanoparticles (Figure 2.15). The authors obtained biotinylated ferritin-conjugated CNTs (*b*Fn–[CNT]) in the first step by addition of *b*Fn to a dispersion of CNTs in buffer at 70 °C followed by rapid cooling over ice. The authors observed that heating the CNT dispersion for adsorption of *b*Fn was critical since attractive hydrophobic interactions between

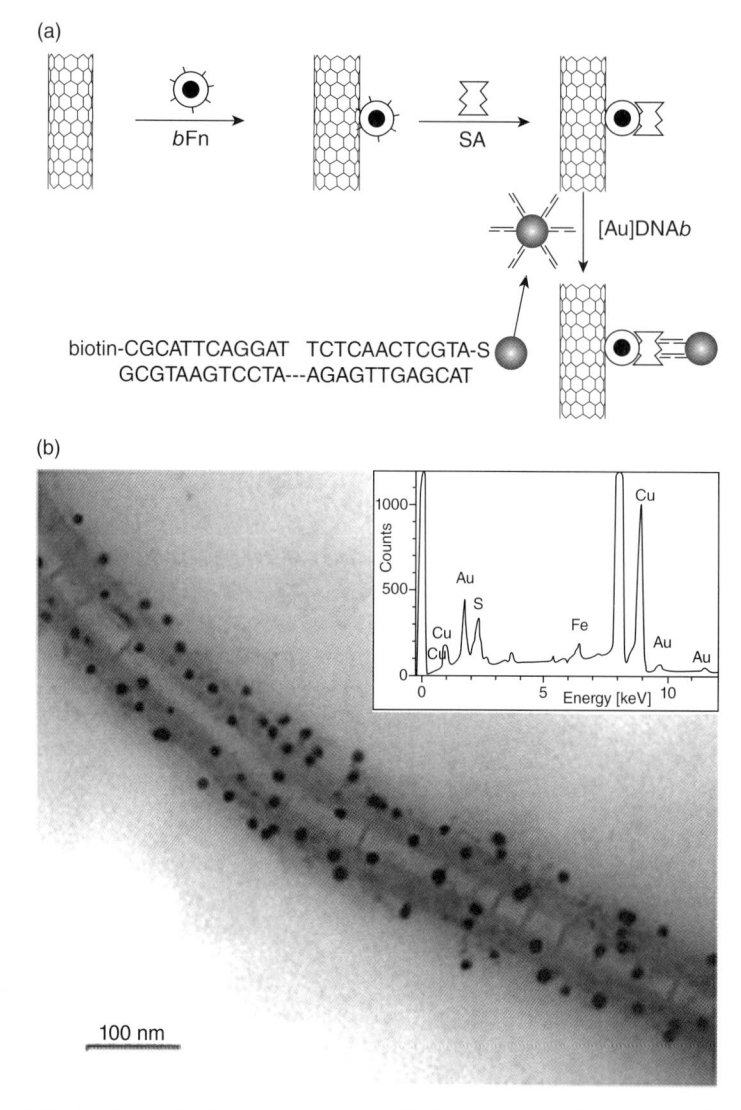

Figure 2.15 (a) Schematic showing stepwise construction of multilayered protein/nanoparticle CNT conjugates. (b) Unstained TEM image of a [Au]–DNAb–SA–bFn–[CNT] hybrid nanoconjugate showing single CNT with assembled Au and ferritin nanoparticles. Inset, EDAX spectrum corresponding to the TEM image showing the presence of Au, Fe, and S (background peaks for Cu are also observed). M. Li, E. Dujardinb, S. Mann, *Chem. Commun.* **2005**, *39*, 4952. Reproduced by permission of The Royal Society of Chemistry.

2.6 LbL on CNTs for Template Development

bFn and the CNT surface were enhanced by heating to 70 °C, and that rapid cooling of the surface-adsorbed protein molecules produced irreversible attachment. Improved hydrophilicity and dispersibility of the CNTs was observed after bFn deposition. The authors obtained protein-coated nanotubes coated with SA in the top layer by addition of SA to a dispersion of bFn-CNT. Delamination of bFn from CNTs was not observed upon formation of SA-bFn-CNT conjugates. On the other hand, the authors observed that the conjugated SA molecules were oriented such that biotin-terminated DNA-functionalized Au nanoparticles ([Au]–DNAb) could be subsequently coupled to residual binding sites exposed on the nanotube surface to generate a [Au]–DNAb–SA–bFn–[CNT] assembly. TEM was utilized for determining the structure of the assembly and the authors observed that the DNA-modified Au nanoparticles were specifically assembled on the protein-functionalized CNT surface (Figure 2.15). Iron oxide cores of conjugated bFn molecules were also observed at a lower contrast in association with the Au nanoparticles. The structure of the nanohybrid was further confirmed using EDAX analysis. The authors also reported disassembly of Au nanoparticles on the surface of the nanohybrids by heating them at 70 °C to cause the denaturation of the 24 base DNA complexes. Reassembly of Au nanoparticles was reported when the nanohybrids were cooled to room temperature.

2.6
LbL on CNTs for Template Development

So far in this chapter, a description of LbL assembly and the application of the modified CNTs are shown, but this section is focused on developing hollow nanotubes of other materials (indium oxide (In_2O_3), nickel oxide (NiO), copper oxide (CuO), etc.) by using LbL assembly on CNT templates. Sol–gel process with CNTs in combination with calcinations was first used to synthesize the metal-oxide nanotubes [73, 74]. This approach did not have a great impact because of toxicity, long reaction time, and the relatively high cost of materials, and hence was not pursued much. In contrast to this, a recent report by Du et al. showed the successful and relatively easy synthesis of porous metal-oxide nanotubes by coating CNT with LbL assembly of respective ionic salt of the metal (Figure 2.16a), followed by calcinations (at 550 °C) to get rid of the CNT template [75]. Confirmation of the porous metal-oxide tubes was done by TEM (Figure 2.16b) and XPS (results not shown here). XRD was also performed to match the peaks with already-published literature of respective metal-oxides and to confirm that no trace of CNTs or metal is left after calcinations.

Some of the In_2O_3 nanotubes synthesized by this method ought to be broken due to high concentration of nanoparticles of size 5 nm around the periphery of the tubes. The primary coating layer of these amorphous nanotubes is thin and the presence of nanoparticles tends to develop some cracks, and hence also significantly enhances the gas sensitivity due to high surface-to-volume ratio. Figure 2.17 shows the gas sensor performance of four different kinds of sensors at room

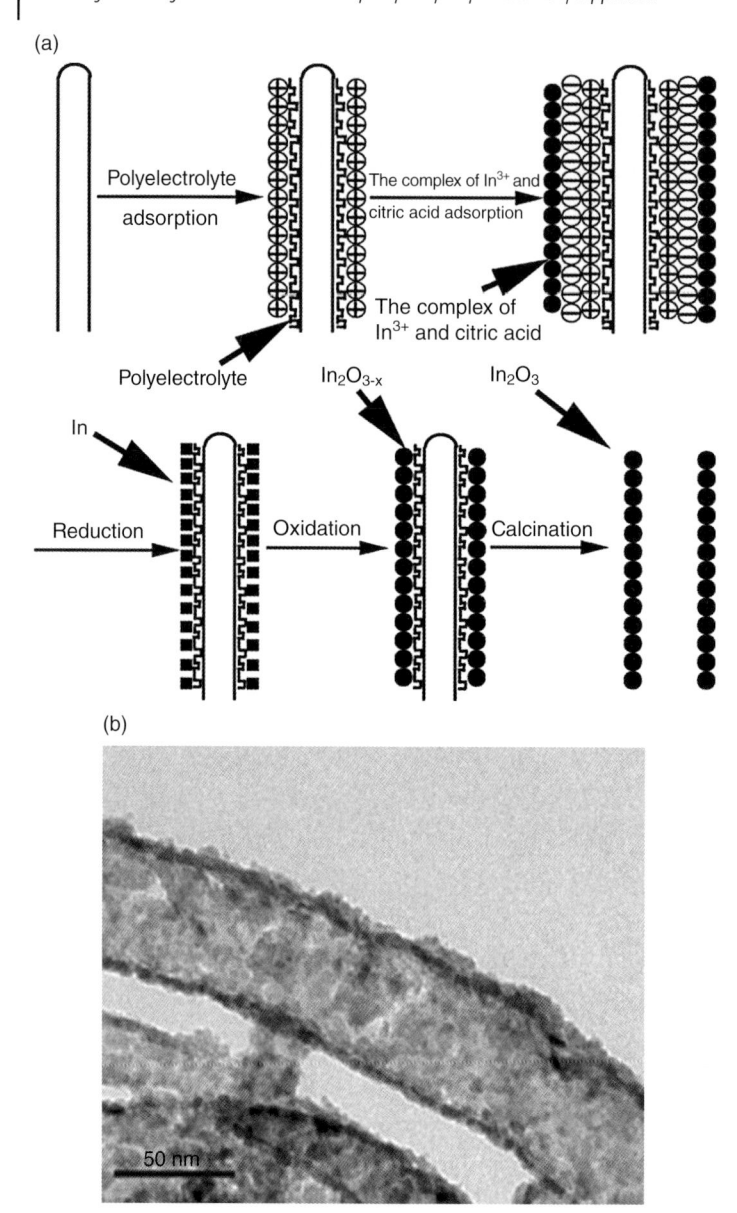

Figure 2.16 (a) Schematic diagram for the growth process of In_2O_3 nanotubes, (b) Hr-TEM images of porous In_2O_3 nanotubes after calcinations that also composed of nanoparticles of about 5 nm. N. Du, H. Zhang, X. Ma, Z. Liu, J. Wu, D. Wang, Porous Indium Oxide Nanotubes: Layer-by-Layer Assembly on Carbon Nanotubes Templates and Application for Room-Temperature NH_3 gas sensors *Adv. Mater.* **2007**, *19*, 1641. Copyright Wiley-VCH Verlag GmbH & Co. KGaA. Reproduced with permission.

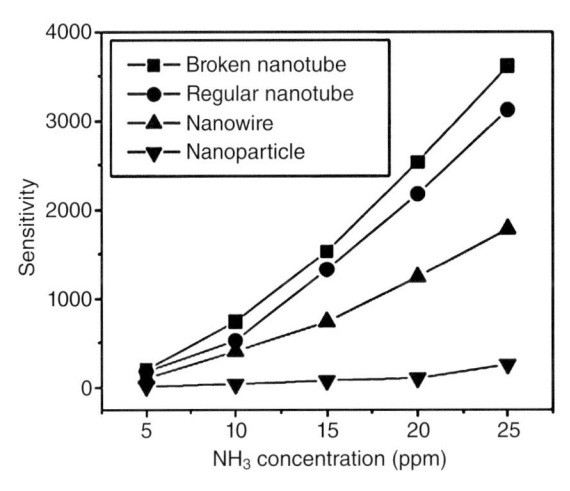

Figure 2.17 Sensitivity response of four different types of gas sensors based on In_2O_3 nanostructure including broken In_2O_3 nanotubes, regular In_2O_3 nanotubes, In_2O_3 nanowires, and In_2O_3 nanoparticles versus ammonia concentration at room temperature. N. Du, H. Zhang, X. Ma, Z. Liu, J. Wu, D. Wang, Porous Indium Oxide Nanotubes : Layer-by-Layer Assembly on Carbon Nanotubes Templates and Application for Room-Temperature NH_3 gas sensors *Adv. Mater.* **2007**, *19*, 1641. Copyright Wiley-VCH Verlag GmbH & Co. KGaA. Reproduced with permission.

temperature based on In_2O_3 nanostructure with ammonia concentration, which clearly show the high sensitivity of broken In_2O_3 nanotubes as compared to nanowires or nanotubes. Broken In_2O_3 nanotubes' sensitivity for NH_3 concentration relatively showed 10% better performance as compared to regular nanotubes. Du *et al.* have also reported the performance of gold (Au) and SnO_2 coating on CNT by LbL assembly for carbon monoxide (CO) gas sensor and showed that the sensor developed by the LbL technique is twice as sensitive to a well-studied [76, 77] Au/SnO_2 nanocrystal hybrid sensor [78].

2.7
Applications of LbL-Modified CNTs

2.7.1
Donor Acceptor Assemblies

Guldi and coworkers recently reported improved photocurrent generation in nanohybrid noncovalent assemblies of polycation-grafted CNT and anionic porphyrins [79]. The authors covalently grafted water-soluble poly-[(vinylbenzyl)trimethylammonium chloride (PVBTA^{n+})] on SWNTs by the free radical polymerization of the monomer in the presence of SWNTs. The authors then assembled polyanionic

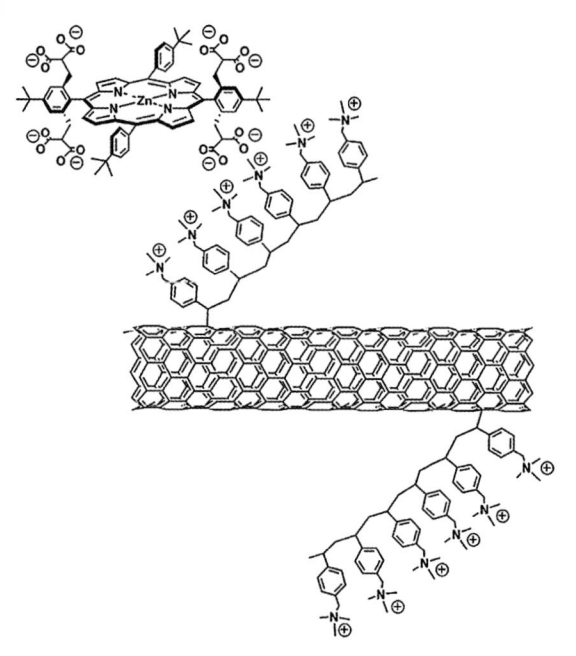

Figure 2.18 Schematic of CNT-PVBTA-ZnP^{8-} nanohybrid. Reprinted with permission from G. M. A. Rahman, A. Troeger, V. Sgobba, D. M. Guldi, N. Jux, D. Balbino, M. N. Tchoul, W. T. Ford, A. Mateo-Alonso, M. Prato, Improving Photocurrent Generation : Supramolecularly and Covalently Functionalized Single-Wall Carbon Nanotubes – Polymer/Porphyrin Donor-Acceptor Nanohybrids. *Chem. Eur. J.* **2008**, *14*, 8837. Copyright Wiley-VCH Verlag GmbH & Co. KGaA. Reproduced with permission.

porphyrins (viz. H$_2$P^{8-} and ZnP^{8-}) on the cationic CNT-PVBTA (Figure 2.18). Soret band (400–450 nm) of the porphyrins was monitored as a function of the added amount of CNT-PVBTA. A bathochromic shift and broadening of the Soret band of the porphyrin ZnP^{8-} was reported by the authors, which suggested the immobilization of the porphyrins (and possible stacking) on the cationic CNTs. Quenching of the porphyrin fluorescence was also reported with increased concentration of the nanotubes. The authors have previously reported electron transfer from enzymes to the carbon nanotubes (*vide supra*). In this case, the authors have also observed this phenomenon upon photoexcitation affording long-lived radical ion pairs with lifetimes as long as 2.2 µs. Photocurrent measurements performed on the noncovalent ZnP^{8-}-CNT-PVBTA displayed high photon-to-current efficiencies of 9.90% for such systems, when a potential of 0.5 V was applied. Such nanohybrids have potential for applications in the ever-important field of photovoltaics. Guldi and coworkers also reported electrostatic assembly of CdTe nanoparticles on SWNTs and MWNTs to afford photoactive nanohybrids [80]. The CdTe nanoparticles act as electron donors, while the SWNTs and MWNTs act as electron acceptors. Internal photon conversion efficiencies (IPCE) of up to 2.3% were reported for such nanohybrids.

2.7.2
Bioapplications

The higher loading of enzymes using LbL assembly on CNTs can be used for a variety of biological applications. Wang and coauthors reported a novel method for dramatically amplifying the electrochemical detection of proteins or DNA based on a stepwise LbL assembly of multilayer enzyme films on a CNT template [81]. The authors reported LbL assembly of PDDA and the enzyme alkaline phosphatase (ALP) on oxidized CNTs. A variety of proteins, antibody, and antigens were deposited on the nanohybrid using electrostatic interactions. By utilizing the coupling of CNT carriers and LbL loading, the authors report an increased ratio of enzyme tags per binding event, which results in a significant amplification factor. Utilizing such amplified bioelectronic assays, the authors reported detection of DNA and proteins down to 80 copies (5.4 aM) and 2000 protein molecules (67 aM), respectively. Due to the increased amplification, the CNT-LbL biolables are suited for the ultrasensitive detection of infectious agents and disease markers.

The CNT-based electrochemical glucose biosensors are typically based on glucose oxidase (GOx), which catalyzes the oxidation of glucose to gluconolactone with evolution of H_2O_2. The amount of glucose can be determined using the electrocatalytic redox measurement of the generated hydrogen peroxide. Since the detection of glucose is inherently dependent on the enzymatic oxidation reaction, immobilization of GOx on CNT is a critical factor for the biosensor development. A flow injection amperometric glucose biosensor based on electrostatic self-assembling GOx on a CNT-modified GC transducer was reported by Liu and Lin [82]. The enzyme GOx was deposited on the negatively charged oxidized CNT by alternatively assembling PDDA and GOx. The optimum potential for the biosensor operation was determined using hydrodynamic voltammetric studies with 5 mM glucose under batch conditions, and to improve the signal-to-noise ratio was chosen to be −100 mV. Typical amperometric responses of the PDDA/GOx/PDDA/CNT/GC electrode for the successive injection of various concentrations of glucose were studied at an applied potential of −100 mV in a flow injection system (Figure 2.19). The signal was expressed as the area under the peak and the resulting calibration curve was generated. A linear calibration curve was observed for glucose concentrations ranging in between 15 µM and 6 mM with a correlation coefficient of 0.992 (Figure 2.20a). Even for low concentrations between 15 and 120 µM, the authors reported good linear relationship (correlation coefficient, 0.997) between the peak area and glucose concentration (Figure 2.20b). The measurements were performed at a pH of 7.4, as the authors observed a decrease in the signal strength with decrease in pH, which they attributed to the leaching out of GOx into the solution. The sensors also displayed a reproducible response and were highly selective.

In a novel application, Gao and coauthors prepared magnetic MWNTs utilizing the LbL self-assembly process [83]. They first grafted poly(2-diethylaminoethyl methacrylate) (PDEAEMA) on MWNTs by MWNT-Br initiated *in situ* ATRP of 2-diethylaminoethyl methacrylate (DEAEMA). The PDEAEMA-grafted MWNTs

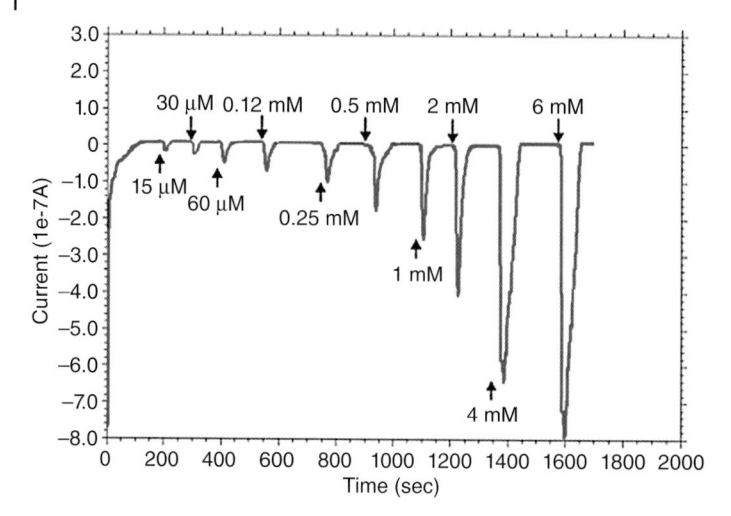

Figure 2.19 Amperometric response of the PDDA/GOx/PDDA/CNT/GC electrode to various concentrations of glucose in the flow injection system. Working potential −100 mV; flow rate 250 μL/min. Reprinted with permission from G. Liu, Y. Lin, *Electrochem. Commun.* **2006**, *8*, 251. Copyright 2006 Elsevier B.V.

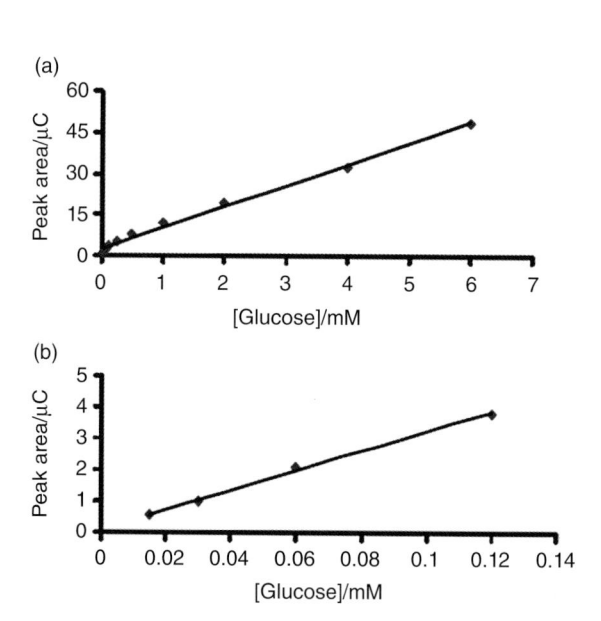

Figure 2.20 (a) The resulting calibration curve in Figure 2.15 at a concentration range from 15 μM to 6 mM and (b) the calibration curve at low concentration range from 15 to 120 μM. The data were obtained from Figure 2.15. Reprinted with permission from G. Liu, Y. Lin, *Electrochem. Commun.* **2006**, *8*, 251. Copyright 2006 Elsevier B.V.

were then quaternized with methyl iodide (CH_3I), resulting in cationic PE-grafted MWNTs (MWNT-PAmI). Magnetic iron oxide (Fe_3O_4) nanoparticles (~10 nm) were assembled onto the MWNT surfaces by electrostatic interactions between MWNT-PAmI and Fe_3O_4, affording magnetic MWNTs. The concentration of the nanoparticles assembled on the MWNTs was adjusted by changing the feed ratio of Fe_3O_4 to MWNT-PAmI. The dispersibility of the magnetic MWNTs was dependent on the final loading of the Fe_3O_4 nanoparticles. For low loading (<8 wt% of iron), the magnetic nanotubes remained dispersed; however, at high loading (>25 wt.% of iron), the magnetic nanotubes precipitated and could not be redispersed. The TEM analysis displayed that the Fe_3O_4 nanoparticles were deposited on the nanotubes as aggregates and uniform surface coverage was not observed.

Magnetic nanomaterials can find important applications in the field of biotechnology and the authors utilized the magnetic nanotubes as a magnetic handle to manipulate blood cells in a magnetic field. The authors employed sheep erythrocytes for the manipulation studies. Magnetic nanotubes were first dispersed via sonication in a phosphate-buffer saline (PBS) solution at a pH of 7.4. The sheep erythrocytes were also dispersed in a separate PBS solution, the dispersed magnetic nanotubes were added, and the mixture shaken for a predetermined time. Using a high-speed CCD camera, the authors reported that in a 2D magnetic field the cells that were attached to the magnetic nanotubes could be rotated at high speeds. These blood cells could also be aligned or moved from one spot to another by utilizing an appropriate magnetic field. The rotation of a sheep erythrocyte attached to a magnetic nanotube bundle is displayed in Figure 2.21. The authors report that this blood cell was rotated clockwise with a speed of 1 cycle per 3 s under the 0.5 Hz and 12.7 kA/m magnetic field. Such techniques introduce possibilities for noninvasive manipulation of cells in the fields of biotechnology and medicine.

2.8
Conclusions

Even though LbL technique has been utilized for obtaining thin films of CNTs, LbL assembly on CNTs as a functionalization protocol has received little attention. CNTs possess unique electronic properties due to their extended π-system, which can be affected upon covalent functionalization of the nanotubes. Noncovalent functionalization of the CNTs such as LbL assembly is an important approach as it preserves the sp^2 hybridization of the carbon atoms and in the process maintains the unique electronic properties of the CNTs. PEs wrap around the CNTs during the LbL process; hence tuning ionic concentration of their solution to alter their persistence length can be critical. The charge density on the PE is also critical in the LbL process as observed in the case of the hyperbranched PE. The LbL assemblies provide the freedom for utilization for a variety of functional materials including polymers, biopolymers, enzymes, nanoparticles, etc. Further study and utilization of LbL assembly on CNTs thus can have a major impact for applications

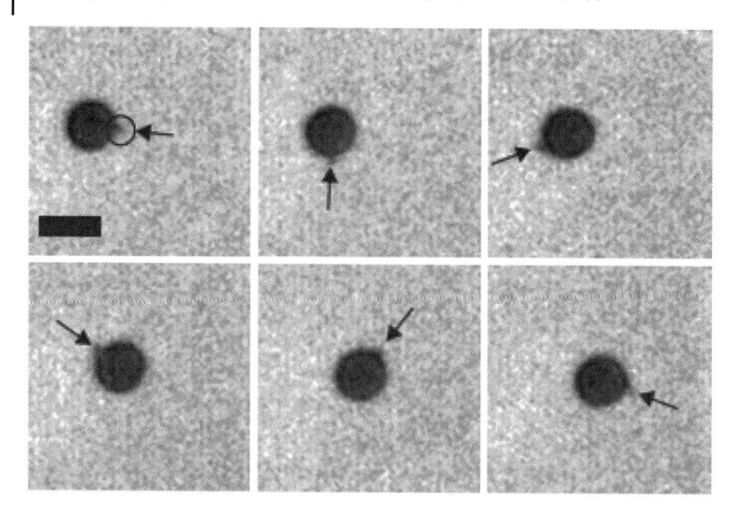

Figure 2.21 Rotational motion of sheep erythrocytes that are attached to magnetic nanotubes. The frequency and intensity of the magnetic field are 0.5 Hz and 12.7 kA/m, respectively. The snapshots correspond to 0th, 0.2th, 0.4th, 0.6th, 0.8th, and 1.0th cycle of the rotational magnetic field. The blood cells are rotated in the clockwise direction. The scale bars are 5 μm. Reprinted with permission from C. Gao, W. Li, H. Morimoto, Y. Nagaoka, T. Maekawa, *J. Phys. Chem. B* **2006**, *110*, 7213. Copyright 2006 American Chemical Society.

in the diverse fields such as materials science, biotechnology, sensing, energy, drug delivery, etc.

References

1 Iijima, S. (1991) *Nature*, **354**, 56.
2 Baughman, R.H., Zakhidov, A.A., and De Heer, W.A. (2002) *Science*, **297**, 787.
3 Hu, T.J., Odom, T.W., and Lieber, C.M. (1999) *Acc. Chem. Res.*, **32**, 435.
4 Kroto, H.W., Heath, J.R., O'Brien, S.C., Curl, R.F., and Smalley, R.E. (1985) *Nature*, **318**, 162.
5 Ajayan, P.M. and Zhou, O.Z. (2001) *Carbon Nanotubes Topics Appl. Phys.*, **80**, 391.
6 Decher, G., Hong, J.D., and Schmitt, J. (1992) *Thin Solid Films*, **210/211**, 831.
7 Haas, H., Decher, G., Mohwald, H., and Kalacbevt, A. (1993) *J. Phys. Chem.*, **97**, 12835.
8 Decher, G., and Hong, J.D. (1991) *Makromol. Chem. Macromol. Symp.*, **95**, 321.
9 Decher, G. (1997) *Science*, **277**, 1232.
10 Ariga, K., Hill, J.P., and Ji, Q. (2007) *Phys. Chem. Chem. Phys.*, **9**, 2319.
11 Brent, P., Kurihara, K., and Kunitake, T. (1992) *Langmuir*, **8**, 2486.
12 Feldheim, D.L., Grabar, K.C., Natan, M.J., and Mallouk, T.C. (1996) *J. Am. Chem. Soc.*, **118**, 7640.
13 Ariga, K., Lvov, Y., Onda, M., Ichinose, I., and Kunitake, T. (1997) *Chem. Lett.*, 125.
14 Lvov, Y., Ariga, K., Onda, M., Ichinose, I., and Kunitake, T. (1997) *Langmuir*, **13**, 6195.

15 Kleinfeld, E.R., and Ferguson, G.S. (1994) *Science*, **265**, 370.
16 Lvov, Y., Decher, G., and Sukhorukov, G. (1993) *Macromolecules*, **26**, 5396.
17 Lvov, Y., Lu, Z., Schenkman, J.B., and Rusling, J.F. (1998) *J. Am. Chem. Soc.*, **120**, 4073.
18 Lvov, Y., Haas, J., Decher, G., Möhwald, H., Mikhailov, A., Mtchedlishvily, B., Morgunova, E., and Vainshtein, B. (1994) *Langmuir*, **10**, 4232.
19 Watanabe, S. and Regan, S.L. (1994) *J. Am. Chem. Soc.*, **116**, 8855.
20 Cheung, J.H., Fou, A.F., Ferreira, M., and Rubner, M. (1993) *Polym. Prepr. ACS Proc.*, **34**, 757.
21 Ferreira, M., Cheung, J.H., and Rubner, M.F. (1994) *Thin Solid Films*, **244**, 806.
22 Cheung, J.H., Fou, A.F., and Rubner, M.F. (1994) *Thin Solid Films*, **244**, 985.
23 Van Cott, K., Guzy, M., Neyman, P., Brands, C., Heflin, J.R., Gibson, H.W., and Davis, R.M. (2002) *Angew. Chem. Int. Ed.*, **41**, 3236.
24 Heflin, J.R., Guzy, M.T., Neyman, P.J., Gaskins, K.J., Brands, C., Wang, Z., Gibson, H.W., Davis, R.M., and Van Cott, K.E. (2006) *Langmuir*, **22**, 5723.
25 Hammond, P.T., and Whitesides, G.M. (1995) *Macromolecules*, **28**, 7569.
26 (a)Sun, Y., Zhang, X., Sun, C., Wang, B., and Shen, J. (1996) *Macromol. Chem. Phys.*, **197**, 147; (b)Wang, X., Kim, Y.G., Drew, C., Ku, B.C., Kumar, J., and Samuelson, L.A. (2004) *Nano Lett.*, **4**, 331.
27 Laurent, D. and Schlenoff, J.B. (1997) *Langmuir*, **13**, 1552.
28 Kurth, D.G., Lopez, J.P., and Dong, W. (2005) *Chem. Commun.*, 2119.
29 Stepp, J. and Schlenoff, J.B. (1997) *J. Electrochem. Soc.*, **144**, L155.
30 Onoda, M. and Yoshino, K. (1995) *Jpn. J. Appl. Phys.*, **34**, L260.
31 Yancey, S.E., Zhong, W., Heflin, J.R., and Ritter, A.L. (2006) *J. Appl. Phys.*, **99**, 1.
32 Onda, M., Lvov, Y., Ariga, K., and Kunitake, T. (1996) *Biotechnol. Bioeng.*, **51**, 163.
33 Yoo, D., Shiratori, S.S., and Rubner, M.F. (1998) *Macromolecules*, **31**, 4309.
34 Shiratori, S.S., and Rubner, M.F. (2000) *Macromolecules*, **33**, 4213.
35 Krozer, A., Nordin, S.A., and Kasemo, B. (1995) *J. Colloid Interface Sci.*, **176**, 479.
36 Lvov, Y., Decher, G., and Mohwald, H. (1993) *Langmuir*, **9**, 481.
37 Glinel, K., Moussa, A., Jonas, A.M., and Laschewsky, A. (2002) *Langmuir*, **18**, 1408.
38 Stuart, M.A.C. and Tamai, H. (1988) *Langmuir*, **4**, 1184.
39 Schlenoff, J.B., Ly, H., and Li, M. (1998) *J. Am. Chem. Soc.*, **120**, 7626.
40 Yang, S.Y. and Rubner, M.F. (2002) *J. Am. Chem. Soc.*, **124**, 2100.
41 Sukhishvili, S.A. and Garnick, S. (2000) *J. Am. Chem. Soc.*, **122**, 9550.
42 Podsiadloa, P., Shima, B.S., and Kotov, N.A. (2009) *Coord. Chem. Rev.*, **253**, 2835.
43 Taylor, A.D., Michel, M., Sekol, R.C., Kizuka, J.M., Kotov, N.A., and Thompson, L.T. (2008) *Adv. Funct. Mater.*, **18**, 3354.
44 Mamedov, A.A., Kotov, N.A., Prato, M., Guldi, D.M., Wicksted, J.P., and Hirsch, A. (2002) *Nat. Mater.*, **1**, 190.
45 Zykwinska, A., Radji-Taleb, S., and Cuenot, S. (2010) *Langmuir*, **26**, 2779.
46 Rivadulla, F., Mateo-Mateo, C., and Correa-Duarte, M.A. (2010) *J. Am. Chem. Soc.*, **132**, 3751.
47 Artyukhin, A.B., Bakajin, O., Strove, P., and Noy, A. (2004) *Langmuir*, **20**, 1442.
48 Chen, R.J., Zhan, Y.G., Wang, D.W., and Dai, H.J. (2001) *J. Am. Chem. Soc.*, **123**, 3838.
49 Mayya, K.S., Schoeler, B., and Caruso, F. (2003) *Adv. Funct. Mater.*, **13**, 183.
50 Huang, S.C.J., Artyukhin, A.B., Wang, Y., Ju, J.W., Stroeve, P., and Noy, A. (2005) *J. Am. Chem. Soc.*, **127**, 14176.
51 Ruths, J., Essler, F., Decher, G., and Riegler, H. (2000) *Langmui*, **16**, 8871.
52 Nierlich, M., Boue, F., Lapp, A., and Oberthur, R. (1985) *J. Physique*, **46**, 649.
53 Spiteri, M.N., Boue, F., Lapp, A., and Cotton, J.P. (1996) *Phys. Rev. Lett.*, **77**, 5218.
54 Artyukhin, A.B., Shestakov, A., Harper, J., Bakajin, O., Stroeve, P., and Noy, A. (2005) *J. Am. Chem. Soc.*, **127**, 7538.
55 Kong, H., Luo, P., Gao, C., and Yan, D. (2005) *Polymer*, **46**, 2472.
56 O'Connell, M.J., Boul, P., Ericson, L.M., Huffman, C., Wang, Y., Haroz, E., Kuper, C., Tour, J., Ausman, K.D., and

Smalley, R.E. (2001) *Chem. Phys. Lett.*, **342**, 265.
57 Zhang, Y., He, H., Gao, C., and Wu, J. (2009) *Langmuir*, **25**, 5814.
58 Kolb, C.H., Finn, G.M., and Sharpless, B.K. (2001) *Angew. Chem. Int. Ed.*, **40**, 2004.
59 Rostovtsev, V.V., Green, G.L., Fokin, V.V., and Sharpless, B.K. (2002) *Angew. Chem. Int. Ed.*, **41**, 2596.
60 Kolb, C.H. and Sharpless, B.K. (2003) *Drug Discov. Today*, **8**, 1128.
61 Tron, G.C., Pirali, T., Billington, R.A., Canonico, P.L., Sorba, G., and Genazzani, A. (2008) *Med. Res. Rev.*, **28**, 278.
62 Lutz, J.F. and Zarafshani, Z. (2008) *Adv. Drug Deliv. Rev.*, **60**, 958.
63 Binder, W.H., and Kluger, C. (2006) *Curr. Org. Chem.*, **10**, 1791.
64 Zhang, Y., He, H.K., and Gao, C. (2008) *Macromolecules*, **41**, 9581.
65 Cui, H., Cui, Y., Sun, Y., Zhang, K., and Zhang, W.D. (2010) *Nanotechnology*, **21**, 215601.
66 Zhang, W.D., Wen, Y., Lin, S.M., Tiju, W.C., Xu, G.Q., and Gan, L.M. (2002) *Carbon*, **40**, 1981.
67 Zhang, W.D., Wen, Y., Tiju, W.C., Xu, G.Q., and Gan, L.M. (2002) *Appl. Phys. A*, **74**, 419.
68 Dresselhaus, M.S. (1992) *Nature*, **358**, 195.
69 Luo, H.X., Shi, Z.J., Li, N.Q., Gu, Z.N., and Zhuang, Q.K. (2001) *Anal. Chem.*, **73**, 915.
70 Guldi, D.M., Rahman, G.M.A., Jux, N., Balbinot, D., Hartnagel, U., Tagmatarchis, N., and Prato, M. (2005) *J. Am. Chem. Soc.*, **127**, 9830.
71 Guldi, D.M., Rahman, G.M.A., Qin, S., Tchoul, M., Ford, W.T., Marcaccio, M., Paolucci, D., Paolucci, F., Campidelli, S., and Prato, M. (2006) *Chem. Eur. J.*, **12**, 2152.
72 Li, M., Dujardinb, E., and Mann, S. (2005) *Chem. Commun.*, **39**, 4952.
73 Satishkumar, B.C., Govindraj, A., Vogel, E.M., Basumalick, L., and Rao, C. (1997) *J. Mater. Res.*, **12**, 604.
74 Satishkumar, B.C., Govindraj, A., Nath, M., and Rao, C.N.R. (2000) *J. Mater. Chem.*, **10**, 2115.
75 Du, N., Zhang, H., Ma, X., Liu, Z., Wu, J., and Wang, D. (2007) *Adv. Mater.*, **19**, 1641.
76 Nelli, P., Fagila, G., Sberveglieri, G., Cereda, E., Gabetta, G., Dieguez, A., Romano-Rodriguez, A., and Morante, J. (2000) *Thin Sold Films*, **371**, 249.
77 Ramgir, N.S., Hwang, Y.K., Jhung, S.H., Kim, H., Hwang, J., Mulla, I.S., and Chang, J. (2006) *Appl. Surf. Sci.*, **252**, 4928.
78 Du, N., Zhang, H., Ma, Z., and Yang, D. (2008) *Chem. Commun.*, 6182.
79 Rahman, G.M.A., Troeger, A., Sgobba, V., Guldi, D.M., Jux, N., Balbino, D., Tchoul, M.N., Ford, W.T., Mateo-Alonso, A., and Prato, M. (2008) *Chem. Eur. J.*, **14**, 8837.
80 Guldi, D.M., Rahman, G.M.A., Sgobba, V., Kotov, N.A., Bonifazi, D., and Prato, M. (2006) *J. Am. Chem. Soc.*, **128**, 2315.
81 Munge, B., Liu, G., Collins, G., and Wang, J. (2005) *Anal. Chem.*, **77**, 4662.
82 Liu, G. and Lin, Y. (2006) *Electrochem. Commun.*, **8**, 251.
83 Gao, C., Li, W., Morimoto, H., Nagaoka, Y., and Maekawa, T. (2006) *J. Phys. Chem. B*, **110**, 7213.

3
Noncovalent Functionalization of Electrically Conductive Nanotubes by Multiple Ionic or π–π Stacking Interactions with Block Copolymers

Petar Petrov and Levon Terlemezyan

In the past decades, both carbon nanotubes (CNTs) [1] and nanotubes of electrically conductive polymers (CPNTs) [2] have attracted considerable attention due to their numerous exiting potential applications in various fields such as advanced materials with electronic and optical properties, molecular devices, sensors, energy storage and transformation, and so on [2–5]. The unique structure, topology, and size of CNTs make their properties exciting compared to the parent, planar graphite-related structures [5]. In particular, the helicity in the arrangement of the carbon atoms in hexagonal arrays on the surface honeycomb lattices, along with the diameter, introduces significant changes in the electronic density of states, and hence provides specific electronic properties of CNTs – they can be either metallic or semiconducting [4].

Similarly, the growing interest in nanotubes based on electrically conductive polymers mainly originates from the fact that their electrical properties are usually different from those of the bulk materials. These phenomena arise from the quantum chemical effects including quantum confinement and finite size effect as well as the nano-sized filler effect. The ability to selectively tune defects, electronic states, and surface chemistry has motivated the development of a variety of methods to fabricate conducting polymeric nanomaterials. By using these fabrication techniques, nanotubes of polypyrrole, polyaniline, polythiophene, poly(3,4-ethyelenedioxythiophene), and poly(p-phenylene vinylene) have been obtained by different research groups and discussed in several review articles [2, 6–8].

The key problem related to the application of CNTs and CPNTs is their pure solubility and processability. In the case of conducting polymers a general procedure of rendering rigid-chain macromolecules soluble is to attach covalently flexible side groups to the stiff polymer backbone. Such approach resulted in the preparation of soluble derivatives, for example, of poly(thiophene), poly(p-phenylene vinylene), and poly(aniline). The modified polyconjugated polymers, however, usually dissolve only in their neutral, that is, undoped state. Upon doping they become insoluble. The idea of preparation of conductive polymers processible in their doped state consists in introducing of processing improving moieties not as side groups attached to the polymer by a covalent bond but rather as an inherent

---

Surface Modification of Nanotube Fillers, First Edition. Edited by Vikas Mittal.
© 2011 Wiley-VCH Verlag GmbH & Co. KGaA. Published 2011 by Wiley-VCH Verlag GmbH & Co. KGaA.

part of the doping anions. In such a case the polymer intractable in its neutral undoped state becomes processable after doping [9].

The surface modification of CNTs with small organic molecules and macromolecules is the commonly applied methodology to enhance their solubility and processability [10, 11]. Apart from improving the chemical affinity of CNTs to polymer matrices, the grafting of polymers is expected to have more benefits because the long polymer chains help to dissolve the tubes into a wide range of solvents even at a low degree of functionalization compared to the low molecular weight functionalities [12]. The modification of CNTs by polymers can be classified into two main categories based on whether the bonding to the nanotube surface is covalent or not. The covalent attachment of macromolecules can be achieved by (i) growing polymers from CNT surfaces via *in situ* polymerization of monomers initiated by chemical species grafted on the CNT sidewalls and CNT edges [13–17]; or (ii) grafting preformed polymers with a defined molecular weight and specific terminal reactive groups or radical precursor by addition reactions [14, 18–21]. The main drawback of covalent functionalization chemistry is that the formation of covalent bonds inevitably causes damages in the original graphene structure and may alter CNT properties, especially their strength and conductivity.

Although the noncovalent CNT modification is based on relatively weak interactions, the adsorption and/or wrapping of polymers [22, 23] to the outer surface of the CNTs effectively disrupt nanotube bundles and provide stable dispersions of individual CNTs. Such nondestructive strategy received increasing attention, since it allows modification at mild conditions without generation of any defects on CNT structure.

3.1
Noncovalent Functionalization of CNTs by π–π Stacking with Block Copolymers Bearing Pyrene Groups

The noncovalent sidewall functionalization of CNTs by π-stacking interactions is one of the gentlest routes to modify their surface properties without introduction of defect sites within the sp^2-hybridized nanotube structure. It has been reported that the stacking between aromatic organic molecules and CNTs via their π-electrons do not significantly perturb the original structure of CNTs, and thus their electronic characteristics [24]. Among various polycyclic aromatic hydrocarbons, pyrene (Py) and its derivatives have shown a remarkable affinity for complexation with CNTs, thus paving the way for immobilization of variety of low molecular weight pyrene derivatives [25–27] as well as pyrene-containing polymers [28–37]. The concept to explore pyrene-bearing molecules for modification of CNTs has several valuable benefits. Since the pyrene molecule can be depicted as a piece of graphene sheet, its electronic structure is quite similar to those of CNTs. Thus, the interaction between pyrene and CNTs strongly resemble the interaction between CNTs themselves and does not alter their electronic properties. On the other hand, one may use the diversity of organic and polymer chemistry to obtain huge variety of pyrene-containing (macro)molecules with desired properties able to bind to both single-walled and multiwalled CNTs. By modification of CNTs with polymers containing pendant pyrene groups, one can preserve the original electronic properties of CNTs and, at the same time, to impart a remarkable stability of CNTs dispersion in organic and aqueous media and to facilitate the preparation of homogeneous CNT-based polymer composites. The method allows incorporation of pyrene moiety into polymers of different nature, thus offering the possibility to finely tune the surface properties of CNTs with respect to their specific affinity to solvents, reactivity, polarity, charge, etc.

Similarly to the covalent modification of CNTs with macromolecules, two general approaches, "grafting from" and "grafting to," have been exploited for noncovalent sidewall functionalization of CNTs via π-stacking interactions. The first method involves adsorption of pyrene-based initiator onto CNT surface and subsequent polymerization. For example, Gómez et al. attached ring-opening metathesis polymerization initiators to SWNTs surfaces and initiated the polymerization of norbornene [28]. Primary limitation of this strategy is that only one pyrene unit can be attached at the chain-end of each macromolecule which, as described below, may be not sufficient to ensure strong binding of the long polymer chains to CNTs and consequently long-term stability of CNTs dispersion. The "grafting to" approach offers more possibilities to design and synthesize (co)polymers of desired composition, architecture, and functionality. Hence, our discussion will be oriented to the utilization of preformed copolymer modifiers, especially the benefit of block copolymer architecture versus other pyrene-carrying (co)polymers.

In an early work Jérôme and coworkers [29] published the concept that various pyrene-containing (co)polymers of different nature can be synthesized and subsequently used for modification of multiwalled CNTs (Figure 3.1). Thus, random poly(methyl methacrylate) (PMMA)-based copolymers (Figure 3.1a) impart high stability to dispersions of MWNTs in toluene, THF, and chloroform, while nanotubes modified with poly(2-(dimethylamino)ethyl methacrylate) (PDMAEMA) (Figure 3.1b) are well dispersed in water. Noteworthy, an excess of MMA (DMAEMA) to the Py-carrying monomer, (1-pyrene)methyl 2-methyl-2-propenoate (PyMMP), was necessary to form a fragment between two adjacent Py units that makes a loop responsible for the stability of CNTs dispersion and, in addition, to ensure certain mobility of anchor molecules for optimal π-stacking with CNTs surface.

The PMMA polymers bearing pyrene moieties randomly distributed along the chain of nearly the same composition were synthesized either by conventional free radical polymerization [30] or by controlled radical polymerization such as atom transfer radical polymerization (ATRP) [29]. The advantages of conventional technique are low capital outlay, simplified procedure and polymers free of metal residues typical for ATRP process. However, the living/controlled polymerization techniques allow preparation of (co)polymers with well-defined molecular weight, tailored polymer architecture, content and position of pyrene, etc., which determines the overall efficiency of modification as well as the potential for further applications.

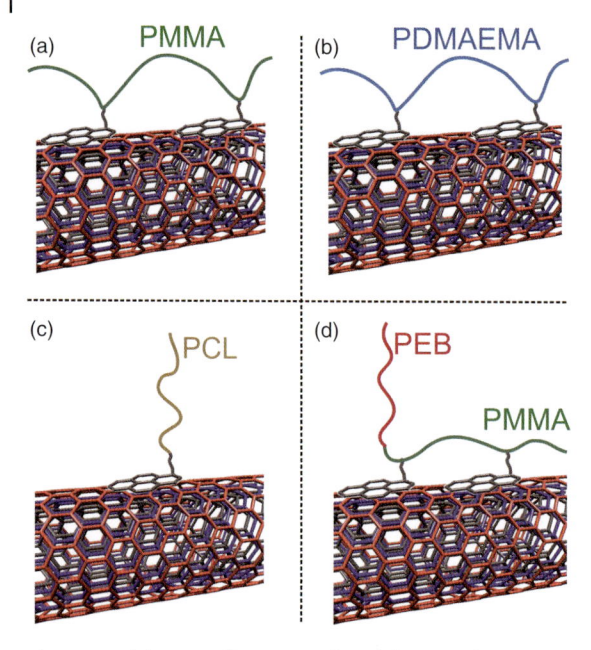

Figure 3.1 Schematic illustration of modification of multiwalled carbon nanotubes by π-interactions with pyrene-bearing polymers of different nature and molecular architecture.

In consequence, pyrene end-capped poly(ε-caprolactone) (PCL) polymers (Figure 3.1c) obtained by ring-opening polymerization of ε-caprolactone from pyrene-based initiator were also found to improve the dispersibility of MWNTs in toluene and THF [29]. The complexation with PCL-Py avoided sedimentation of MWNTs for several days. Interestingly, when a poly(ethylene-co-butylene) chain (PEB) was associated to PMMA-co-PyMMP in a diblock structure (Figure 3.1d; Table 3.1), the corresponding functionalized MWNTs remained individually dispersed for a period as long as several months (Figure 3.2). In this case, a relatively short PMMA-co-PyMMP block containing a few Py units adsorbs onto MWNTs surface, while the longer PEB chains stay extended from the nanotubes surface into solution, providing the steric repulsion and the overall stability of dispersion, respectively.

Among all (co)polymers studied, the use of block copolymer modifiers offers significant benefits concerning the preparation of long-term stable dispersions with exclusively low amount of grafted copolymer (mass of adsorbed polymer/mass of CNTs, calculated by TGA is ca. 0.11) [30].

One should mention that whichever polymer carrying Py group(s) is used for modification, stable dispersions were obtained at very low initial polymer:MWNTs mass ratio (ranging from 2 to 10 (w/w)). Notably, the copolymer excess (nongrafted polymer) can be separated by filtration, recovered and reused for the same purpose. For instance, Choi *et al.* [33] demonstrated by means of TGA analysis that three

Table 3.1 Poly(ethylene-co-butylene) precursor and poly(ethylene-co-butylene)-block-PMMA-co-PyMMP copolymers synthesized by ATRP and used for modification of MWNTs.

Entry	$M_n^{a)}$ (g/mol)	$M_n^{b)}$ (g/mol)	M_w/M_n	Molar composition of PMMA-co-PyMMP block[a]	Py content groups/chain
1 – PEB(Kraton L1203)	4200[c]	–	–	–	–
2 – PEB-b-P(MMA-co-PyMMP)	5460	5650	1.2	0.77:0.23	2
3 – PEB-b-P(MMA-co-PyMMP)	7480	7190	1.1	0.76:0.24	5

a) Calculated from ^{1}H NMR.
b) Measured by GPC.
c) Provided by supplier.

Figure 3.2 Dispersions in toluene of MWNTs modified by poly(ethylene-co-butylene) precursor (1) and poly(ethylene-co-butylene)-block-PMMA-co-PyMMP copolymers (2 and 3), given in Table 3.1.

washing cycles are enough for complete removal of free Py-functionalized poly(styrene-co-maleic anhydride)-*block*-polystyrene (PMAS) from the PMAS-MWNTs dispersion in THF (Figure 3.3). From a practical point of view, the use of Py-copolymer modifiers is a cost-effective method and can be easily adapted for a large-scale production.

A detailed comparison between the adsorption affinity of alpha-pyrene functionalized PMMA and PMMA-*block*-poly(methyl methacrylamide) containing pyrene units (Figure 3.4) for MWNTs has been made by Meuer et al. [35]. The polymers were synthesized via reversible addition–fragmentation chain transfer (RAFT) polymerization and in case of block copolymers, additional polymer analogous reaction to attach pyrene moiety was employed.

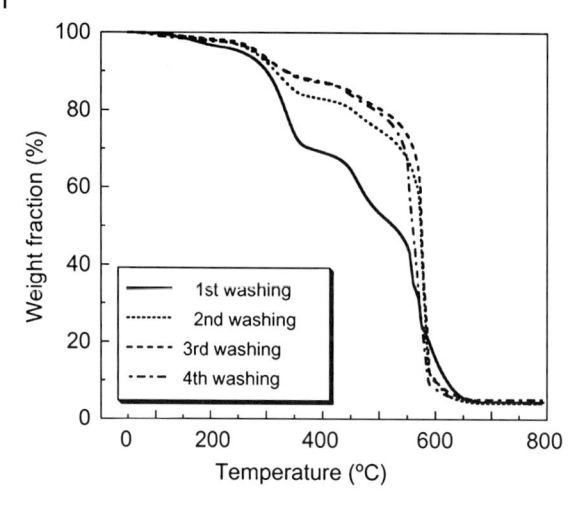

Figure 3.3 Thermogravimetric analyses of carbon nanotubes modified with pyrene-functionalized poly(styrene-co-maleic anhydride)-block-polystyrene after repeated washing procedures. Reproduced from Ref. [33] with permission from Elsevier.

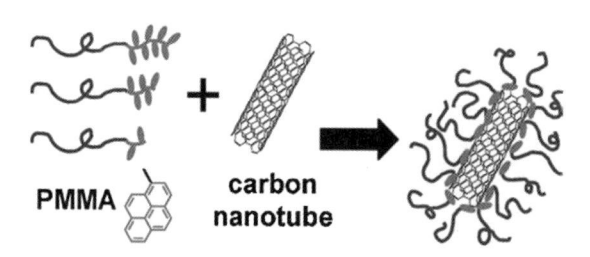

Figure 3.4 Modification of MWNTs with poly(methyl methacrylate)-block-poly(pyren-1-yl-methyl methacrylamide) comprising anchor blocks of different lengths. Reproduced from Ref. [35] with permission from John Wiley and Sons.

These techniques allowed a precise control over the length of both the soluble PMMA and anchor blocks as well as the position of pyrene unit(s) either at the chain-end or at each repeating unit of anchor block. Several important features were emphasized. The amount of the block copolymers adsorbed onto MWNTs is remarkably larger compared to alpha-pyrene functionalized PMMA of similar molecular weight. Obviously, the use of an anchor block due to the multiple binding of the pyrene groups increases the overall binding efficacy. In addition, the amount of adsorbed polymer increases with the increase of anchor block length up to certain extent, and then a decrease is observed. Certainly, for relatively short anchor blocks the entropy loss to fit the surface requirements as a result of the deformation of three-dimensional polymer coils does not contribute too much to the free energy of the system, while with the increase of anchor blocks the entropy factor becomes dominating. Another important tendency, observed with the alpha-pyrene-functionalized PMMA is that the length of the soluble PMMA block plays an important role on the adsorption capability. As a rule, the longer the soluble block, the smaller the grafting density. This phenomenon results from the fact that macromolecules of larger hydrodynamic radii decrease the access to the nanotubes surface and, therefore, diminish the number of chains bound to a single MWNT. On the other hand, the desorption of macrochains, if any, is more likely to happen for longer soluble blocks, as the binding energy of a single anchor unit can be overruled by the entropy win of a free polymer chain in solution.

Specifically for anchor blocks comprising Py at each repeating unit, the insertion of spacer between the polymer backbone and Py group can facilitate the adsorption of block copolymer. Namely, enlarging the spacing of the pyrene from the backbone from one to four C atoms enhanced the adsorption of copolymer by a factor of 1.5 to 3.

Although the prevailing number of studies have been oriented to sidewall functionalization of MWNTs, the "grafting to" method exploring Py-containing copolymers is applicable for modification of single-walled CNTs too [31]. Specifically, Adronov et al. found that a successful adsorption of copolymers can be achieved only onto SWNTs subjected to oxidation/shortening procedure [31, 37] since the as-prepared SWNTs did not exhibit any solubility both in the presence and absence of Py-functionalized polymers. In contrast, when pristine MWNTs were substituted for SWNTs a stable dispersion was formed [37]. Such behavior is attributed to the greater diameter of MWNTs and the decreased surface curvature, respectively, which improves the interaction between the flat pyrene units and nanotubes surface.

Generally, the sidewall functionalization of shortened SWNTs by Py-copolymers of different architecture follows similar regularity as for MWNTs. In particular, the effect of incorporation and distribution of Py within series of well-defined polystyrene-based copolymers on the copolymer/SWNTs interactions were studied [31]. Copolymers having pyrene groups segregated in one block have superior efficacy in dispersing SWNTs when compared to polymers containing an equal amount of pyrene randomly distributed throughout the polystyrene backbone (Figure 3.5).

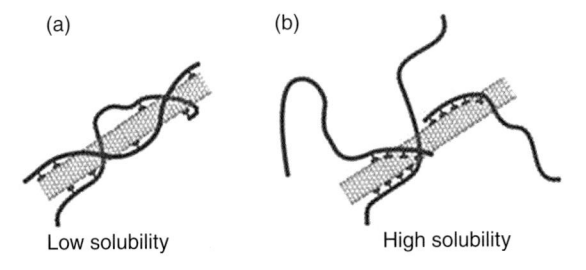

Figure 3.5 Schematic diagram of nanotube interactions with (a) a pyrene-containing random copolymer versus (b) a block copolymer. Reproduced from Ref. [31] with permission from John Wiley and Sons.

The solubility difference imparted by the random copolymers versus the block copolymers should be attributed to the different position of pyrene functionalities being able to π-stack to the nanotube surface. In the case of random PS copolymers, Py triggers anchoring of polymer chains onto nanotubes at random points along the polymer's entire length, as schematically depicted in Figure 3.5a. This pyrene distribution presumably allows relatively short fragments of the polymer chain to form loops that stretch out into THF and provide certain stability of PS–SWNTs dispersion. However, the presence of a long PS block that is free to extend into solution (Figure 3.5b) is considered to play a crucial role in enhancing the dispersibility of SWNTs. Another two factors, the length of pyrene-containing block and the pyrene concentration, have been found to influence notably the capacity of diblock copolymers to adsorb onto SWNTs. Interestingly, the polymer series having shorter pyrene-containing blocks (Mn in the 3000–5000 Da range) have the ability to solubilize larger amount of SWNTs as compared to the polymers of 5–10 times longer blocks of similar composition. Assuming full nanotube surface coverage by the anchor blocks, it is supposed that the number of shorter polymer chains that have an access and bind to the nanotube surface is higher than the number of longer chains (Figure 3.6).

The influence of pyrene concentration in the anchor block is evaluated by means of polymers having equal PS blocks and anchor blocks of different Py/styrene composition. The most concentrated SWNTs dispersion (57 mg/L) was obtained with copolymers bearing Py at each repeating unit, while the incorporation of styrene in between two Py-functionalized units progressively decreased the concentration of SWNTs in THF. It should be mentioned that in this study 1-pyrene butyric acid was used for the synthesis of polymerizable pyrene derivative. This means that the four C atoms between Py-group and polymer backbone act as a spacer and, as proposed by Meuer and coauthors [35], help to overcome the entropy cost for optimal π-stacking with respect to the CNT surface.

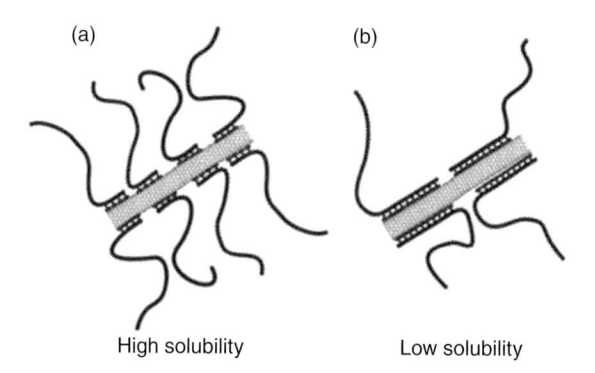

Figure 3.6 Schematic representation of the interaction between short pyrene-functionalized blocks (a) and long pyrene-functionalized blocks (b) in block copolymers. Reproduced from Ref. [31] with permission from John Wiley and Sons.

Interesting is the fact that the interaction between SWNTs and Py-carrying PS copolymers is a reversible process and some nanotube-bound polymer was degrafted during the washing procedure with THF [31]. Taking into account a few reports that Py-functionalized block copolymers cannot be detached from the MWNTs surface by simple washing [29], one may suggest that the strength of complexes of Py-copolymers with single- and multiwalled CNTs are not of the same magnitude. Despite the contribution of polymer chains, it seems that the π-stacking of planar pyrene molecules with the more flattened MWNTs conformation favors stronger interactions as compared to the more curved SWNTs.

Another original work by Bahur and Adronov [37] demonstrated the advantage of blocky structure of linear-dendritic polymer hybrids over the dendrones themselves for enhancing the dispersibility of both SWNTs and MWNTs. The affinity to CNTs of various hybrids comprising linear PS blocks associated to aliphatic polyester dendrons (from G0 to G4 generation), peripherally functionalized with pyrene (from 1 to 16 units), was studied. The pyrene groups at the dendrimer periphery are relatively mobile, even at the higher generations, and are supposed to be capable of orienting themselves around the wall of a CNT for optimal π-stacking interactions (Figure 3.7).

The hybrids bearing one or two pyrene units did not impart any appreciable solubility to the nanotubes. The increased number of pyrene units per polymer chain to four or higher resulted in the formation of stable SWNTs dispersions with concentrations up to 64 mg/mL. Similar trend was found for MWNTs. Thus, for such systems minimum four Py-groups per chain are needed to overcome the tendency for the polymer to dissociate into solution and for CNTs to re-bundle, resulting in predominant immobilization of polymer chains on the nanotube surface. The capability of dendrons themselves to impart solubility to the SWNTs was studied as well. However, relatively dark nanotube dispersion (36 mg/L) was obtained only in the case of G4 dendron, which seems to be large enough to allow a part of the pyrene-functionalized branches to bind to the nanotube surface while

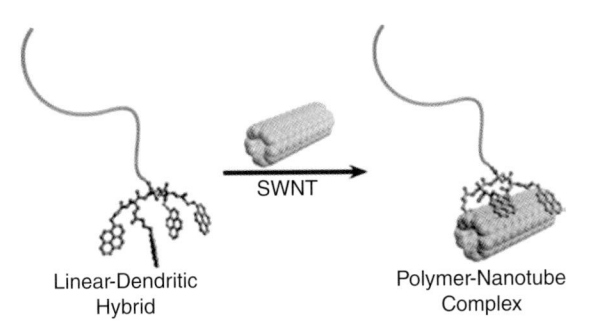

Figure 3.7 Schematic illustration of the interaction between a linear-dendritic hybrid polymer (second-generation dendron) and a SWNT. Reproduced from Ref. [37] with permission from John Wiley and Sons.

others extend into solution, imparting solubility to the dendron–nanotube complex. Indeed, when MWNTs were mixed with the G4 dendron, no nanotube solubility was observed. In this case, it is likely that the larger diameter of the MWNTs, relative to SWNTs, allowed all pyrene units to bind to the nanotube surface. Therefore, the presence of linear PS block, attached to dendritic moiety, was essential for dispersing MWNTs in THF.

3.2
Noncovalent Functionalization of Electrically Conducting Nanotubes by Multiple Ionic Interactions with Double-Hydrophilic Block Copolymers

Among the conducting polymers, polyaniline (PANI) is one of the most promising for electrical and optical applications because of its simple preparation, low cost, good thermal and environmental stability, structure versatility, and controllable chemical and physical properties relating to its oxidation and protonation states [38]. However, like most conducting polymers, PANI is somewhat intractable and is only slightly soluble in a limited number of solvents. The preparation of various forms of colloidal nanoparticles has been used to improve the processability of PANI. Most of the processible forms of PANI involved organic solvents. However, the increasing environmental concerns about these solvents require nontoxic aqueous solution or dispersion processing. Typically, aqueous colloidal dispersions of submicrometer PANI particles have been produced by chemical oxidative polymerization of aniline (ANI) in the presence of various polymeric stabilizers such as poly(ethylene oxide) (PEO), poly(vinyl alcohol), poly(N-vinyl pyrrolidone), poly(vinyl pyridine), poly(vinyl methyl ether), etc. [38–43]. There are a few reports describing synthesis of stable PANI dispersions by employing amphiphilic block copolymer micelles, such as polystyrene-block-poly(ethylene oxide), as the reaction medium [44]. Compared with the PANI nanoparticles, one-dimensional nanostructured PANI including nanofibers, nanowires, nanorods, nanotubes, nanobelts, and nanoribbons presents several advantages in fabricating nanodevices and in preparing nanoscale electrical connections in highly conducting polymer composites [6]. It has been shown that the tubular morphology of the polyaniline is very attractive and plays an important role for the enhanced charge transport across the electrode/electrolyte interface and conductivity, as compared to the conventionally synthesized polyaniline.

To make PANI nanotubes dispersible in water, a strategy based on their surface modification with double-hydrophilic poly(ethylene oxide)-*block*-poly(acrylic acid) copolymer (PEO-*b*-PAA) was reported [45]. Briefly, PANI nanotubes used in this study were prepared by oxidative polymerization of aniline in the presence of two structure-directing agents – salicylic acid (SA) and sodium dodecyl sulfate (SDS). It has been previously shown [46] that, in oxidative polymerization of ANI in the presence only of SA, changing synthesis conditions, especially the molar ratio SA to ANI, the morphology of the PANI obtained can be changed from one-dimensional nanotubes to three-dimensional hollow spheres. When the SA to ANI molar ratio is equal or lower than 0.1, the formation of PANI nanotubes reaches 90%. Moreover, it was found recently [45] that the addition of a very small amount of the surfactant SDS (0.35 mM), much less than the critical micelle concentration, resulted in substantial reduction of the aggregated nanotube morphology on the account of isolated PANI nanotubes, which is the favorable structure from the point of view of further modification. The as-prepared PANI in its "emeraldine salt" form (PANI-ES) was firstly treated with NH_4OH (aq) to obtain PANI in its "emeraldine base" form (PANI-EB) and then subjected to complexation with PEO-*b*-PAA. The complex is formed between the strong basic imine/amine centers of PANI and acidic groups of PAA [47] which resulted in the attachment of PAA blocks onto the PANI nanotubes surface, while the PEO blocks stand water soluble and provide the stability of dispersion. It should be mentioned that PEO has been reported to stabilize PANI dispersions in many cases and to be inefficient in the others due to the inefficient adsorption. PAA has not been found to be a good stabilizer because of the formation of an intermolecular complex between the polyacid and the conducting polymer polycation [48]. In case of PEO-*b*-PAA, the attached copolymer chains improve the stability of the sonicated PANI dispersion as seen from Figure 3.8.

While the PANI-EB dispersion has been completely separated in a 24-h period, the sonicated dispersion of PANI-PEO$_{113}$PAA$_{19}$ is still stable. This was also confirmed by the UV-vis spectra shown in Figure 3.9b. Next to the 2-day storage, the absorbance of the dispersion of PANI-PEO$_{113}$PAA$_{19}$ is much higher as compared to the PANI-EB dispersion. Interestingly, when the PEO$_{113}$PAA$_{19}$ copolymer was added to the aqueous dispersion of PANI-EB, its color immediately changed to green, obviously because of the doping of PANI chains with the carboxylic acid groups of the copolymer. Notably, the content of AA constitutional units in the copolymer is quite low, and the carboxylic acid groups are confined on the copolymer chains. Thus, some of these groups are unable to participate in the doping

Figure 3.8 Visual observation of aqueous dispersions of PANI-PEO$_{113}$ PAA$_{19}$ and PANI-EB next to 24-h storage. Reproduced from Ref. [45] with permission from Springer.

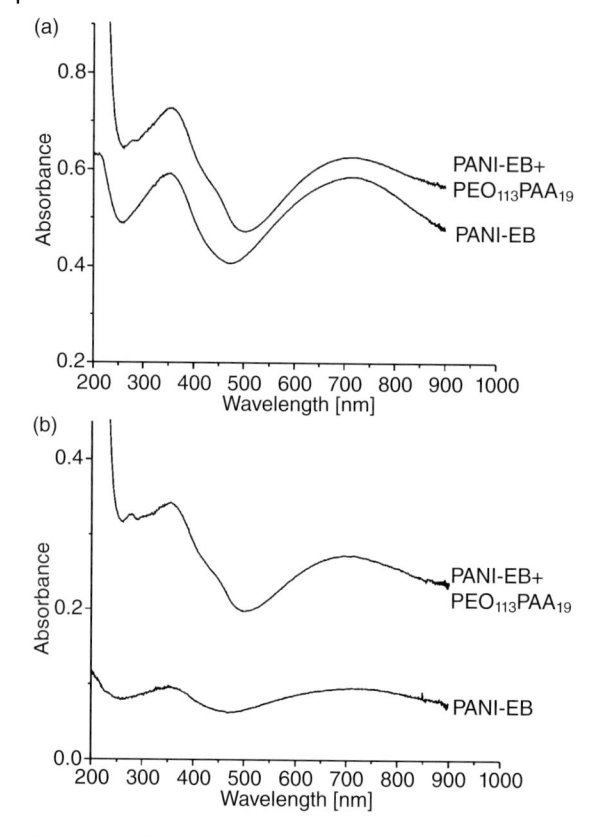

Figure 3.9 UV-vis absorption spectra of aqueous dispersions of PANI-EB (pH = 5) and PANI-PEO$_{113}$PAA$_{19}$ (pH = 3) (a) immediately after preparation; and (b) next to 24-h storage. Reproduced from Ref. [45] with permission from Springer.

[47], and the respective effect is relatively weak, distinctive polaron band at wavelengths beyond 800 nm in the electronic absorption spectrum [49] being hardly visible (Figure 3.9).

Nevertheless, comparing the spectra of PANI-EB and PANI-PEO$_{113}$PAA$_{19}$ shown in Figure 3.9a, some obvious differences can be noted: (i) a shoulder at ca. 440 nm is visible for PANI-PEO$_{113}$PAA$_{19}$, while it is absent in PANI-EB spectrum; and (ii) the absorbance at wavelengths beyond 750 nm is higher for PANI-PEO$_{113}$PAA$_{19}$ as compared to PANI-EB. These two bands are related to the doped protonated form (polaron structure) of PANI [49, 50].

The doping of PANI nanotubes with PEO$_{113}$PAA$_{19}$ was confirmed by conductivity measurements as well. Four-point probe conductivity of pelletized as-prepared PANI-ES obtained by the oxidative polymerization of aniline using simultaneously SA and SDS as structure-directing agents was measured to be 2.1×10^{-3} S cm^{-1}. PANI-EB obtained by the treatment of as-prepared PANI with NH$_4$OH (aq) revealed conductivity ca. 10^{-9} S cm^{-1} (as measured by two-disk method since it

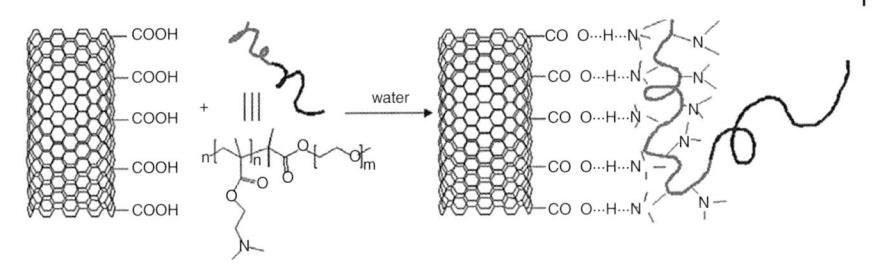

Figure 3.10 Functionalization procedure of MWNTs with PEO-b-PDMAEMA through multisite ionic interactions. Reproduced from Ref. [52] with permission from Elsevier.

is too low to be measured by four-point method). The following treatment of PANI-EB with $PEO_{113}PAA_{19}$ resulted in a product with four-point probe conductivity of 1.2×10^{-3} S cm^{-1}. Incidentally, these bulk conductivity values should be only taken as relative, not as absolute values, since it has been shown that contact resistance between nanotubes in compressed pellets decreases the conductivity by a factor of approximately 10^4 times when compared to an individually probed nanotubes [51].

Double-hydrophilic block copolymers have been successfully used for noncovalent functionalization of MWNTs via ionic interactions [52]. This modification strategy involved an oxidation procedure to generate carboxylic groups onto nanotubes surface and subsequent mixing of oxidized MWNTs with well-defined PEO-b-PDMAEMA block copolymers. The PDMAEMA segments were anchored onto the surface of the oxidized MWNTs through multisite ionic interactions between the amino and the carboxyl groups while the PEO segments formed corona (Figure 3.10) to promote the solubility of MWNTs in different solvents.

The modified tubes form stable dispersions in water, ethanol, and chloroform without any precipitation for at least several months. To confirm that the functionalization is driven by ionic interactions rather than physical adsorption or covalent bonding, the pH of the solution was decreased by adding aqueous solution of hydrochloric acid. Expectedly, the decrease of pH triggered the precipitation of MWNTs. This behavior is ascribed to the neutralization of the amino groups of PDMAEMA blocks, which destroys the ionic interaction between the copolymer and the MWNT resulting in desorption of macromolecules from the nanotube surface.

In conclusion to this chapter, the noncovalent surface modification of nanotubes with copolymers by complex formation effectively disrupts the tendency of as-synthesized nanotubes to aggregate into bundles and provides stable dispersions of individual objects. Such nondestructive strategy has the advantages of modification performed at mild conditions without alteration the native structure of NTs. The use of block copolymer modifiers comprising an anchor block that interact with the tubes through multiple binding and another inert block that stay extended into solution significantly improves the absorption phenomenon and long-term stability of dispersions.

Acknowledgment

The authors thank Mrs. P. Mokreva for the polyaniline-related experiments.

References

1 Iijima, S. (1991) *Nature*, **354**, 56.
2 Jang, J.S. (2006) *Adv. Polym. Sci.*, **199**, 189.
3 Baughman, R.H., Zakhidov, A.A., and De Heer, W.A. (2002) *Science*, **297**, 787.
4 Sgobba, V. and Guldi, D.M. (2009) *Chem. Soc. Rev.*, **38**, 165.
5 Ajayan, P.M. and Zhou, O.Z. (2001) *Carbon Nanotubes. Topics in Applied Physics* (eds M.S. Dresselhaus, G. Dresselhaus, and P. Avouris), Springer, Berlin, p. 391.
6 Zhang, D. and Wang, Y. (2006) *Mater. Sci. Eng. B*, **134**, 9.
7 Wan, M. (2009) *Macromol. Rapid Commun.*, **30**, 963.
8 Sapurina, I. and Stejskal, J. (2008) *Polym. Int.*, **57**, 1295.
9 Pron, A. and Rannou, P. (2002) *Prog. Polym. Sci.*, **27**, 135.
10 Hirsch, A. (2002) *Angew. Chem. Int. Ed.*, **41**, 1853.
11 Bahr, J.L. and Tour, J.M. (2002) *J. Mater. Chem.*, **12**, 1952.
12 Kharisov, B.I., Kharissova, O.V., Gutierrez, H.L., and Méndez, U.O. (2009) *Ind. Eng. Chem. Res.*, **48**, 572.
13 Bahr, J.L., Yang, J., Kosynkin, D.V., and Bronikowski, M.J. (2001) *J. Am. Chem. Soc.*, **123**, 6536.
14 Tasis, D., Tagmatarchis, N., Bianco, A., and Prato, M. (2006) *Chem. Rev.*, **106**, 1105.
15 Viswanathan, G., Chakrapani, N., Yang, H., Wei, B., Chung, H., Cho, K., Ryu, C.Y., and Ajayan, P.M. (2003) *J. Am. Chem. Soc.*, **125**, 9258.
16 Kong, H., Gao, C., and Yan, D. (2004) *J. Am. Chem. Soc.*, **126**, 412.
17 Baskaran, D., Mays, J.W., and Bratcher, M.S. (2004) *Angew. Chem. Int. Ed.*, **43**, 2138.
18 Baker, S.E., Cai, W., Lasseter, T.L., Weidkamp, K.P., and Hamer, R.J. (2002) *Nano Lett.*, **2**, 1413.
19 Blake, R., Gun'ko, Y.K., Coleman, J., Cadek, M., Fonseca, A., Nagy, J.B., and Blau, W.J. (2004) *J. Am. Chem. Soc.*, **126**, 10226.
20 Shaffer, M.S.P. and Koziol, K. (2002) *Chem. Commun.*, 2074.
21 Liu, Y., Yao, Z., and Adronov, A. (2005) *Macromolecules*, **38**, 1172.
22 O'Connell, M.J., Boul, P., Ericson, L.M., Huffman, C., Wang, Y., Haroz, E., Kuper, C., Tour, J., Ausman, K.D., and Smalley, R.E. (2001) *Chem. Phys. Lett.*, **342**, 265.
23 Grunlan, J.C., Liu, L., and Kim, Y. (2006) *Nano Lett.*, **6**, 911.
24 Tournus, F., Latil, S., Heggie, M.I., and Charlier, J.-C. (2005) *Phys. Rev. B*, **72**, art. no. 075431.
25 Chen, R.J., Zhang, Y., Wang, D., and Dai, H. (2001) *J. Am. Chem. Soc.*, **123**, 3838.
26 Nakashima, N., Tomonari, Y., and Murakami, H. (2002) *Chem. Lett.*, **6**, 638.
27 Paloniemi, H., Ääritalo, T., Laiho, T., Liuke, H., Kocharova, N., Haapakka, K., Terzi, F., Seeber, R., and Lukkari, J. (2005) *J. Phys. Chem. B*, **109**, 8634.
28 Gómez, F.J., Chen, R.J., Wang, D., Waymouth, R.M., and Dai, H. (2003) *Chem. Commun.*, 190.
29 Petrov, P., Stassin, F., Pagnoulle, C., and Jérôme, R. (2003) *Chem. Commun.*, 2904.
30 Lou, X., Daussin, R., Cuenot, S., Duwez, A.-S., Pagnoulle, C., Detrembleur, C., Bailly, C., and Jérôme, R. (2004) *Chem. Mater.*, **16**, 4005.
31 Bahun, G.J., Wang, C., and Adronov, A. (2006) *J. Polym. Sci. Part A: Polym. Chem.*, **44**, 1941.
32 Yuan, W.Z., Sun, J.Z., Dong, Y., Häussier, M., Yang, F., Xu, H.P., Qin, A., Lam, J.W.Y., Zheng, Q., and Tang, B. (2006) *Macromolecules*, **39**, 8011.

33. Choi, I.H., Park, M., Lee, S.-S., and Hong, S.C. (2008) *Eur. Polym. J.*, **44**, 3087.
34. Meuer, S., Braun, L., and Zentel, R. (2008) *Chem. Commun.*, 3166.
35. Meuer, S., Braun, L., and Zentel, R. (2009) *Macromol. Chem. Phys.*, **210**, 1528.
36. Morishita, T., Matsushita, M., Katagiri, Y., and Fukumori, K. (2009) *Carbon*, **47**, 2716.
37. Bahun, G.J. and Adronov, A. (2010) *J. Polym. Sci. Part A: Polym. Chem.*, **48**, 1016.
38. Gospodinova, N. and Terlemezyan, L. (1998) *Prog. Polym. Sci.*, **23**, 1443.
39. Vincent, B. and Waterson, J. (1990) *J. Chem. Soc. Chem. Commun.*, 683.
40. Gospodinova, N., Mokreva, P., and Terlemezyan, L. (1992) *J. Chem. Soc. Chem. Commun.*, 923.
41. Stejskal, J., Kratochvil, P., and Helmstedt, M. (1996) *Langmuir*, **12**, 3389.
42. Armes, S.P., Aldissi, M., Agnew, S., and Gottesfeld, S. (1990) *Langmuir*, **6**, 1745.
43. Banerjee, P., Bhattacharyya, S.N., and Mandal, B.M. (1995) *Langmuir*, **11**, 2414.
44. Sapurina, I., Stejskal, J., and Tuzar, Z. (2001) *Colloids Surf. A*, **180**, 193.
45. Petrov, P., Mokreva, P., Tsvetanov, C.B., and Terlemezyan, L. (2008) *Colloid Polym. Sci.*, **286**, 691.
46. Zhang, L.J. and Wan, M.X. (2003) *Adv. Funct. Mater.*, **13**, 815.
47. Chen, S.-A. and Lee, H.-T. (1995) *Macromolecules*, **28**, 2858.
48. Stejskal, J. (2001) *J. Polym. Mater.*, **18**, 225.
49. Stejskal, J., Kratochvíl, P., and Radhakrishnan, N. (1993) *Synth. Metals*, **61**, 225.
50. Nekrasov, A.A., Ivanov, V.F., and Vannikov, A.V. (2000) *J. Electroanal. Chem.*, **482**, 11.
51. Long, Y., Chen, Z., Wang, N., Ma, Y., Zhang, Z., Zhang, L., and Wan, M. (2003) *Appl. Phys. Lett.*, **83**, 1863.
52. Wang, Z., Liu, Q., Zhu, H., Liu, H., Chen, Y., and Yang, M. (2007) *Carbon*, **45**, 285.

4
Modification of Nanotubes with Conjugated Block Copolymers

Jianhua Zou, Jianhua Liu, Binh Tran, Qun Huo, and Lei Zhai

4.1
Introduction

Since their discovery in the early 1990s [1], carbon nanotubes (CNTs) have been extensively studied because of their intriguing electrical, mechanical, optical, thermal, and optical properties. For example, single-walled carbon nanotubes (SWCNTs) can carry an electrical current with a density exceeding 10^9 A cm^{-2}, which is three orders of magnitude greater than that of copper [2]. The density normalized modulus and strength of SWCNTs is ~19 and ~56 times, respectively, higher than those of steel wires [3]. Such exceptional properties make CNTs promising candidates for applications varying from nanoelectronics to sensors, energy conversion and storage devices (fuel cells, photovoltaics, batteries, and supercapacitors), and conductive nanocomposites [4–13]. However, the lack of solubility and processability of CNTs in common solvents and matrices presents one of the major barriers along the path to the CNT commercial products. As-produced CNTs form bundles or aggregates due to the strong inter-CNT van der Waals interactions [14]. Significant effort has been devoted to address this issue with three major approaches being developed to disperse CNTs into solvents or matrices. The first approach is mechanical exfoliation where CNTs are dispersed into a solvent or matrix by mechanical forces such as sonication or mechanical stirring [8, 9, 15, 16]. This method represents the most convenient way to produce CNT dispersions. However, the obtained CNT dispersions are not stable and CNTs precipitate shortly after removing mechanical exfoliation force. As a result, this approach is not suitable for the applications that require a stable CNT dispersion in solution. The second approach involves chemical surface modification of CNTs, which has been proved to be very effective to achieve a stable CNT dispersion. Various methods to covalently modify CNTs have been developed [17], among which the acid oxidation (HNO_3 or $KMnO_4/H_2SO_4$) method is mostly widely applied [18, 19]. Acid oxidation leaves oxygenated functional groups of carbonyl and carboxylic acid on CNT sidewalls and ends [20]. These functional groups may be further reacted, leading to CNTs functionalized with esters, amides, zwitterions, or polymer chains [21–26]. The major disadvantage of chemical modification approach is that the

---

Surface Modification of Nanotube Fillers, First Edition. Edited by Vikas Mittal.
© 2011 Wiley-VCH Verlag GmbH & Co. KGaA. Published 2011 by Wiley-VCH Verlag GmbH & Co. KGaA.

electrical and mechanical properties of CNTs are deteriorated due to the disruption of the conjugated electronic structure and shortening of the CNTs. The third approach utilizes a third component (dispersant) to disperse CNTs. In this approach, CNTs are mechanically debundled through ultrasonication or other methods with dispersants including surfactants [27, 28], polymeric electrolytes [29, 30], proteins [31, 32], DNAs [33], and block copolymers [34–40]. The dispersant binds to the debundled CNTs surface through certain noncovalent interactions (e.g., van der Walls interaction and π–π interaction) and prevents CNTs from aggregating. The third component approach has demonstrated the most promising potential for its high efficiency to disperse CNTs without deteriorating CNT properties. Recently, conjugated polymers including poly(*m*-phenylene vinylene) [41, 42], poly(3-alkylthiophene) [43, 44], poly(arylene ethynylene) [45, 46], polyfluorene [47, 48], and poly(Zn-porphyrin) [49, 50] have emerged as a new class of CNT dispersant. Conjugated polymers strongly interact with CNTs through π–π interactions, which allow conjugated polymers to bind to the CNT surface either through helical wrapping or nonhelical adsorption along CNTs depending on the stiffness of the polymer chain. The advantages of conjugated polymers as CNT dispersants are obvious. First, the long polymer chain helps to disperse CNTs even at low degrees of modification. Second, conjugated polymers with extended conjugation have been proved to strongly interact with CNTs with high interaction energy [51, 52]. Therefore, compared to small molecular dispersants, conjugated polymers help to achieve more homogenous and stable CNT dispersions [49]. Despite their outstanding performance, there is plenty of room to improve the dispersing efficiency of conjugated polymers. Strong π–π intermolecular interactions among the conjugated polymers cause limited solubility of the dispersed CNTs. Furthermore, conjugated polymer dispersed CNTs normally are less practical with deficient functional groups on the CNT surface. To optimize their functionality as CNT dispersants, conjugated polymers need to be modified to increase their solubility and introduce functional groups. To achieve this goal, the most convenient way is to add one block of functional polymer to conjugated polymer to form a conjugated block copolymer. When CNTs are sonicated with a conjugated block copolymer, the conjugated polymer block binds to the CNT surface through π–π interactions and the functional polymer block orients outward from the surface of CNTs and provides the dispersed CNTs with good solubility and stability in a variety of solvents and host polymer matrices. Additionally, the functional block may also introduce various functional groups onto CNT surfaces, providing a noninvasive approach to functionalize CNTs. In this chapter, we present the recent progress in dispersing and functionalizing CNTs using conjugated block copolymers [53–55]. Due to its high binding efficiency to CNT and the well documented synthetic procedure, we focus our discussion on poly(3-hexylthiophene) (P3HT) block copolymers.

The schematic illustration of dispersing and functionalizing CNTs using conjugated block copolymers is depicted in Scheme 4.1. It is believed that the dispersing of CNTs relies on the binding of the conjugated block to the debundled CNT surface through π–π interactions, which disrupts the strong van der Waals interac-

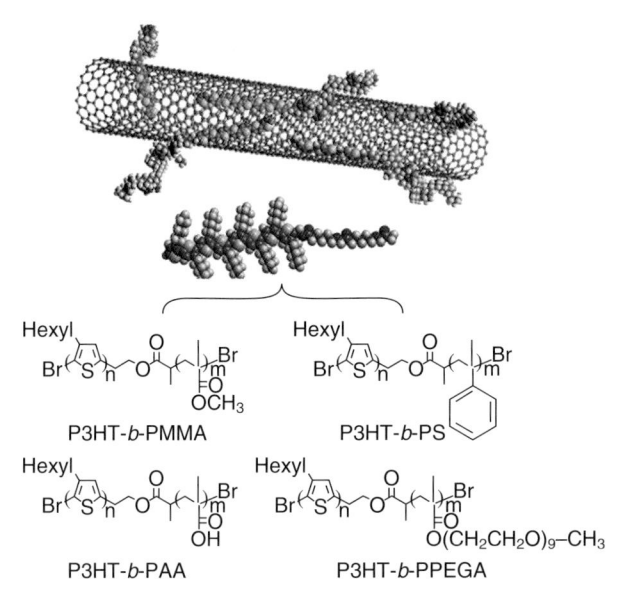

Scheme 4.1 Schematic illustration of dispersing CNTs using P3HT block copolymers. Reproduced from Ref. [55] with permission from John Wiley.

tions between CNTs and prevents them from aggregating. With the conjugated block binding to CNT surface, the functional block with variable composition as listed in Scheme 4.1, will orient outward from CNT surfaces and endow the dispersed CNTs with solubility in a wide variety of solvents, compatibility with host polymer matrices, and desired functionality. The process to disperse and functionalize CNTs using conjugated block copolymers is simple and straightforward, which is described as following. Certain amount of CNTs and block copolymer are mixed in a solvent, and a brief sonication is then applied to the mixture to obtain a homogenous CNT dispersion while the dispersion temperature is controlled below 20 °C. Besides the simple procedure, another advantage of this method is that both the structure and properties of the dispersed and functionalized CNTs remain the same as the pristine CNTs attributed to the noncovalent nature of the method.

4.2
Synthesis of P3HT Block Copolymer

P3HT atom transfer radical polymerization (ATRP) macroinitiator (P3HT-ATRP) was synthesized according to the literature recorded procedure as shown in Scheme 4.2 [56]. The ATRP polymerization of various monomers using P3HT-ATRP as the initiator was then carried out using the detailed procedure described as below [57].

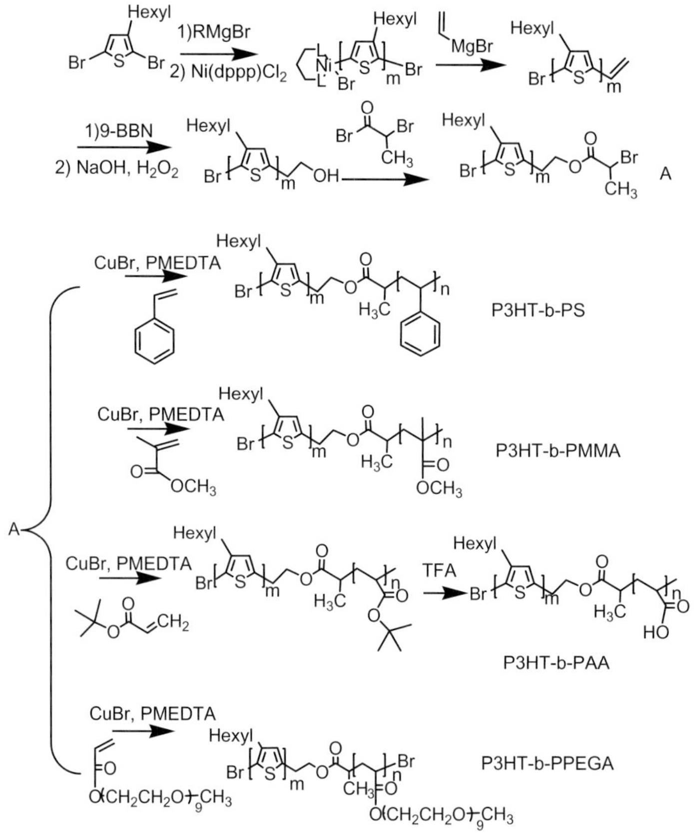

Scheme 4.2 Synthetic scheme of P3HT block copolymers.

4.2.1
Poly(3-hexylthiophene)-b-polystyrene (P3HT-b-PS)

P3HT-ATRP (300 mg, 0.065 mmol), CuBr (9.3 mg, 0.065 mmol), pentamethyldiethylenetriamine (PMDETA) (22.5 mg, 0.13 mmol), and styrene (1.95 g, 19.5 mmol) were dissolved in 3 mL toluene. After being degassed by three cycles of freeze-pump-thaw, the polymerization was carried out at 95 °C for 5 h. Gel permeation chromatography (GPC) measurement of the obtained P3HT-*b*-PS using polystyrene as standard indicated a number average molecular weight (M_n) of 23 000 Da with a molecular weight distribution (M_w/M_n) of 1.3. The molar ratio of 3-hexylthiophene unit to styrene unit was determined to be 1 : 3.3 by ^{1}H NMR spectral analysis.

4.2.2
Poly(3-hexylthiophene)-b-poly(methyl methacrylate) (P3HT-b-PMMA)

P3HT-ATRP (300 mg, 0.065 mmol of ATRP initiator), CuBr (9.3 mg, 0.065 mmol), PMDETA (22.5 mg, 0.13 mmol), and methyl methacrylate (2.6 g, 26 mmol) were dissolved in 3 mL toluene. After being degassed by three cycles of freeze-pump-thaw, the polymerization was carried out at 95 °C for 10 h. The M_n of the P3HT-b-PMMA was determined by GPC to be 15 900 Da with a molecular weight distribution of 1.4. The molar ratio of 3-hexylthiophene unit to methyl methacrylate unit was determined to be 1 : 2.49 by ^{1}H NMR.

4.2.3
Poly(3-hexylthiophene)-b-poly(acrylic acid) (P3HT-b-PAA)

P3HT-b-PAA was synthesized through a hydrolysis of poly(3-hexylthiophene)-b-poly(tert-butyl acrylate) (P3HT-b-PtBA), which was synthesized by following procedure. P3HT-ATRP (300 mg, 0.065 mmol of ATRP initiator), CuBr (9.3 mg, 0.065 mmol), PMDETA (22.5 mg, 0.13 mmol), and tert-butyl acrylate (3.3 g, 26 mmol) were dissolved in 4 mL toluene. After being degassed by three cycles of freeze-pump-thaw, the polymerization was carried out at 105 °C for 20 h. The M_n of the obtained P3HT-b-PtBA was determined to be 27 300 Da with a molecular weight distribution of 1.6. The molar ratio of 3-hexylthiophene unit to tert-butyl acrylate unit was determined to be 1 : 6.8 P3HT-b-PtBA was then hydrolyzed to obtain P3HT-b-PAA. 0.5 mL of trifluoroacetic acid (TFA) was added to 10 mL CH_2Cl_2 solution of 32 mg P3HT-b-PtBA. The hydrolysis was conducted at room temperature with shaking for 24 h. P3HT-b-PAA was collected by centrifugation and purified by washing three times with CH_2Cl_2, and dried under vacuum. The obtained P3HT-b-PAA is not soluble in chloroform, but well dissolved in THF, methanol, and NaOH aqueous solutions.

4.2.4
P3HT-b-poly(poly(ethylene glycol) methyl ether acrylate) (P3HT-b-PPEGA)

P3HT-ATRP (300 mg, 0.065 mmol of ATRP initiator), CuBr (9.3 mg, 0.065 mmol), PMDETA (22.5 mg, 0.13 mmol), and poly(ethylene glycol) methyl ether acrylate (PEGA, MW = 452, Sigma Aldrich) (3 g, 6.5 mmol) were dissolved in 4 mL toluene. After being degassed by three cycles of freeze-pump-thaw, the polymerization was carried out at 100 °C for 10 h. The M_n of P3HT-b-PPEGA was determined by GPC to be 18 200 Da with a molecular weight distribution of 1.3. The molar ratio of 3-hexylthiophene unit to PEGA unit was determined to be 1 : 0.78. P3HT-b-PPEGA is soluble in a wide variety of solvents, such as chloroform, toluene, THF, DMF, acetonitrile, methanol, and ethanol.

4.3
Dispersion of CNTs

4.3.1
Dispersing CNTs using P3HT-b-PS

Both SWCNTs and MWCNTs can be well dispersed by P3HT-b-PS in chloroform. The superiority of P3HT-b-PS over other dispersants was demonstrated by the low polymer to CNT mass ratio to achieve a good CNT dispersion (0.6:1 and 0.5:1 for SWCNTs and MWCNTs, respectively, in chloroform). The highest CNT concentration is 2.5 and 3.0 mg/mL of SWCNTs (2:1 polymer to SWCNTs mass ratio) and MWCNTs (1:1 polymer to MWCNTs mass ratio), respectively, which is significantly higher than most reported CNT dispersions. The CNTs are stable in the dispersion for over one year without precipitating out. The CNT dispersion was testified by centrifuging at 13 200 rpm for 30 min. After centrifugation, most SWCNTs remained in the dispersion, while most MWCNTs sedimentated due to their larger dimensions. The sedimented MWCNTs can be collected, which enables us to remove the free P3HT-b-PS in the dispersion. The collected MWCNTs can be redispersed in chloroform, toluene, and THF. The obtained dispersion is as stable as the dispersion before purification, indicating that the binding of P3HT-b-PS to CNT surface is not degraded by the centrifugation and redispersing process. The stability of the CNT dispersion is attributed to the presence of PS block on CNT surface, which dissolves well in the solvent and stabilizes the CNTs. As a control experiment, P3HT homopolymer was applied to disperse CNTs. The results indicated that more P3HT homopolymer was required to disperse CNTs. The obtained CNT dispersion was not as stable as P3HT-b-PS/CNT dispersion. The sedimentation of the CNTs happened in a few days. The control experiment clearly demonstrated that the PS block played a critical role to obtain a homogenous and stable CNT dispersion.

P3HT-b-PS dispersed MWCNTs (1:1 mass ratio) were purified by three cycles of centrifuge and redispersion to remove the free P3HT-b-PS and then examined by transmission electron microscopy (TEM) and energy dispersive X-ray (EDX) analysis. As shown in Figure 4.1a, MWCNTs exist as individual tubes instead of aggregations. The high resolution TEM image of a MWCNT segment (Figure 4.1c) clearly reveals the presence of a 2 nm thick amorphous coating on MWCNT surface in contrast with the clean surface of pristine MWCNTs (Figure 4.1d). The thin layer coating is believed to be P3HT-b-PS attached on CNT surface since the EDX spectrum in Figure 4.2 clearly shows the sulfur signal arising from P3HT-b-PS.

P3HT-b-PS dispersed SWCNTs were also investigated by TEM. As shown in Figure 4.1b, most SWCNTs have a diameter of 5–6 nm, which are the completely debundled individual SWCNTs coated with P3HT-b-PS since the diameter corresponds well with the thickness of the copolymer coating (about 2 nm) and the diameter of the individual SWCNTs (0.8–1.2 nm). Another group of SWCNTs, which are the minor component, have a diameter of 10–15 nm and are believed to be SWCNT bundles that are not completely dispersed. The P3HT-b-PS dispersed SWCNTs were further examined by atomic force microscope (AFM). The AFM image and the height analysis clearly suggest that most SWCNTs were dispersed into individual tubes (Figure 4.3).

The proposed mechanism of dispersing CNTs by P3HT-b-PS was investigated by ^{1}H NMR. Previous studies revealed that interaction between polymer and CNTs led to the broadening and reduced intensity of peaks in the ^{1}H NMR spectrum of the polymer. The closer the proton is to the surface of CNTs, the broader and

(a)   (b)

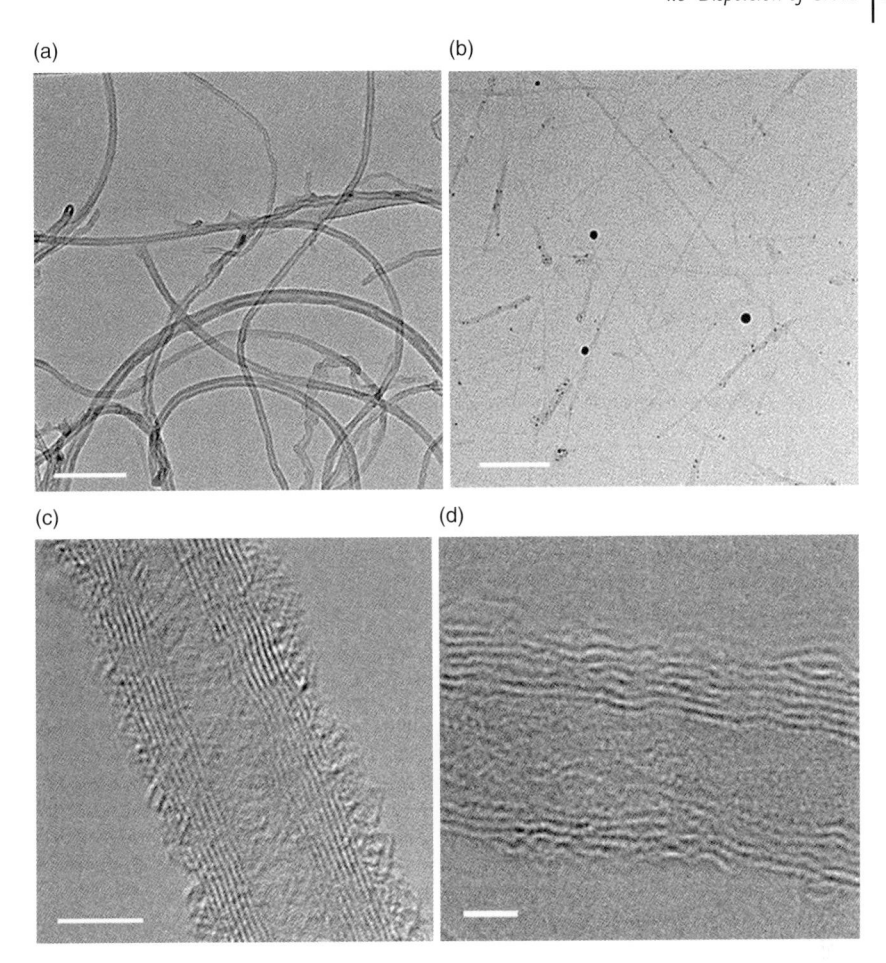

(c)   (d)

Figure 4.1 TEM images of (a) P3HT-*b*-PS/MWCNT (1:1, mass ratio); (b) P3HT-*b*-PS/SWCNT (3:1); HRTEM images of (c) MWCNT covered by a thin layer of P3HT-*b*-PS; (d) pristine MWCNT. The scale bars in a, b, c, and d are 100, 100, 5, and 2 nm, respectively. Reproduced from Ref. [53] with permission from John Wiley.

less intensity the peak will be [41, 45]. Figure 4.4 shows the ^{1}H NMR spectra of P3HT-*b*-PS and P3HT-*b*-PS/MWCNTs with proton assignment. The signal of the proton on the thiophene ring (7) of the P3HT block, which is observed in P3HT-*b*-PS spectrum disappears in the spectrum of P3HT-*b*-PS/SWCNT due to the broadening and overlapping with signals from the PS block. The broadening of the signals of proton 1 and 6 on P3HT block is also observed as indicated by the dash line. In contrast, the signals from PS block (8) are only slightly broadened indicating that P3HT-*b*-PS assembles on the CNT surface in such a way that the P3HT block interacts closely with CNTs, while the PS block orients away from the surface of CNTs. The conclusion is further quantitatively supported by the relative signal intensity of proton 1, 6, and 8. For P3HT-*b*-PS, the ratio of 1, 6, and 8 is

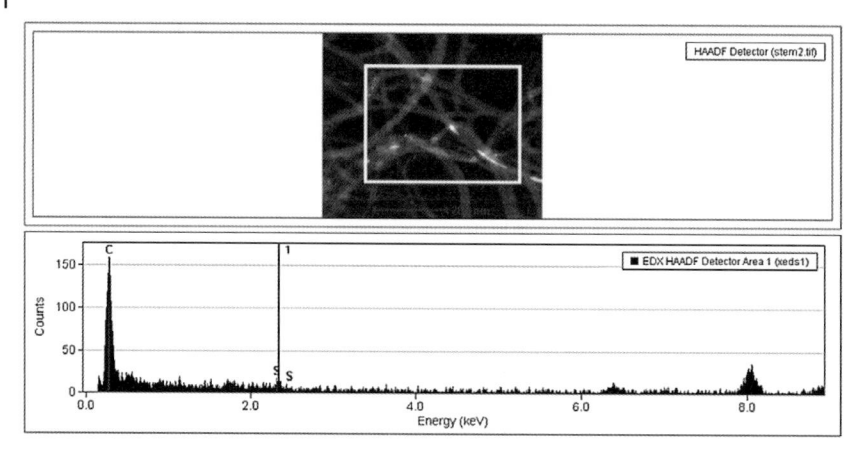

Figure 4.2 Dark field TEM image of P3HT-*b*-PS dispersed MWCNTs and EDX spectrum from the area of the TEM image marked by rectangle.

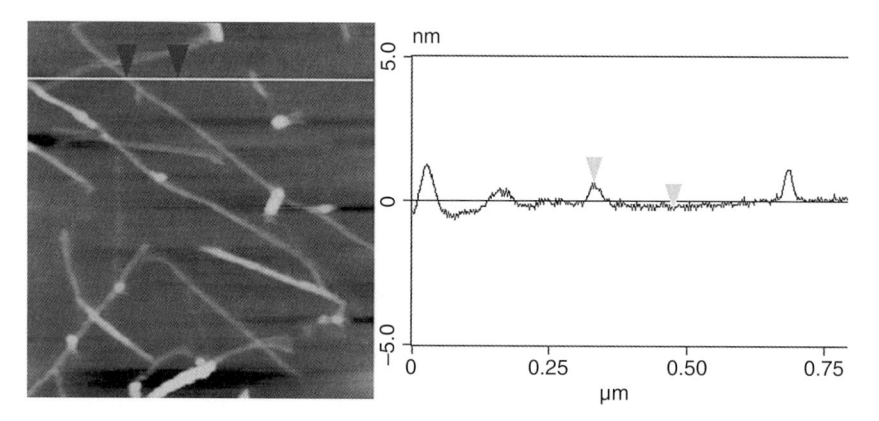

Figure 4.3 AFM image of P3HT-*b*-PS dispersed SWCNTs ($1.0 \times 1.0\,\mu m^2$, 5 nm z-scale) with the section analysis.

2:3:15.5, while the ratio is 1.6:2.9:15.5 for P3HT-*b*-PS/SWCNT. The signals intensities of protons 1 and 6, especially 1, which is the closest one to thiophene ring are greatly reduced, suggesting that the π–π interaction between the P3HT block and CNTs leads to the binding of P3HT-*b*-PS to CNT surface, and the PS block provides the dispersed CNTs with solubility and stability in solvents.

The assembly structure of P3HT on the CNT surface was also studied by investigating the UV-Vis spectrum of the P3HT-*b*-PS/SWCNT dispersion. As shown in Figure 4.5a, P3HT-*b*-PS exhibits an absorption band at 454 nm, revealing the regioregular nature of the P3HT block in the copolymer [58]. The UV-Vis absorption band of a solid-state P3HT-*b*-PS film shifts to longer wavelength, and splits into three peaks (513, 560, and 604 nm) (Figure 4.5a), which is due to the increased coplanarity and extended π-conjugation length of the P3HT block in the solid state [58, 59]. The UV-Vis absorption spectrum of P3HT-*b*-PS/SWCNT dispersion

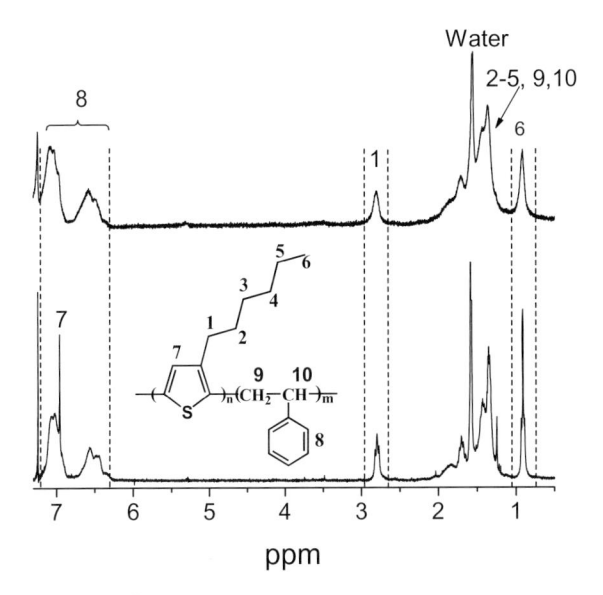

Figure 4.4 ¹H NMR spectra of (a) P3HT-b-PS/SWCNT dispersion and (b) P3HT-b-PS. Reproduced from Ref. [53] with permission from John Wiley.

exhibits similar features to that of the solid-state P3HT-b-PS with three main absorption peaks at 513, 560, and 604 nm (Figure 4.5b). This interesting result indicates that the P3HT block binds to the CNT surface with a coplanar geometry. Therefore, the P3HT block nonhelically binds along instead of helically wrapping around the CNTs. The UV-Vis spectrum of the P3HT-b-PS/SWCNT solid thin film is identical to that of P3HT-b-PS/SWCNT in solution, indicating that the assembly structure of P3HT block on SWCNTs surface is not disrupted by solvent drying due to the strong π–π interaction. Actually, the dispersed CNTs can be repeatedly dried and redispersed back to solution by a brief sonication without noticeable decrease of the dispersion quality.

The π–π interaction between P3HT block and CNTs is further evident from the substantial fluorescence quenching of P3HT-b-PS after being dispersed with CNTs. Figure 4.6a shows the picture of P3HT-b-PS, P3HT-b-PS/SWCNT, and P3HT-b-PS/MWCNT (all three samples have same P3HT-b-PS concentration) in chloroform under an irradiation of UV light. Both SWCNTs and MWCNTs substantially quench the fluorescence of P3HT-b-PS. Quantitatively, 97% fluorescence emission of P3HT-b-PS is quenched after being dispersed with SWCNTs as shown in Figure 4.6b. The fluorescence quenching is due to the electron transfer and/or energy transfer between P3HT-b-PS and CNTs [60]. A control experiment of dispersing SWCNTs by P3HT homopolymer with the same P3HT to SWCNT weight ratio shows that 75% of the fluorescence emission is quenched, indicating stronger interaction of P3HT-b-PS with CNTs than P3HT homopolymer. The stronger interaction also contributes to the higher efficiency of P3HT-b-PS in dispersing CNTs compared to P3HT.

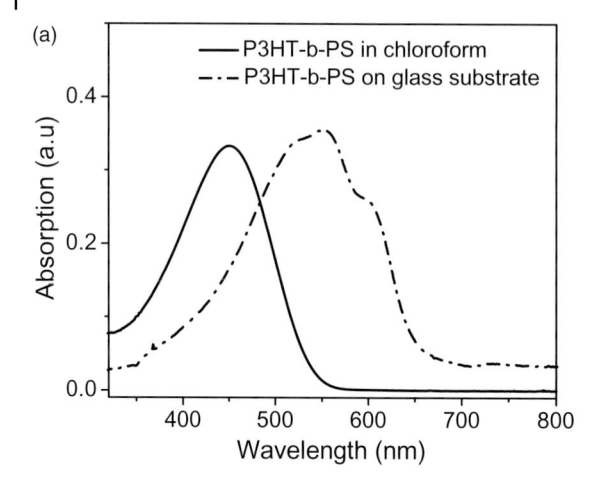

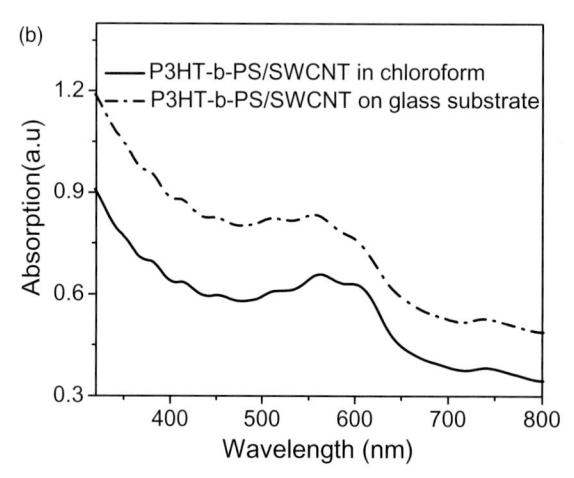

Figure 4.5 UV-Vis spectra of (a) P3HT-*b*-PS in chloroform and a cast film on a glass substrate; (b) P3HT-*b*-PS/SWCNT (2:1) in chloroform and a cast film on a glass substrate. Reproduced from Ref. [53] with permission from John Wiley.

Raman spectroscopy was applied to investigate the interaction between P3HT-*b*-PS and SWCNTs since Raman modes of the SWCNTs is affected by the adsorption of foreign molecules. As shown in Figure 4.7, the tangential vibrational mode (G band) of SWCNTs shifts significantly from 1579.9 (pristine SWCNT) to 1595.6 cm^{-1} and the radical breathing modes (RBMs) shifts slightly from 179.8 to 181.7 cm^{-1} after being dispersed by P3HT-*b*-PS. A previous study indicated that the G band and RBM band of SWCNTs shifted to higher frequency when SWCNTs lost electrons [61]. Therefore, the observed band shift suggests a molecular-level interaction between CNTs and the block copolymers.

It has been clearly demonstrated that the superior dispersing ability of P3HT conjugated block polymer is due to that the P3HT block binds to CNTs surface

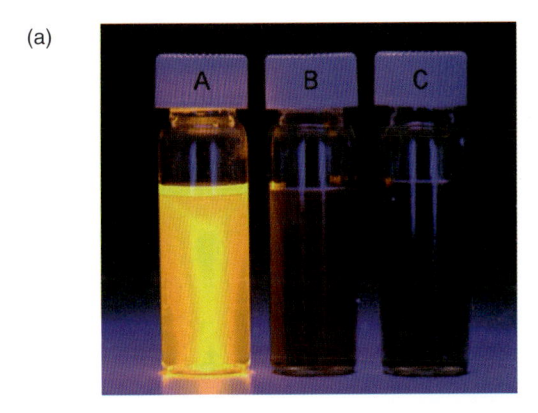

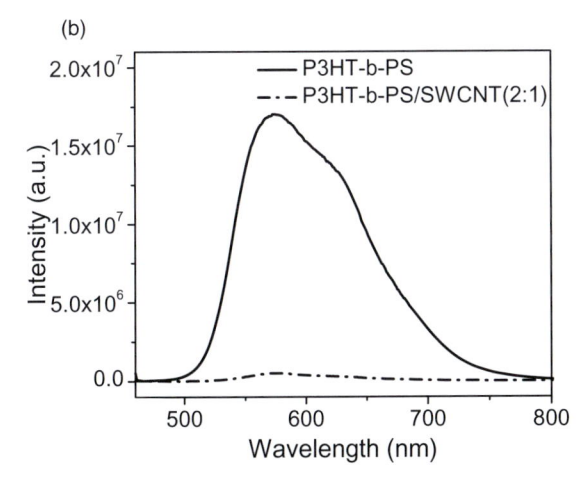

Figure 4.6 (a) Pictures of P3HT-b-PS (A), P3HT-b-PS/SWCNT (2:1) (B), and P3HT-b-PS/MWCNT (1:1) (C) in chloroform under UV light irradiation and (b) fluorescence spectra of P3HT-b-PS and P3HT-b-PS/SWCNT (2:1) dispersion in chloroform. Reproduced from Ref. [53] with permission from John Wiley.

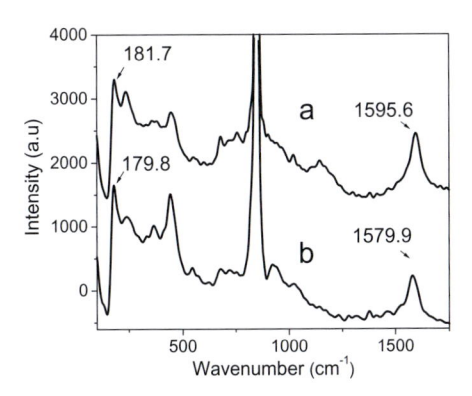

Figure 4.7 Raman spectra of (a) SWCNTs dispersed by P3HT-b-PS and (b) pristine SWCNTs. Reproduced from Ref. [53] with permission from John Wiley.

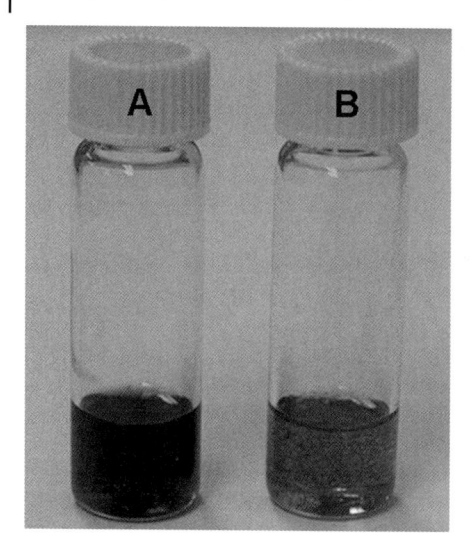

Figure 4.8 P3HT-*b*-PS dispersed SWCNTs (A) and P3HT dispersed SWCNTs (B) in polystyrene chloroform solution (10 wt.%).

through π–π interaction and the functional block (e.g., PS) provides CNTs with a good solubility and stability in the solvents. Additionally, the functional block on the surface of CNTs is expected to provide CNTs with excellent miscibility and compatibility in a compatible polymer matrix, offering a versatile approach to fabricate uniform CNT/polymer nanocomposites. To verify this concept, we tested the stability and compatibility of P3HT-*b*-PS dispersed SWCNTs with PS by adding the dispersed SWCNTs into PS chloroform solution (10 wt.%). As a control experiment, P3HT dispersed SWCNTs were also tested. As shown in Figure 4.8, P3HT-*b*-PS dispersed SWCNTs are stable in the solution because the PS block on the SWCNT surface increases the compatibility of SWCNTs with PS. In contrast, P3HT dispersed SWCNTs precipitate out of the solution after overnight standing due to the incompatibility between CNTs and PS.

Percolation threshold, defined as the lowest filler loading to form a three-dimensional filler network in a matrix, has been used as an indicator to evaluate the CNT dispersing quality in polymer matrices. Namely, a better dispersing quality leads to a lower percolation threshold. Decreasing the percolation threshold is also a practical method to produce cost-effective CNT/polymer composites for charge dissipation and electrical conductivity applications. The SWCNT/PS nanocomposites were fabricated to determine the percolation threshold of P3HT-*b*-PS dispersed SWCNTs in a PS matrix. The P3HT-*b*-PS dispersed SWCNTs were mixed with PS dissolved in chloroform (10 wt.%). After being stirred for half an hour, the mixture was cast on a glass substrate by a drawdown bar to obtain SWCNT/PS nanocomposite films with a thickness of ~30 μm. The electrical conductivity of the nanocomposite thin films was measured by Keithley 2400 using a four-probe setup. As a control experiment, P3HT dispersed SWCNT/PS nanocom-

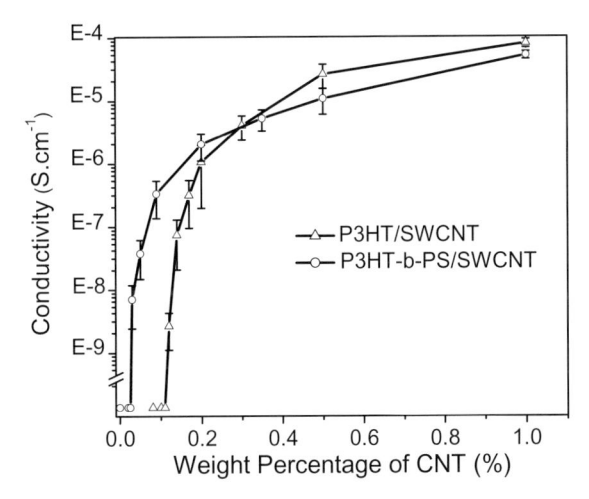

Figure 4.9 Conductivity of SWCNT/PS nanocomposites prepared from SWCNTs dispersed by P3HT-*b*-PS and P3HT. Reproduced from Ref. [53] with permission from John Wiley.

posite films were also prepared and their conductivity were measured. The percolation threshold of SWCNTs in the PS matrix was determined from the conductivity/SWCNT concentration plot as shown in Figure 4.9. The percolation threshold of P3HT-*b*-PS dispersed SWCNTs was determined to be 0.03 wt.%, which is significantly lower than that of P3HT dispersed SWCNTs (0.12 wt.%). The result indicates that P3HT-*b*-PS dispersed SWCNTs are better dispersed in PS matrix than P3HT dispersed SWCNTs due to the compatibility of the PS block with the PS matrix. We also observed that when the SWCNT loading amount in the nanocomposite was higher than 0.3 wt.%, the P3HT composites are more conductive than P3HT-*b*-PS composites. We believe that it is because P3HT-*b*-PS dispersed SWCNTs have lower conductivity due to the insulating PS block covering on their surface.

The compatibility of conjugated block copolymer dispersed CNTs with the host polymer matrix improves not only the dispersing quality but also the stability of CNTs in the matrix. The stability of P3HT-*b*-PMMA dispersed SWCNTs in a PMMA matrix was examined and compared with that of P3HT dispersed SWCNTs. We prepared the SWCNT/PMMA composite films (1 wt.% SWCNT in PMMA) with a thickness of 30 μm following the same procedure described above except that the composite films were cast right after the mixing of CNT dispersion with PMMA solution. The images of the composite films were taken by an optical microscope under the transmission mode. As shown in Figure 4.10a and c, both P3HT-*b*-PMMA and P3HT dispersed SWCNTs are dispersed uniformly in the PMMA matrix with only a few SWCNT aggregations observed in the P3HT composite. The composite thin films were then annealed at 140 °C under vacuum for 10 h. Because the annealing temperature is higher than the glass transition temperature (T_g) of PMMA, SWCNTs can move freely in the matrix. Such movement

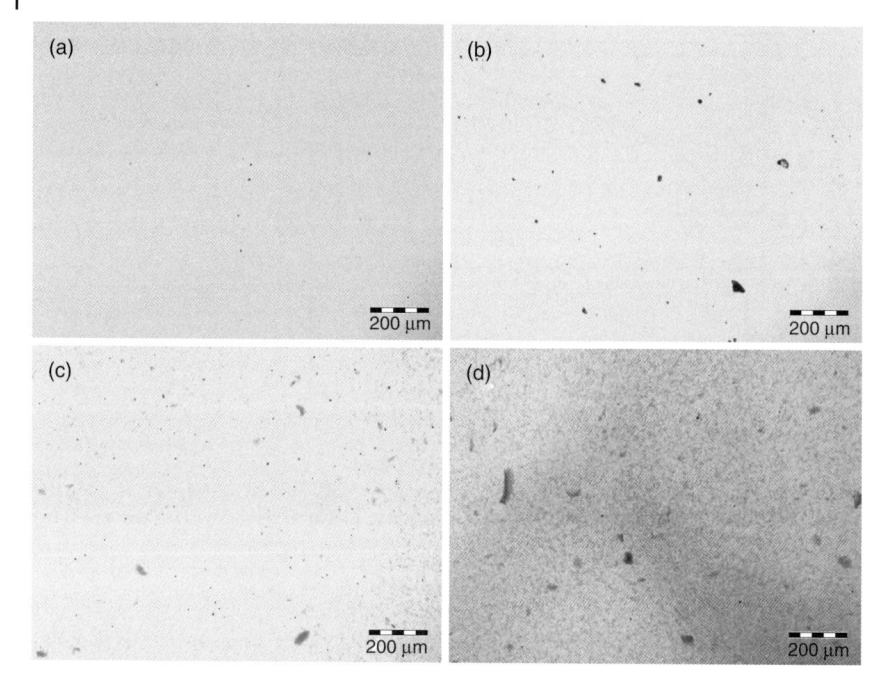

Figure 4.10 Optical microscope images of P3HT-*b*-PMMA/SWCNT-PMMA composite film (a) before and (b) after annealing, P3HT /SWCNT-PMMA composite film (c) before and (d) after annealing. Reproduced from Ref. [55] with permission from John Wiley.

will cause SWCNTs to phase separate from the PMMA matrix and form aggregations unless the composite films are thermodynamically stable. Figure 4.10b shows that P3HT-*b*-PMMA dispersed SWCNTs remain stable in the PMMA matrix after annealing without forming aggregations, indicating that the P3HT-*b*-PMMA composite is thermodynamically stable due to the compatibility of SWCNTs with PMMA matrix. However, P3HT dispersed SWCNTs aggregate substantially after annealing (Figure 4.10d), indicating a thermodynamically unstable system due to the incompatibility of the SWCNTs with the PMMA matrix. The result demonstrates that the compatibility of the functional block with the polymer matrix plays a crucial role in stabilizing CNTs in a CNT/polymer composite. It is worth to note that various functional blocks can be introduced into conjugated block copolymers through well establish synthetic routes to provide versatile dispersants of CNT for different polymer matrices.

P3HT-*b*-PS is a rod-coil copolymer, whose self-assembly behavior has been widely studied [62]. When the rod-coil block copolymer in a highly volatile solvent (e.g., chloroform, carbon disulfide, and toluene) is cast on a solid substrate, porous honeycomb structures will be formed after evaporation of the solvent [63–67]. Therefore, it is interesting to investigate the self-assembly behavior of P3HT-*b*-PS dispersed CNTs. The SEM image (Figure 4.11a) of the thin film formed by drop

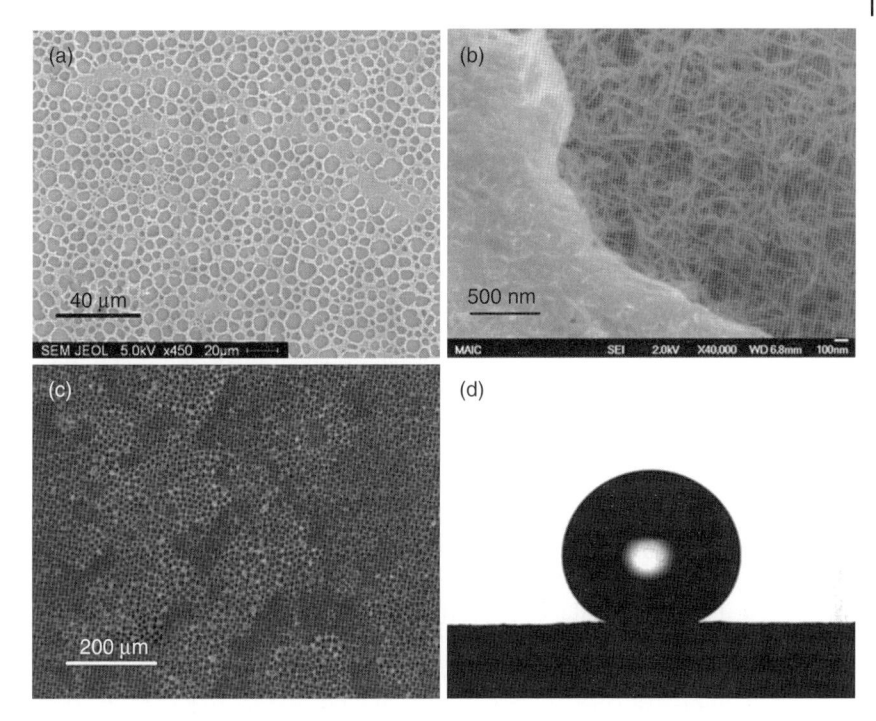

Figure 4.11 (a) SEM image of the honeycomb structures formed by solvent casting of P3HT-*b*-PS/MWCNT dispersions. (b) High-resolution SEM image of the honeycomb structure showing P3HT-*b*-PS rich ridge and MWCNT rich base. (c) Fluorescence image of the honeycomb structure. (d) A digital photo image of a 10 μL water droplet on superhydrophobic surface with a contact angle of 159.6°. Reproduced from Ref. [54] with permission from John Wiley.

casting the P3HT-*b*-PS/MWCNT chloroform dispersion onto a glass substrate reveals a honeycomb like morphology. Actually, the honeycomb morphology was observed on various substrates including glass, silicone wafer, paper, fabric, aluminum foil, and gold. The diameter of the honeycomb pore is in the range of 2–15 μm and the height of the honeycomb ridge is from 200 to 500 nm. The high-resolution SEM image (Figure 4.11b) reveals that the ridge of the honeycomb structure is mainly composed of P3HT-*b*-PS decorated with a small amount of MWCNTs, while the base appears to be entangled CNTs. The different composition of the ridge and base makes the honeycomb structure observable under fluorescence microscope. As shown in Figure 4.11c, the honeycomb ridge appears to be bright green and the base is dark. The honeycomb structure generates microscale roughness on the thin film surface. The microscale roughness combined with the low surface energy of the surface makes the thin film superhydrophobic [68–70]. The water contact angle of the thin film was characterized to be 154°–160° with the sliding angle less than 5° (Figure 4.11d). With the conductivity in the range of 30–100 S/cm, the superhydrophobic film is of particular interest for applications as both electromagnetic interference (EMI) shielding and moisture/vapor barrier material in electronic devices.

Conjugated polymers, with their unique electrical properties, have been widely investigated as volatile organic compound (VOC) sensing materials [71–74]. However, such applications are limited by low sensitivity, long response time (tens of seconds to minutes), and the interference of moisture. The low sensitivity and long response time are caused by the slow diffusion of gas into the conjugated polymers, which can be improved by using nanoscale structured conjugated polymer with large surface area. The interference of moisture is due to the water condensed at the interface between conducting polymers and substrates which leads to a cooperative sensing response of both organic vapor and water. Such interference can be eliminated or reduced by making the substrate hydrophobic to reduce the penetration of moisture into the interface between the substrate and VOC sensing materials [75]. The superhydrophobic P3HT-b-PS/MWCNT thin film is expected to improve the sensitivity and response time in VOC sensing applications by utilizing its micro- and nanostructures with large surface area, as well as to eliminate the interference of moisture because of its superhydrophobicity.

The VOC sensing property of a superhydrophobic P3HT-b-PS/MWCNT film was evaluated by exposing the film to various VOCs. The change of film resistance with exposure time was recorded in Figure 4.12. The film resistance quickly decreases and reaches a steady value within 5 s upon the exposure to toluene, chloroform and tetrahydrofuran (THF). After the film is removed from the VOC atmosphere, its resistance recovers to the original value in 3 s. The repeat doses of VOCs show identical response of film resistance, suggesting a good stability and sensitivity of the sensors. On the other hand, the film is completely inert to moisture due to its superhydrophobic nature. The response of film resistance

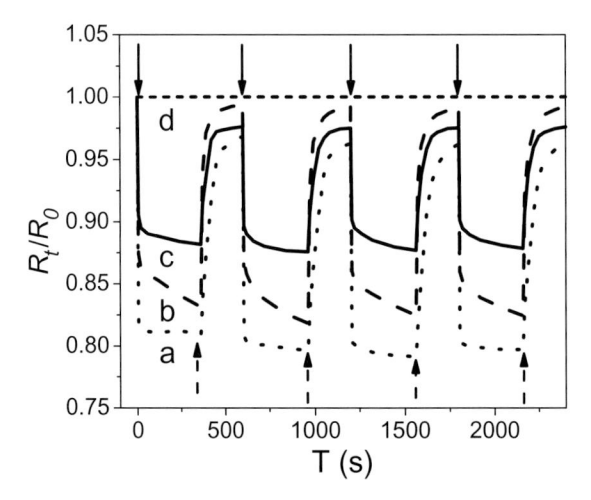

Figure 4.12 The normalized resistance of the superhydrophobic film exposed to organic vapor: (a) toluene; (b) chloroform; (c) tetrahydrofuran; (d) moisture. Solid arrows and dash arrows indicate the moment that the sample was exposed to and removed from the vapor, respectively. Reproduced from Ref. [54] with permission from John Wiley.

toward moisture-rich organic vapors of toluene, chloroform, and THF was exactly the same as what was observed from the pure organic vapors, indicating no moisture interference with the film.

4.3.2
Dispersing and Functionalizing CNTs by P3HT-b-PAA

CNTs bearing carboxylic groups on their surface (CNT-COOH) are important and versatile intermediates toward many applications. For example, ruthenium oxide nanoparticles were attached on CNT-COOH as supercapacitor electrode materials [76]. Platinum nanoparticles on CNT-COOH demonstrated high electrocatalytic activity for methanol oxidation, which is crucial for fuel cell applications [77]. CNT-COOH grafted with ATRP initiator can be used to initiate ATRP polymerization to produce CNT/polymer composites [78]. Some biological applications of CNTs require the attachment of functional molecules or particles onto CNTs through the reaction with carboxylic groups [79–81]. Although CNT-COOH can be obtained by the well-established approach of treating CNTs with HNO_3/H_2SO_4 [11], the CNT-COOH has poor electrical and mechanical properties due to the serious damage of the CNTs. Additionally, carboxylic groups mainly locate at the defects of CNT fragments and its density cannot be precisely controlled. To develop a more practical approach of functionalizing CNT with carboxylic groups, we synthesized the conjugated block copolymer of P3HT-b-PAA. Pristine MWCNTs can be homogenously dispersed in THF by P3HT-b-PAA with a brief sonication. P3HT block binds to the dispersed MWCNTs through π–π interaction and the PAA block orients at the CNT surface and functionalize MWCNTs with carboxylic groups. TEM image (Figure 4.13a) of the obtained MWCNT-COOH shows that MWCNTs were dispersed into individual tubes with a large aspect ratio in contrast to the short CNT fragment obtained by the HNO_3/H_2SO_4 treatment. This is because of the noncovalent nature of the process, during which only a short time of sonication was applied. Thus, both the structure and properties of CNTs are maintained. High-resolution TEM image of the MWCNT-COOH (Figure 4.13b) reveals that the MWCNT is covered homogenously by an amorphous thin layer of the block copolymers, indicating homogenous and nonspecific binding of P3HT-b-PAA on CNT surfaces, which allows a uniform distribution of carboxylic groups. Another advantage of the method is that the carboxylic group density on CNT surface can be controlled by the composition of P3HT-b-PAA. P3HT-b-PAA with a longer PAA block will lead to a higher density of carboxylic group on CNTs. To investigate the relationship between the block copolymer composition and the acid group density on CNTs. Two P3HT-b-PAAs with the P3HT to PAA molar ratio of 1:6.8 and 1:0.9 were synthesized and used to disperse MWCNTs in THF. The P3HT-b-PAA to MWCNT ratio was kept in such a way that the P3HT block to MWCNT weight ratio is 0.5:1. The dispersion was centrifuged and redispersed in THF for three times to remove free P3HT-b-PAA. The purified MWCNT-COOH was collected by centrifugation and dried in vacuum. The weight ratio of P3HT-b-PAA to MWCNTs in the obtained MWCNT-COOH was characterized by thermogravimetric analysis (TGA) [47] to be 1.72:1 and 0.63:1 for P3HT-b-PAA (1:6.8) and P3HT-b-PAA

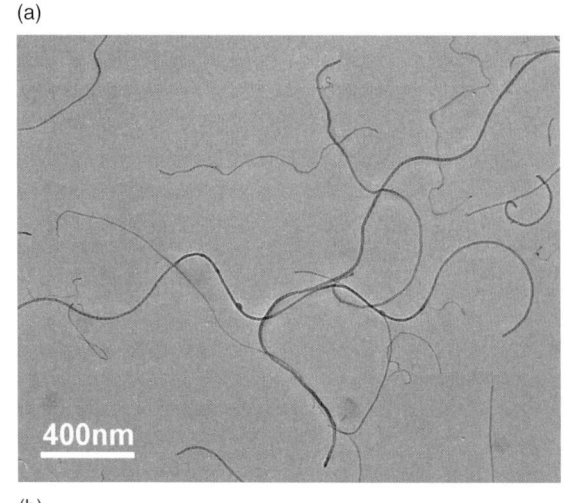

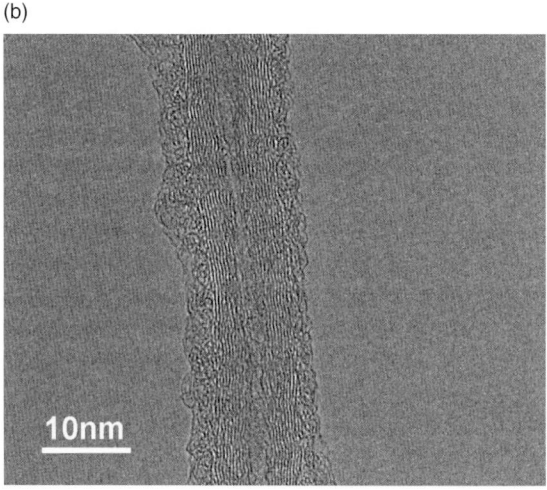

Figure 4.13 (a) A TEM image of MWCNT dispersed by P3HT-*b*-PAA, (b) A HRTEM image indicating homogenous coverage of CNT by P3HT-*b*-PAA. Reproduced from Ref. [55] with permission from John Wiley.

(1:0.9), respectively. P3HT-*b*-PAA (1:6.8) leads to higher polymer content in the MWCNT-COOH indicating higher carboxylic group density on MWCNT surface. The carboxylic group density on the obtained MWCNT-COOH was quantitatively determined by a titration (NaOH, 10^{-3} M) [82] to be 7.5 and 1.2 mmol/g for P3HT-*b*-PAA (1:6.8) and P3HT-*b*-PAA (1:0.9), respectively. Therefore, P3HT-*b*-PAA can be used to disperse and functionalize CNTs with carboxylic groups without degrading the structure and properties of CNTs. This method provides uniform distribution of carboxylic groups on CNT surfaces and precisely controlled carboxylic group density by varying the P3HT-*b*-PAA composition.

4.3.3
Dispersing and Functionalizing CNTs by P3HT-b-PPEGA

We have demonstrated conjugated block copolymer dispersed CNTs with good solubility in certain solvents. CNTs soluble in a variety of solvents including both polar and nonpolar solvents are also important. Such CNTs can be obtained by dispersing CNTs with conjugated block copolymer having an amphiphilic functional block. The amphiphilic functional block which is soluble both in polar and nonpolar solvents will orient on the CNT surface and endow CNTs with solubility in various solvents. To demonstrate this idea, we synthesized P3HT-b-PPEGA, in which the functional block of PPEGA is an amphiphilic polymer which is soluble in a wide variety of solvents. Pristine MWCNTs was first dispersed homogenously by P3HT-b-PPEGA in chloroform with 30 min of sonication. The obtained dispersion was then dried and the solubility of the dried MWCNTs was tested by redispersing them into various solvents with sonication. The solubility of P3HT dispersed MWCNTs in various solvents was also investigated and compared with P3HT-b-PPEGA dispersed MWCNTs. We tested a variety of solvents including chloroform, toluene, methanol, ethanol, dimethylformamide (DMF), and acetonitrile, which are all good solvents of PPEGA. The results showed that P3HT-b-PPEGA dispersed MWCNTs were soluble in all these solvents. However, P3HT dispersed MWCNTs were only soluble in chloroform. These results indicated that the solubility of conjugated block copolymer dispersed CNTs was mainly determined by the functional block which orients at CNT surface. Therefore, the solubility of CNTs can be manipulated by the functional block. The photographs of P3HT-b-PPEGA dispersed MWCNTs in different solvents (Figure 4.14) indicate that MWCNTs are well dispersed in all the solvents except that the dispersions in methanol and DMF display a reddish color compared with the dispersions in chloroform and toluene. The UV-Vis spectra also reveal the difference by showing

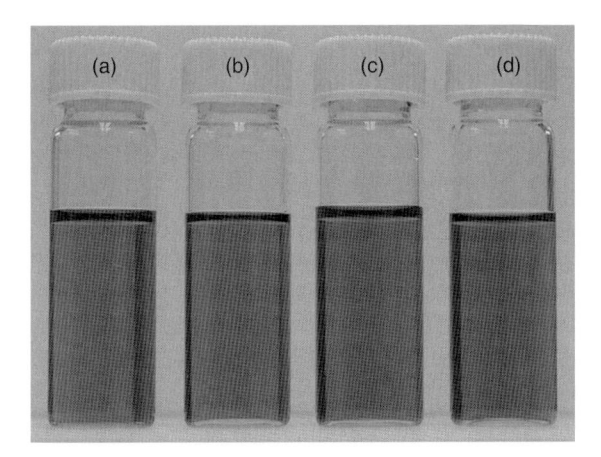

Figure 4.14 The photographs of P3HT-b-PPEGA dispersed MWCNTs in (a) chloroform, (b) toluene, (c) methanol, (d) DMF. Reproduced from Ref. [55] with permission from John Wiley.

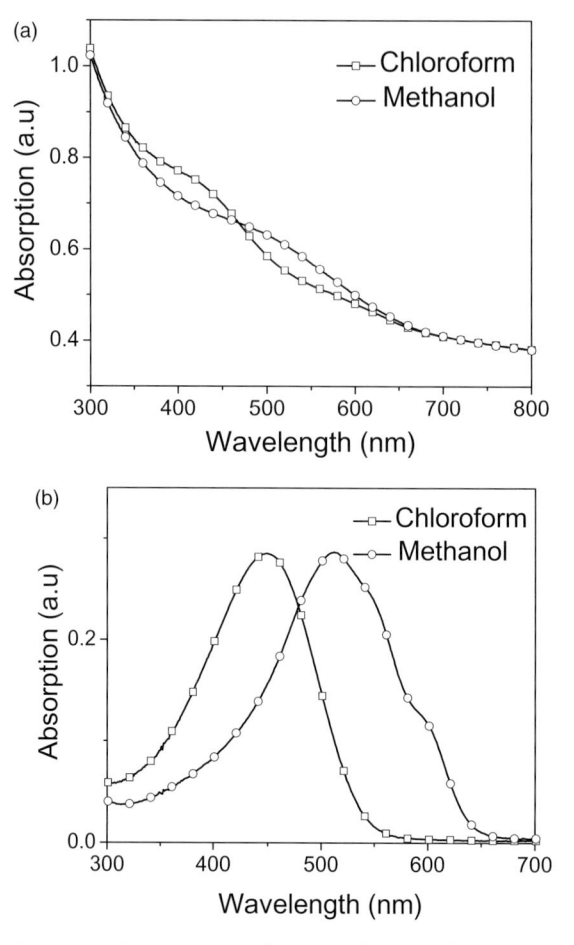

Figure 4.15 UV-Vis spectra of (a) P3HT-*b*-PPEGA dispersed MWCNTs, (b) P3HT-*b*-PPEGA in chloroform and methanol. Reproduced from Ref. [55] with permission from John Wiley.

an absorption at 500 nm in methanol comparing with the absorption at 430 nm in the chloroform (Figure 4.15a). Such a difference is due to the presence of free P3HT-*b*-PPEGA in the dispersion. In chloroform and toluene, P3HT-*b*-PPEGA has a UV-Vis absorption peak at 450 nm, which is assigned to individual P3HT-*b*-PPEGA chains. While in methanol, ethanol, DMF, and acetonitrile, P3HT-*b*-PPEGA has a UV-Vis absorption peak at 512 nm due to the aggregation of P3HT block (Figure 4.15b).

4.4
Conclusions and Perspective

In this chapter, we described a noncovalent approach to disperse and functionalize pristine CNTs through a simple sonication process. The dispersant used in this approach is the conjugated block copolymers, which are composed of a conjugated block of poly(3-hexylthiophene) and a functional block. When pristine CNTs are sonicated in the presence of the conjugated block copolymer, the P3HT block will bind to the debundled CNT surfaces through π–π interactions and prevent CNTs from aggregating. The functional block will orient on the CNT surface and introduce both solubility and functional groups to CNTs. This approach represents a general method to disperse and functionalize pristine CNTs since both the composition and content of the functional block can be varied to endow CNTs with various functionality and solubility in different solvents. The existing work shows the success of this approach and calls for the synthesis of more conjugated block copolymers with different nonconjugated blocks in order to obtain dispersed CNTs with different functionalities. A further research of this approach relies on using more than one conjugated block copolymers to disperse and functionalize CNTs in one batch in order to fabricate CNTs with multiple functionalities.

Acknowledgments

The financial supports from National Science Foundation (CAREER Award DMR 0746499 and CBET 0608870) are greatly appreciated.

References

1 Iijima, S. (1991) *Nature*, **354**, 56.
2 Yao, Z., Kane, C.L., and Dekker, C. (2000) *Phys. Rev. Lett.*, **84**, 2941.
3 Wong, E.W., Sheehan, P.E., and Lieber, C.M. (1997) *Science*, **277**, 1971.
4 O'Connell, M.J. (2006) *Carbon Nanotubes: Properties and Applications*, CRC Press and Taylor & Francis Group, London.
5 Endo, M., Iijima, S., and Dresselhaus, M. (1997) *Carbon Nanotubes*, Elsevier, Amsterdam.
6 Reich, S., Thomson, C., and Maultzsch, J. (2004) *Carbon Nanotubes: Basic Concepts and Physical Properties*, Wiley-VCH Verlag GmbH, Weinheim.
7 Dai, L. (2006) *Carbon Nanotechnology: Recent Developments in Chemistry, Physics, Materials Science and Device Applications*, Elsevier, Amsterdam.
8 Sandler, J.K.W., Kirk, J.E., Kinloch, I.A., Shaffer, M.S.P., and Windle, A.H. (2003) *Polymer*, **44**, 5893.
9 Sandler, J., Shaffer, M.S.P., Prasse, T., Bauhofer, W., Schulte, K., and Windle, A. (1999) *Polymer*, **40**, 5967.
10 Dujardin, E., Ebbesen, T.W., Krishnan, A.E., and Treacy, M.M.J. (1998) *Adv. Mater.*, **10**, 611.
11 Liu, J., Rinzler, A.G., Dai, H., Hafner, J.H., Bradley, R.K., Boul, P.J., Lu, A., Iverson, T., Shelimov, K., Huffman, C.B., Rodriguez-Macias, F., Shon, Y.S., Lee, T.R., Colbert, D.T., and Smalley, R.E. (1998) *Science*, **280**, 1253.
12 Hiura, H., Ebbesen, T.W., and Tanigaki, K. (1995) *Adv. Mater.*, **7**, 275.
13 Hamon, M.A., Hui, H., Bhowmik, P., Itkis, H.M.E., and Haddon, R.C. (2002) *Appl. Phys. A*, **74**, 333.
14 Thess, A., Lee, R., Nikolaev, P., Dai, H., Petit, P., Robert, J., Xu, C., Lee, Y.H., Kim, S.G., Rinzler, A.G., Colbert, D.T., Scuseria, G.E., Tománek, D., Fischer, J.E., and Smalley, R.E. (1996) *Science*, **273**, 483.
15 Yang, D.-Q., Rochette, J.-F., and Sacher, E. (2005) *J. Phys. Chem. B*, **109**, 7788.
16 Curtzwiler, G., Singh, J., Miltz, J., Doi, J., and Vorst, K. (2008) *J. Appl. Polym. Sci.*, **109**, 218.

17 Tasis, D., Tagmatarchis, N., Bianco, A., and Prato, M. (2006) *Chem. Rev.*, **106**, 1105.

18 Osorio, A.G., Silverira, I.C.L., Bueno, V.L., and Bergmann, C.P. (2008) *Appl. Surf. Sci.*, **255**, 2485.

19 Kim, M., Hong, C.K., Choe, S., and Shim, S.E. (2007) *J. Polym. Sci. Part A: Polym. Chem.*, **45**, 4413.

20 Monthioux, M., Smith, B.W., Burteaux, B., Claye, A., Fischer, J.E., and Luzzi, D.E. (2001) *Carbon*, **39**, 1251.

21 Alvaro, M., Atienzar, P., Cruz, P., Delgado, J.L., Garcia, H., and Langa, F. (2004) *Chem. Phys. Lett.*, **386**, 342.

22 Sun, Y.P., Huang, W., Lin, Y., Fu, K., Kitaygorodskiy, A., Riddle, L.A., Yu, Y.J., and Carroll, D.L. (2001) *Chem. Mater.*, **13**, 2864.

23 Chattopadhyay, D., Lastella, S., Kim, S., and Papadimitrakopoulos, F. (2002) *J. Am. Chem. Soc.*, **124**, 728.

24 Sung, J.H., Kim, H.S., Jin, H.J., Choi, H.J., and Chin, I.J. (2004) *Macromolecules*, **37**, 9899.

25 Shaffer, M.S.P., and Koziol, K. (2002) *Chem. Commun.*, 2074.

26 Lin, Y., Meziani, M.J., and Sun, Y. (2007) *J. Mater. Chem.*, **17**, 1143.

27 Barraza, H.J., Pompeo, F., O'Rear, E.A., and Resasco, D.E. (2002) *Nano Lett.*, **2**, 797.

28 Islam, M.F., Rojas, E., Bergey, D.M., Johnson, A.T., and Yodh, A.G. (2003) *Nano Lett.*, **3**, 269.

29 Bandhyopadhyaya, R., Nativ-Roth, E., Regev, O., and Yerushalmi-Rozen, R. (2002) *Nano Lett.*, **2**, 25.

30 Grunlan, J.C., Liu, L., and Kim, Y. (2006) *Nano Lett.*, **6**, 911.

31 Nepal, D. and Geckeler, K. (2007) *Small*, **3**, 1259.

32 Karajanagi, S.S., Yang, H., Asuri, P., Sellitto, E., Dordick, J.S., and Kane, R.S. (2006) *Langmuir*, **22**, 1392.

33 Zheng, M., Jagota, A., Semke, E.D., Diner, B.A., Mclean, R.S., Lustig, S.R., Richardson, R.E., and Tassi, N.G. (2003) *Nat. Mater.*, **2**, 338.

34 Shin, H., Min, B.G., Jeong, W., and Park, C. (2005) *Macromol. Rapid Commun.*, **26**, 1451.

35 Shvartzman-Cohen, R., Levi-Kalisman, Y., Nativ-Roth, E., and Yerushalmi-Rozen, R. (2004) *Langmuir*, **20**, 6085.

36 Nativ-Roth, E., Shvartzman-Cohen, R., Bounioux, C., Florent, M., Zhang, D., Szleifer, I., and Yerushalmi-Rozen, R. (2007) *Macromolecules*, **40**, 3676.

37 Park, I., Lee, W., Kim, J., Park, M., and Lee, H. (2007) *Sens. Actuators B*, **126**, 301.

38 Mountrichas, G., Tagmatarchis, N., and Pispas, S. (2007) *J. Phys. Chem. B*, **111**, 8369.

39 Cotiuga, I., Picchioni, F., Agarwal, U.S., Wouters, D., Loos, J., and Lemstra, P. (2006) *Macromol. Rapid Commun.*, **27**, 1073.

40 Sinani, V.A., Gheith, M.K., Yaroslavov, A.A., Rakhnyanskaya, A.A., Sun, K., Mamedov, A.A., Wicksted, J.P., and Kotov, N.A. (2005) *J. Am. Chem. Soc.*, **127**, 3463.

41 Star, A., Stoddart, J.F., Steuerman, D., Diehl, M., Boukai, A., Wong, E.W., Yang, X., Chung, S.-W., Choi, H., and Heath, J. (2001) *Angew. Chem. Int. Ed.*, **40**, 1721.

42 Dalton, A.B., Stephan, C., Coleman, J.N., McCarthy, B., Ajayan, P.M., Lefrant, S., Bernier, P., Blau, W.J., and Byrne, H.J. (2000) *J. Phys. Chem. B*, **104**, 10012.

43 Roland, G.S.G., Nunzio, M., John, M.B., and Eric, R.W. (2006) *Appl. Phys. Lett.*, **88**, 053101.

44 Ikeda, A., Nobusawa, K., Hamano, T., and Kikuchi, J. (2006) *Org. Lett.*, **8**, 5489.

45 Chen, J., Liu, H., Weimer, W.A., Halls, M.D., Waldeck, D.H., and Walker, G.C. (2002) *J. Am. Chem. Soc.*, **124**, 9034.

46 Rice, N.A., Soper, K., Zhou, N., Merschrod, E., and Zhao, Y. (2006) *Chem. Commun.*, 4937.

47 Cheng, F., Imin, P., Maunders, C., Botton, G., and Adronov, A. (2008) *Macromolecules*, **41**, 2304.

48 Chen, F., Wang, B., Chen, Y., and Li, L.-J. (2007) *Nano Lett.*, **10**, 3013.

49 Cheng, F.Y. and Adronov, A. (2006) *Chem. Eur. J.*, **12**, 5053.

50 Cheng, F.Y. and Adronov, A. (2007) *J. Porphyrins Phthalocyanines*, **11**, 198.

51 Foroutan, M. and Nasrabadi, A.T. (2010) *J. Phy. Chem. B*, **114**, 5320.

52 Yang, M., Koutsos, V., and Zaiser, M. (2005) *J. Phys. Chem. B*, **109**, 10009.
53 Zou, J., Liu, L., Chen, H., Khondaker, S.I., McCullough, R.D., Huo, Q., and Zhai, L. (2008) *Adv. Mater.*, **20**, 2055.
54 Zou, J., Chen, H., Chunder, A., Yu, Y., Huo, Q., and Zhai, L. (2008) *Adv. Mater.*, **20**, 3337.
55 Zou, J., Khondaker, S.I., Huo, Q., and Zhai, L. (2009) *Adv. Funct. Mater.*, **19**, 479.
56 Iovu, M.C., Jeffries-El, M., Sheina, E.E., Cooper, J.R., and McCullough, R.D. (2005) *Polymer*, **46**, 8582.
57 Matyjaszewski, K. and Xia, J. (2001) *Chem. Rev.*, **101**, 2921.
58 Chen, T., Wu, X., and Rieke, R.D. (1995) *J. Am. Chem. Soc.*, **117**, 233.
59 Ai, X., Anderson, N., Guo, J., Kowalik, J., Tolbert, L.M., and Lian, T. (2006) *J. Phys. Chem. B*, **110**, 25496.
60 Fan, C., Wang, S., Hong, J.W., Bazan, G.C., Plaxco, K.W., and Heeger, J.A. (2003) *Proc. Natl Acad. Sci. USA*, **100**, 6297.
61 Rao, A.M., Eklund, P.C., Bandow, S., Thess, A., and Smalley, R. (1997) *Nature*, **388**, 257.
62 Lee, M., Cho, B.-L., and Zin, W.-C. (2001) *Chem. Rev.*, **101**, 3869.
63 Jenekhe, S.A. and Chen, X.L. (1999) *Science*, **283**, 372.
64 Hayakawa, T. and Horiuchi, S. (2003) *Angew. Chem. Int. Ed.*, **42**, 2285.
65 Boer, B.D., Stalmach, U., Nijland, H., and Hadziioannou, G. (2000) *Adv. Mater.*, **12**, 1581.
66 Cheng, C.X., Tian, Y., Shi, Y.Q., Tang, R.P., and Xi, F. (2005) *Langmuir*, **21**, 6576.
67 Yabu, H. and Shimomura, M. (2005) *Chem. Mater.*, **17**, 5231.
68 Erbil, H.Y., Demirel, A.L., Avci, Y., and Mert, O. (2003) *Science*, **299**, 1377.
69 Nakajima, A., Hashimoto, K., Watanabe, T., Takai, K., Yamauchi, G., and Fujishima, A. (2000) *Langmuir*, **16**, 7044.
70 Fürstner, R. and Barthlott, W. (2005) *Langmuir*, **21**, 956.
71 McQuade, D.T., Pullen, A.E., and Swager, T.M. (2000) *Chem. Rev.*, **100**, 2537.
72 Janata, J.` and Josowicz, M. (2003) *Nat. Mater.*, **2**, 19.
73 Li, B., Sauvé, G., Iovu, M.C., Jeffries-El, M., Zhang, R., Cooper, J., Santhanam, S., Schulz, L., Revelli, J.C., Kusne, A.G., Kowalewski, T., Snyder, J.L., Weiss, L.E., Fedder, G.K., McCullough, R.D., and Lambeth, D.N. (2006) *Nano Lett.*, **6**, 1598.
74 Virji, S., Huang, J., Kaner, R.B., and Weiller, B.H. (2004) *Nano Lett.*, **4**, 491.
75 Wang, P.-C., Huang, Z., and MacDiarmid, A.G. (1999) *Synth. Metals*, **101**, 852.
76 Kim, Y.-T., Tadai, K., and Mitani, T. (2005) *J. Mater. Chem.*, **15**, 4914.
77 Halder, A., Sharma, S., Hegde, M.S., and Ravishankar, N. (2009) *J. Phys. Chem. C*, **113**, 1466.
78 Kong, H., Gao, C., and Yan, D. (2004) *J. Am. Chem. Soc.*, **126**, 412.
79 Kam, N.W.S., Jessop, T.C., Wender, P.A., and Dai, H. (2004) *J. Am. Chem. Soc.*, **126**, 6850.
80 Wong, S.S., Joselevich, E., Woolley, A.T., Cheung, C.L., and Lieber, C.M. (1998) *Nature*, **394**, 52.
81 Shi, D., Guo, Y., Dong, Z., Lian, J., Wang, W., Liu, G., Wang, L., and Ewing, R.C. (2007) *Adv. Mater.*, **19**, 4033.
82 Eitan, A., Jiang, K., Dukes, D., Andrews, R., and Schadler, L. (2003) *Chem. Mater.*, **15**, 3198.

5
Theoretical Analysis of Nanotube Functionalization and Polymer Grafting

Kausala Mylvaganam and Liang Chi Zhang

5.1
Introduction

Carbon nanotubes (CNTs) are considered to be promising materials for many applications in the development of nanotechnology. Some applications, however, are hindered by the difficulties in the dispersion, manipulation, and processability of CNTs, which are highly entangled and form bundles due to van der Waals attraction among the tubes. Hence for technical applications, it is necessary to first disperse them without reducing their aspect ratio such that they can be mixed with other substances effectively to form composites, incorporated into fibers or deposited as a film. A variety of mechanical/chemical/physical methods such as ultrasonication, high shear mixing, chemical modification through functionalization, adsorption of charged surfactants and electrolytes, wrapping with polymers, and complex formation by π–π interactions, have been used to disperse CNTs [1–7]. Functionalization of nanotubes has become attractive for synthetic chemists and material scientists as it not only improves the solubility and dispersion of CNTs but also can expand their potential application areas through the change of mechanical and electrical properties.

Meur *et al.* showed that functionalized polymers offer the possibility to disperse CNTs in organic solvents [8]. Polymers have been attached to single-walled carbon nanotubes (SWCNTs) via "grafting from" and "grafting to" techniques [9, 10]. The "grafting from" method involves first the attachment of polymerization initiators to the surface of SWCNTs (i.e., first functionalizing the nanotubes) followed by polymerization of monomers from the resulting nanotube-based macroinitiator. The advantage of this "grafting from" method is that the polymer growth is not limited by steric hindrance, allowing high-molecular-weight polymers to be efficiently grafted. The "grafting to" method involves the attachment of premade polymer chains to the surface of CNTs. In this strategy, architecturally controlled polymers (linear, block, etc.) that have an active end-group can be used as a linker to the nanotube.

Theoretical studies can be very helpful toward a comprehensive understanding of the nature of the functionalization of CNTs. Computational studies based on

---

Surface Modification of Nanotube Fillers, First Edition. Edited by Vikas Mittal.
© 2011 Wiley-VCH Verlag GmbH & Co. KGaA. Published 2011 by Wiley-VCH Verlag GmbH & Co. KGaA.

ab-initio quantum mechanics method using density functional theory (DFT), molecular dynamics simulation, and hybrid quantum mechanics/molecular mechanics (QM/MM) have been carried out to investigate the mechanism behind the functionalization of SWCNTs by molecules or molecular groups.

This chapter will discuss various theoretical considerations in functionalizing CNTs and grafting polymers onto CNTs.

5.2
Theoretical Techniques in Modeling Nanotube Functionalization

5.2.1
Ab-initio Quantum Mechanics Method

Ab-initio methods are based entirely on first principles. Among the theoretical methods, *ab-initio* calculations are complementary to the information obtained by experiments and in some cases can predict hitherto unobserved phenomena. Hence the *ab-initio* methods have been widely used in the design of new materials. In principle, *ab-initio* methods converge to the exact solutions of the underlying equations as the number of approximations is reduced. However, in practice it is impossible to eliminate all the approximations. To treat large molecules, approximate method such as DFT is used.

A variety of functionals have been developed to account for exchange and correlation terms in DFT [11, 12]. The most widely used functional is the local-density approximation [13] (LDA), where the functional depends only on the density at the coordinate where the functional is evaluated. This functional has been shown to be especially appropriate for studying the adsorption of organic molecules on CNTs. However, it is well known that this functional tends to overestimate binding energies and charge transfer [14]. Hybrid functionals such as B3LYP [15–18] that include a component of the exact exchange energy are known to produce reasonable results for bond lengths, bond angles, and bond energies for a wide range of molecules [19]. Sections 5.3–5.5 will describe a few examples where these functionals are used in the study of CNT functionalization and subsequent polymer grafting. Even though the DFT method uses functionals that involve parameters derived from empirical data, they are considered to be *ab-initio* for determining molecular electronic structures of materials. The DFT methods including different functionals have been incorporated in popular commercial quantum mechanical software such as Gaussian 09, DMol3, VASP, SIESTA code, etc.

5.2.2
Molecular Dynamics Modeling

Molecular dynamics simulation has been widely used in understanding and characterizing the properties of CNTs. However, such simulations have to be done carefully to achieve the best representation of reality. It is important to select an appropriate interaction potential that effectively describes the deformation of a nanotube correctly. During a loading process, an improper treatment of the temperature rise can lead to fictitious results. In molecular dynamics, heat conduction is accomplished via the so-called thermostat atoms and various thermostatting methods. For small systems that are practical for molecular dynamics studies, the adiabatic relaxation method often leads to a fluctuation of the vibrational–relaxation rate. In isokinetic thermostatting, the temperature is maintained in different ways. For example, in the Berendsen thermostat scheme with velocity scaling, the velocities of thermostat atoms are scaled to fix the total kinetic energy. In the Gaussian feedback or Evans–Hoover (EH) scheme with force scaling, however, the kinetic energy is monitored and information is fed back into the equations of motion so that the kinetic energy is kept constant to dissipate heat by controlling the thermostatting force. The velocity scaling has been used in general because it is a simpler scheme to implement. For small time steps, the Gaussian isokinetic method and velocity scaling method are identical. However, a very small time step will give an unusually high loading rate. On the other hand, a small displacement step with a small time step will be computationally expensive. The flaw in the isokinetic–thermostatting method is that it is impossible to separate the effects of thermostatting on rate processes. In addition, a system has to be relaxed initially as well as during the simulation so that the velocities of the Newtonian and thermostat atoms reach equilibrium at the specified temperature of simulation; thus appropriate time step and displacement step have to be selected to get a reasonable loading rate.

Hence in the rest of this section, the necessary details such as the potential, number of thermostat atoms, thermostat method, time step, displacement step, and the number of relaxation steps that are central to a reliable simulation of nanotube will be discussed.

In order to examine the reliability of simulations, we will compare and discuss the stress–strain relationship of single-walled zigzag (17,0) and armchair (10,10) nanotubes obtained with different schemes. The simulations were carried out at 300 K with Berendsen (B) and EH thermostats and a time step of 0.5 fs with the interatomic forces described by Tersoff [20, 21] and Brenner [22, 23] potentials under Schemes I and II as described below.

5.2.2.1 Scheme I

In this scheme, the first two layers of atoms on both ends of a CNT were held rigid. The next four layers were taken as thermostat atoms and the remaining were treated as Newtonian atoms. First, the tubes were annealed at the simulation temperature for 5000 time steps. Then the rigid atoms on both ends were pulled along the axial direction with incremental steps of 0.05 Å. Each displacement step was followed by 1000 relaxation steps in order to dissipate the effect of preceding displacement step over the entire length of the tube.

5.2.2.2 Scheme II

In this scheme, all atoms except the rigidly held boundary ones were treated as thermostat atoms. In this case, each displacement step was followed by 50 relaxation steps.

The simulation parameters had a remarkable influence on the results. Different schemes led to significantly different stress–strain curves (Figure 5.1) and necking processes (Figure 5.2). The simulation using Tersoff–Brenner potential and Berendsen thermostat with all atoms as thermostat atoms with 50 relaxations steps after each displacement of 0.008 Å is a reliable and cost-effective method [24].

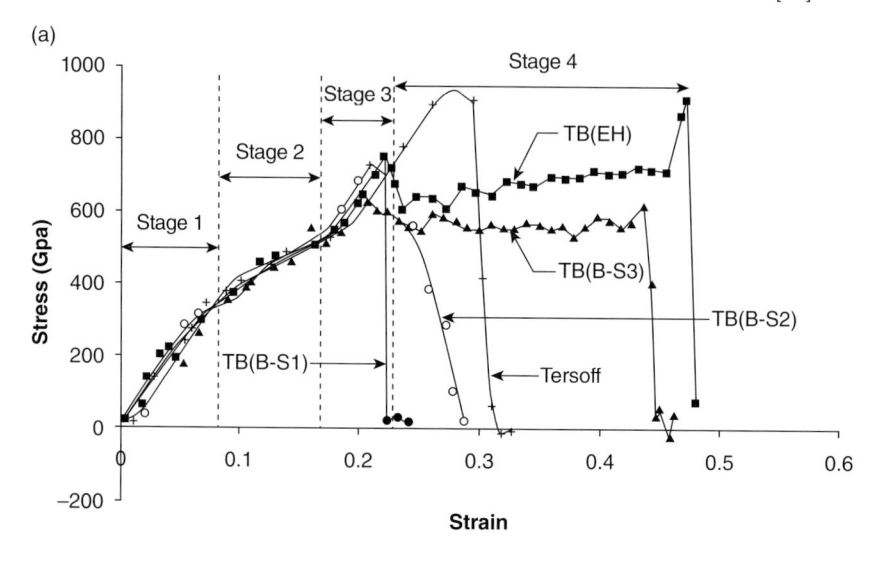

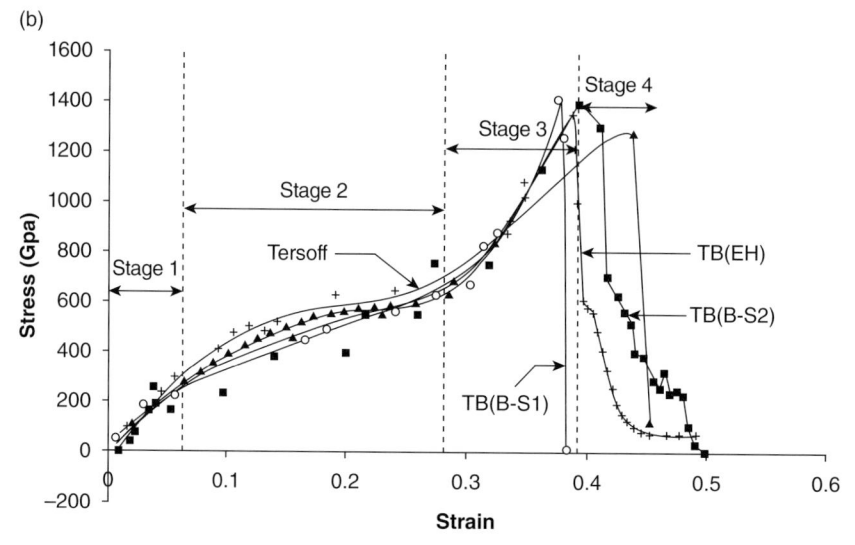

Figure 5.1 The stress–strain curves of (a) a (17,0) zigzag SWCNT and (b) a (10,10) armchair SWCNT using Tersoff and Tersoff–Brenner potentials. In the figure, TB(B-S1) is the calculation with Berendsen thermostat and Scheme 1 simulation details; TB(B-S2) is the calculation with Berendsen thermostat and Scheme 2 simulation details and TB(B-S3) is the calculation as in TB(B-S2) but with a smaller displacement step of 0.008 Å. TB(EH) is the calculation with Evans–Hoover thermostat. Reprinted from [24] with permission.

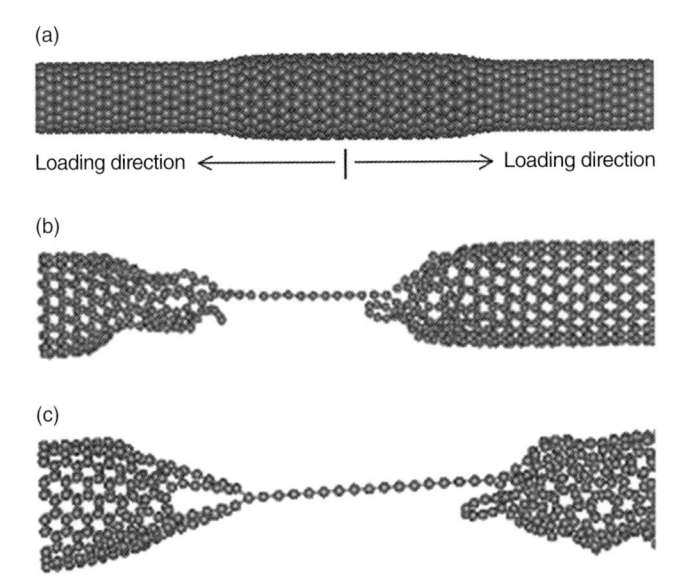

Figure 5.2 (a) Deformation of a zigzag SWCNT, TB(B-S3) calculation showing the propagation of necking. (b) Atomic chain of a zigzag tube when using Berendsen thermostat. (c) Atomic chain of an armchair tube when using Berendsen thermostat. Reprinted from [24] with permission.

5.2.3
Hybrid QM/MM Approach

Ab-initio quantum mechanics (including DFT) procedures are expensive for routine calculations even for a simple system like a CNT. While semiempirical methods, such as molecular dynamics, molecular mechanics, are useful and accessible tools to describe the properties of CNTs semiquantitatively, they are unable to provide reliable estimate for the energetics of reaction mechanisms. A hybrid QM/MM approach solves this problem to a certain extent. It has been implemented in popular commercial software such as Gaussian 09, DMol3, etc. In this approach, a whole molecular system is divided into two or three levels. A relatively small section (usually the reaction center) is treated at a more accurate *ab-initio* level and the rest is treated with a computationally less demanding (molecular mechanics or other semiempirical) method. A couple of examples discussed in Section 5.4 use this approach.

5.3
Functionalizing Carbon Nanotubes through Mechanical Deformation

Tension, compression, torsion, and pure bending are fundamental modes to deform CNTs under certain external conditions of mechanical loading. Theoretical studies have shown that the ridges of a deformed CNT bind functional groups tightly with high binding energy but binding with the deformation-flattened surface is weak – even weaker than that of a nondeformed nanotube. Mylvaganam and Zhang [25] simulated the deformation of (17,0) zigzag nanotubes by molecular dynamics simulation and then revealed the reactivity of the deformed nanotube sections with radicals via optimizing its geometry while keeping the geometry of the deformed tube section fixed, using *ab-initio* DFT with a hybrid functional B3LYP [15–18] and a 3-21G basis set [26]. Before optimizing the geometry, hydrogen atoms had been added to the dangling bonds of the perimeter carbons. The following sections will compare and discuss the reactivity of hydrogen and alkyl radicals with CNT deformed under the above-mentioned fundamental deformation modes.

5.3.1
Tension

Under this type of loading, the nanotube stretched along the loading direction and then necked. The necking propagated over the entire length of the tube and formed one atom chain before breaking into two pieces. Tensile loading does not generate ridges and flat surfaces like other loadings. Hydrogen atom binding with sections of tubes having low strain energy of 2.22 kJ/mol and high strain energies of 15.71 and 37.21 kJ/mol were investigated by quantum mechanics study. The hydrogen atom binding energy with the low strain energy section is 245.9 kJ/mol and with the high strain energy section is 258.5 kJ/mol (Figure 5.3).

5.3.2
Compression

Under compression loading, the tube started buckling once it reaches the strain energy of 21.82 kJ/mol. After this, the strain energy started to decrease and increase

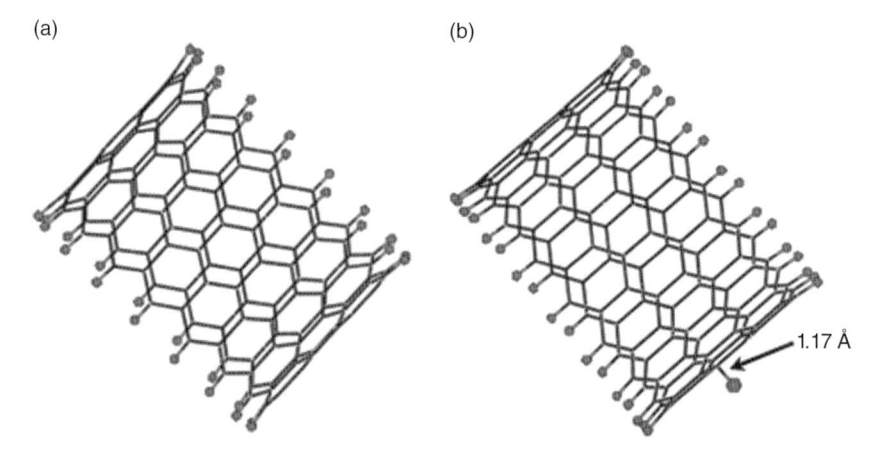

Figure 5.3 (a) Section of the CNT under tensile deformation and (b) optimized structure of the hydrogen radical on the deformed section of the CNT. Reprinted from [25] with permission.

5.3 Functionalizing Carbon Nanotubes through Mechanical Deformation | 97

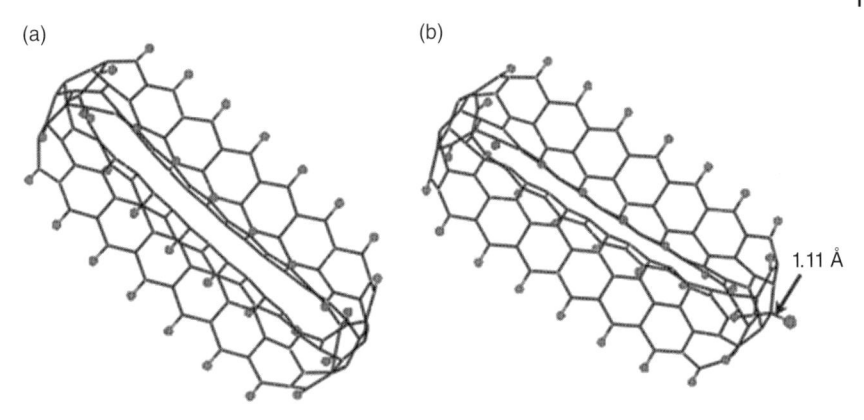

Figure 5.4 (a) Section of the CNT under compressive deformation (strain = 0.135) with the hydrogens at the dangling bonds of the perimeter C atoms optimized at DFT/3-21G level and (b) optimized structure of the hydrogen radical at the ridge. Reprinted from [25] with permission.

as it forms more buckles. Sections of the tube having strain energies of 2.01 kJ/mol (before buckling) and 23.06 kJ/mol (soon after buckling) were used in quantum mechanics study. The section of the compressed tube, with the hydrogen atoms added to the dangling bonds of the perimeter C atoms, is shown in Figure 5.4a. The hydrogen atom placed near the ridge of these sections binds to one of the C atoms with binding energies of 269.5 (before buckling) and 480.4 kJ/mol (after buckling; Figure 5.4b). Before buckling, when the strain energy was 2.01 kJ/mol, the curvature of the tube did not change significantly and hence the binding energy was low.

5.3.3
Torsion

Under this type of deformation mode, the strain energy varied initially cubically with the twisting angle. At the onset of buckling, the energy dropped a little and then changed almost linearly. The buckling started at the strain energy of 7.28 kJ/mol, before which the tube retained its cylindrical shape. Sections of the deformed CNT under the strain energy of 1.97 (before buckling) and 8.51 kJ/mol (soon after buckling) were used in the quantum mechanics study. Figure 5.5a shows a section of the tube after buckling, having clear ridges formed on twisting, with H atoms added and optimized to the dangling bonds of the perimeter C atoms. On placing a hydrogen atom near to those tube sections and optimizing its geometry, the hydrogen atom tightly bound to one of the carbon atoms at a ridge with binding energies 229.9 (before buckling) and 427.1 kJ/mol (after buckling), respectively. The optimized geometry of the hydrogen bound nanotube section after buckling is shown in Figure 5.5b. The shorter bond lengths and higher binding energies compared to the hydrogen radical attached to the nondeformed tube shows that the deformation promotes the reactivity. This could be explained

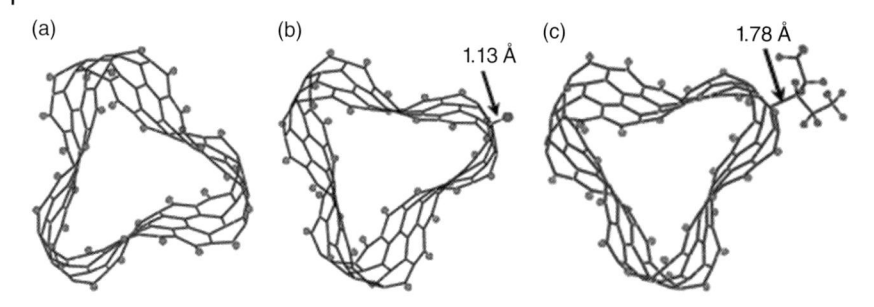

Figure 5.5 (a) Section of the CNT under 40° torsional deformation with the hydrogens at the dangling bonds of the perimeter C atoms optimized at DFT/3-21G level, (b) optimized structure of the hydrogen radical at the ridge, and (c) optimized structure of the C_5H_{11} radical at the ridge. Reprinted from [25] with permission.

by the change in hybridization of C atoms. On twisting the nanotube, some of the sp^2 hybridized carbon atoms become sp^3 and hence one of the sp^3 hybridized orbitals is free to form a covalent bond with an incoming atom or molecule. On a nondeformed tube, the C atoms have only a slight sp^3 character due to its cylindrical shape of the tube, which gives rise to a weak bond with a low binding energy.

On placing an alkyl radical (C_2H_5–CH$^{\cdot}$–C_2H_5) near to one of the C atoms at the ridges of the nanotube section of Figure 5.5a, and optimizing its geometry, the radical bound to the C atom, as shown in Figure 5.5c with a binding energy of 337.6 kJ/mol. On the other hand, the binding energy between the alkyl radical and a deformation-free nanotube is only 4.3 kJ/mol with CNT-alkyl C–C bond length of 3.7 Å, that is, the alkyl radical does not bind to free nanotube.

5.3.4
Bending

Under pure bending, the tube started kinking after reaching a strain energy of 1.88 kJ/mol. Sections of the tube under a strain energy of 1.88 (just before kinking) and 2.95 kJ/mol (soon after kinking) were used in the quantum mechanics study. The center section (kink site) with the hydrogen atoms added to the dangling bonds of the perimeter C atoms is shown in Figure 5.6a. Here again it has the ridges and flattened portion. The hydrogen atom placed near to the ridge bound to one of the C atoms at the ridge with binding energies of 232.0 (before kinking) and 398.9 kJ/mol (after kinking; Figure 5.6b). Although due to ovalization the curvature of the tube changes before kinking, it did not form well-developed ridge and flat surface. Thus the binding energy of radicals was low before kinking.

In summary, quantum mechanics calculations showed that the radicals strongly bind to the C atoms at the ridges; flattened surfaces show even a lower binding

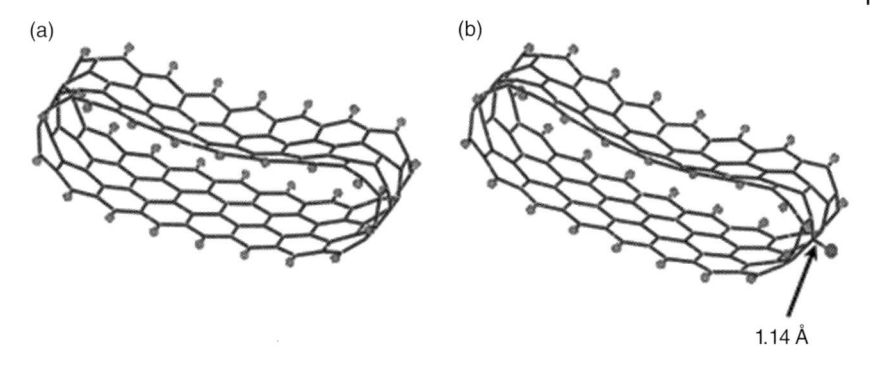

Figure 5.6 (a) Section of the NT under 40° bending deformation with the hydrogens at the dangling bonds of the perimeter C atoms optimized at DFT/3-21G level and (b) optimized structure of the hydrogen radical at the ridge. Reprinted from [25] with permission.

strength compared to the free tube as their behavior approaches that of a graphite sheet. Under torsion, the ridges are formed along the entire length of the tube and this gives a higher binding energy under relatively low strain energy. Thus this appears to be the best deformation mode to promote multiple reactions on nanotubes. Srivastava et al. [27] also predicted that the chemisorption of hydrogen atoms is enhanced by ~1.6 eV (i.e., ~154.4 kJ/mol) at regions of high deformation due to bending and torsion of SWNTs. In the blending of CNT–polymer composites, a CNT is often under a combined loading including many of the loading modes studies above. Furthermore, because of the large aspect ratio of CNTs, multiple kinking/buckling (can be hundreds) may form on a single CNT. These, according to the above discussion, will enhance the formation of chemical bonds. Thus mechanical deformations in composite blending can promote reactivity with nearby chemical species.

5.4 Functionalizing Carbon Nanotubes via Chemical Modification

Chemical functionalization offers an efficient route not only to modify the chemical and physical properties of CNTs but also to further expand their potential application areas. The main approaches for functionalizing CNTs through chemical modification can be grouped into two main categories – covalent attachment of chemical groups and noncovalent adsorption or wrapping of functional molecules onto the tubes.

5.4.1 Covalent Modification

Covalent functionalization of CNTs allows functional groups to be attached to tube ends or sidewalls. When chemical groups are attached to CNTs through covalent bonds, the nanotube carbon atoms are chemically modified. Excessive modification of the CNTs can ruin the tubular framework and thereby reduce its mechanical strength. On functionalizing CNTs by acid treatment, carboxylic acid groups are introduced on the surface of CNTs that leads to stabilization in polar solvents and helps to covalently link amides, amines, polymers, and other groups. However, the acids attack defective sites of the CNTs and cut them into many short nanotubes, thereby decreasing their aspect ratios. Despite this, covalent modification of CNTs has been spotted as a viable route to broaden possible applications. Some details are discussed below.

5.4.1.1 Functionalization with Fluorine

Fluorination of CNTs was one of the first sidewall functionalization reactions in 1996 [28]. Jaffe [29] used *ab-initio* quantum chemistry approach for the chemical modification of CNT sidewalls with fluorine. Owing to the large size of complete segments on CNTs, Jaffe used polycyclic aromatic hydrocarbon molecules with curvature maintained by external constraints to represent pieces of (10,10), (5,5), and (16,0) nanotubes. DFT calculations with a B3LYP functional and a 6-31G(d) basis set showed that the F atoms preferred to bond next to each other. The use of small hydrocarbons to simulate a nanotube sidewall can yield insight into the bonding, but the best model might not always be obvious. For example a complete geometry optimization at the reaction site is important but it would result in the hydrocarbon becoming planar and hence it would not reflect the nanotube sidewall. Bauschlicher [30, 31] used the hybrid QM/MM approach to study the hydrogen and fluorine binding to the sidewall of a (10,0) CNT, where a two-level ONIOM approach (developed by Morokuma and coworkers [32]) in which molecular mechanics is used to treat a sizable segment of the nanotube and DFT(B3LYP) is used to study the area where F or H atom attaches to the sidewall, thus eliminating the constraints used in Jaffe's work. These calculations also showed the F atoms appear to favor bonding next to existing F atoms. Similar to fluorinated fullerenes [33], SWCNTs once fluorinated can serve as a staging point for a wide variety of sidewall chemical functionalizations. By treating fluorinated tubes with strong nucleophiles such as Grignard reagents, alkyl lithium reagents, and metal alkoxides, the fluorine substituents were replaced and derivatized products, for example, methyl-, butyl-, hexyl-, and methoxylnanotubes, were formed [34, 35].

5.4.1.2 Functionalization with Carboxylic Group, Amine Group, and Amide Group

Veloso *et al.* [36] functionalized SWCNTs by investigating the interaction of $-COOH$, $-NH_2$, and $-CONH_2$ groups with an (8,0) SWCNT using total energy *ab-initio* calculations based on DFT adopting a generalized gradient approximation for the exchange and correlation terms. All three functional groups induced local distortions along the radial directions on the tube sidewall due to the local sp^3 hybridization of the C–C or C–N bonding between the C atom of the tube and the C or N atoms of the functional group, demonstrating covalent bonds between the SWCNT and the functional groups as shown in Figure 5.7a–c. These covalently

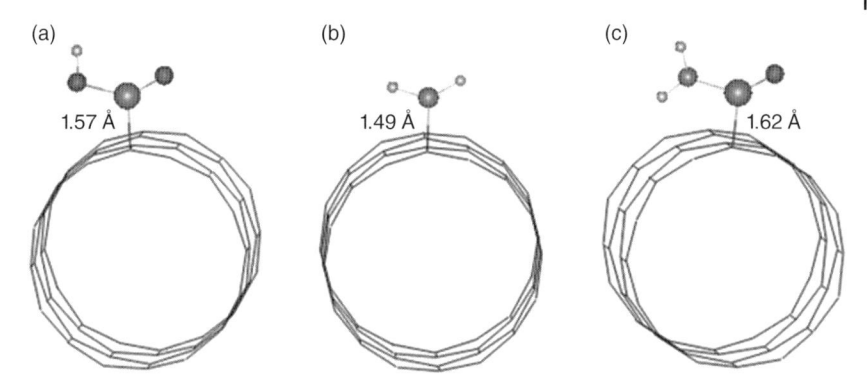

Figure 5.7 Structural configurations of the sidewall functionalization: (a) SWCNT–COOH, (b) SWCNT–NH$_2$, (c) SWCNT–CONH$_2$.

functionalized CNTs are promising materials for catalysis, electronic and optic devices, sensors, gas storage, and making high-performance composites.

5.4.1.3 Functionalization with Nitro Group

Mercuri and Sgamellotti [37] exploited the sidewall functionalization of CNTs with NO$_2$, as a way to develop gas-sensing devices. The adsorption of NO$_2$ on the sidewall of CNT was modeled by performing a hybrid QM/MM calculation. They analyzed the interaction of NO$_2$ (i) with the perfect hexagonal framework of CNT sidewall, (ii) with CNT having Stone–Wales defect constituting two adjacent pentagon–heptagon ring pairs, originating from the rotation of a single C–C bond on the hexagonal network of the sidewall by 180°, and (iii) with a site originating from the vacancy of carbon atom on the sidewall considering the "nitro" attack and "oxo" attack individually. The results indicated a low tendency to adsorption of NO$_2$ for the perfect hexagonal network of the sidewall. Chemisorption on the Stone–Wales defects is slightly favored compared to the defect-free sidewall showing that the defects can partially account for the gas-sensing action. In the "nitro" configuration, the interaction between NO$_2$ to the sidewall at the vacancy indicated a strong chemisorption; for the "oxo" attack, the results suggested the dissociation of NO$_2$ at the vacancy with the formation of a new C–O bond and the release of an NO molecule. Experimental results also indicated a larger tendency to reaction for more defected CNTs and the presence of C–N and carbonyl groups, upon exposure to NO$_2$.

5.4.1.4 Functionalization with Transition Metal Complex

New directions for the functionalizations have been opened by the extension of transition metal chemistry to the interaction with the CNT sidewall. Mercuri and Sgamellotti [38] investigated the interaction of SWCNTs with Vaska's complex [39] (*trans*-Ir(CO)Br(PPh$_3$)$_2$–an electron-rich transition metal complex) by means of hybrid QM/MM calculations in which Vaska's complex and a portion of CNT sidewall (~16–24 carbon atoms) were included in the QM part. For the MM part, the Tripos force field [40] that is known to give good results for π–π interactions and for the QM part, DFT with Becke [15] and Perdew [41] corrections that gives a good description of the metal–ligand interaction were used. Investigation of the interaction between the complex and (i) the defect-free sidewall, (ii) capped-end, and (iii) pentagon–heptagon defect showed that the defect-free sidewall is almost inert to coordination with the transition metal. However, nanotube end-caps or defective sites show higher tendency to coordination with the inorganic fragment, which indicate that a stable adduct is likely to be formed when at least one of the coordinating carbon atoms belongs to a pentagonal ring. Thus CNTs can be functionalized with transition metal complexes. The formation of a stable adduct has been experimentally evidenced by Banerjee and Wong [42].

5.4.1.5 Functionalization via the Introduction of Free Radical

Mylvaganam and Zhang [43] studied the reaction of a simple oxy radical $^{\cdot}OCH_3$, with a (5,0) zigzag nanotube segment having 60 C atoms (length = 11.36 Å) with hydrogen atoms added to the dangling bonds of the perimeter carbons by fully optimizing the geometry using DFT with a hybrid functional B3LYP and a 3-21G basis set. The optimized geometry of the $C_{60}H_{10}$–OCH_3 radical (Figure 5.8) show the newly formed C–O covalent bond having a bond length of 1.463 Å. The three C–C bonds radiating from the substituted sp^3 carbon of the nanotube are lengthened to ~1.5 Å. The newly formed nanotube-bound radical is delocalized over several nanotube carbon atoms and is stabilized by ~168 kJ/mol. This value is likely to increase with the addition of polarization functions to carbon atoms.

5.4.1.6 Functionalization via the Introduction of Anion

Reaction of secondary butyl anion ($CH_3CH^-CH_2CH_3$) with the nanotube model $C_{60}H_{10}$ was studied by fully optimizing the geometry at the DFT(B3LYP)/3-21G level. The optimized geometry of the $C_{60}H_{10}$–C_4H_9 anion is presented in Figure 5.9. This anion is lower in energy compared to its corresponding reactants by

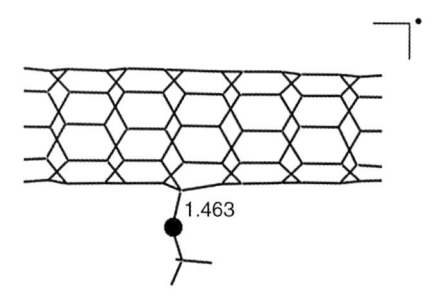

Figure 5.8 DFT(B3LYP)/3-21G optimized geometry of $C_{60}H_{10}$–OCH_3 radical. Note that the solid sphere within the molecular framework represents an oxygen atom. Reprinted from [43] with permission.

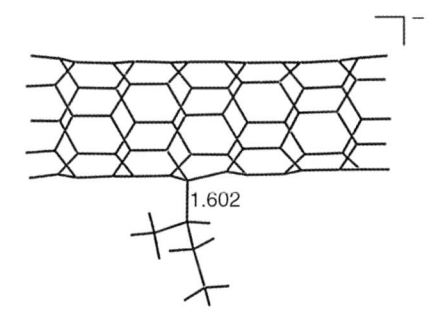

Figure 5.9 DFT(B3LYP)/3-21G optimized geometry of $C_{60}H_{10}$–C_4H_9 anion [43].

~514 kJ/mol. This shows that the CNT can be functionalized with the standard anionic initiator butyl lithium and the resulting CNT anion is very stable.

Nanotubes modified through covalent bonds via the introduction of radicals or anions (methods (e) and (f)) can serve as initiators for the *in-situ* polymerization process that can further lead to polymer-grafted CNT composites.

5.4.2
Noncovalent Modification

A CNT surface can be modified via van der Waals forces and π–π interactions, by adsorption or wrapping of poly-nuclear aromatic compounds, surfactants, polymers, or biomolecules. The main advantage is that chemical functionalities can be introduced to the CNTs without affecting the structure and electronic network of the tubes. Unlike the covalent modifications described above, which introduce defects in the form of sp^3 hybridized carbons that can cause loss of remarkable thermal, electrical, and mechanical properties of CNTs, the non-covalent modification also known as supramolecular functionalization preserves the above properties and enables the utilization of nanotubes in their unmodified state.

5.4.2.1 $CHCl_3$ Adsorption

Girao *et al.* [44] functionalized SWCNTs through chloroform adsorption. As grown, CNTs generally show several defects. Single vacancies are found to be the most important structural defects in CNTs since they significantly promote changes in the physical and chemical properties of the nanotubes. Scientists investigated the reaction of $CHCl_3$ with pristine, and defective SWCNTs in (i) "atom" configuration where the H atom or Cl atom is placed directly above a C atom of the nanotube surface, (ii) "ring" configuration where the H or Cl atom interacts with the SWCNT ring, (iii) "net" configuration where each one of the three Cl atoms is situated on the center of the nanotube ring, (iv) vac-A configuration, where one Cl atom close to the vacancy, and (v) vac-B configuration where three Cl atoms close to the vacancy. Using first principles calculations based on

DFT adopting LDA to account for exchange and correlation terms, they predicted a physisorption regime in all cases. This shows that SWCNTs are promising materials for extracting trihalomethanes from the environment. The stability of the SWCNT–CCl_2 is comparable with SWCNT–COOH system and is in agreement with FTIR and Raman scattering experiments.

5.4.2.2 Adsorption of Metalloporphyrin Complexes

In porphyrins, the flat planar aromatic structures are ideal for π-stacking interactions with the sidewalls of SWCNTs and they play an important role in industry, technology, and biological systems [45]. Compared to porphyrin–SWCNT complexes, the supramolecular metalloporphyrin–SWCNT systems have more interesting magnetic, optic, and spintronic properties, suggesting wider applications such as designing novel biosensors, catalysts, and spintronic molecular devices. Zhao and Ding [46] investigated the interaction of metalloporphyrins (CoP, NiP, CuP, and ZnP) molecules with zigzag (10,0) and armchair (6,6) SWCNTs using DFT within the generalized gradient approximation of the Perdew–Burke–Ernzerhof functional [47], which is proven to give good description for the nature of intermolecular bonding [48]. They found that the adsorption energies of metalloporphyrin functionalized semiconducting (10,0) tubes varied from −0.59 to −0.61 eV and metallic (6,6) tubes varied from −0.43 to −0.51 eV, depending on the identity of the center metal of metalloporphyrin. The closest metal–tube distances were in the range of 3.15–3.3 Å for semiconducting tube and 3.4–3.6 Å for metallic tube. These results showed that the adsorption of metalloporphyrin complexes on semiconducting SWCNTs is much stronger than that on the metallic SWCNTs. In addition, the charge transfer from metalloporphyrin molecules to the (10,0) SWCNT is significantly larger than that to the (6,6) SWCNT. Hence the semiconducting tubes can be separated from metallic tubes through the adsorption of metalloporphyrin. Moreover, irrespective of the tube's chirality, CoP–SWCNT and CuP–SWCNT complexes showed a net magnetic moment because of the presence of the unpaired electrons, making them possess interesting magnetic and optic properties. For the above purposes, it is vital to functionalize the sidewalls of SWCNTs with metalloporphyrin molecules, which is critical to the physical properties of SWCNTs and their applications.

5.4.2.3 Interaction of Aromatic Amino Acids

Noncovalent interactions of biological molecules have potential applications in molecular electronics, as biosensors and chemical sensors. Electrical conductance of CNTs changes appreciably as proteins are immobilized on their sidewalls [49]. Studies on the structure–function–affinity of peptides with CNTs have shown that the aromatic amino acid phenylalanine has an important role in the enhancement of the adsorption properties of CNTs [50]. Rajesh *et al.* [51] investigated the nature of interaction of four aromatic amino acids (phenylalanine, histidine, tyrosine, and tryptophan) with CNTs, which has implications toward developing biosensors, through *ab-initio* calculations. They have used DFT within the generalized gradient approximation as implemented in Vienna *ab-initio* simulation package (VASP) for geometry optimizations and Moller–Plesset second-order perturbation theory (MP2) using a 6-31G* basis set to get more accurate prediction of the interaction between the amino acids and CNTs. These studies showed that the aromatic rings of the amino acids prefer to orient in parallel with respect to the plane of the CNTs, which bears the signature of weak π–π interactions. In addition, the strength of the interaction was found to be in excellent correlation to the polarizability of the aromatic motifs of the amino acids.

5.4.2.4 Encapsulation of Organic Molecules

For molecular electronic applications, it is essential to obtain both p- and n-type air-stable CNTs. SWCNT filed-effect transistors under ambient conditions are p-type due to electron withdrawing by the absorbed oxygen. Thus the production of n-type air-stable CNT transistors is technologically important.

Air-stable amphoteric doping has been realized by encapsulating organic molecules inside SWCNTs. On tuning the electron affinity or ionization potential of the encapsulated molecules, researchers have shown that both p- and n-type doping of CNTs can be realized [52, 53]. Encapsulation of electrophilic organic molecules realizes a controllable p-type doping, whereas nucleophilic organic molecules lead to controllable n-type doping. Lu et al. [52] modeled the doping of a zigzag (16,0) SWCNT with electrophilic organic molecules such as tetracyano-p-quinodimethane and tetraflurocyano-p-quinodimethane and nucleophilic organic molecule such as tetrathiafulvalene using DFT. They find that these molecules are physisorbed inside the nanotube.

Sa et al. [53] investigated the functionalization of an armchair (8,8) SWCNT by tetrathiafulvalene and found that the external absorption is less stable than the encapsulation one. The molecule is physisorbed (i.e., no covalent bonding between the molecule and the CNT) inside the tube. Mulliken population analysis showed that tetrathiafulvalene molecules are electron donors to the armchair nanotube. This charge transfer induces an ambipolar system and the resulting electrostatic interaction may play a role for its stability.

5.5
Polymer Grafting

It is known that the radicals and negative charges generated on fullerene C_{60} can serve as initiators for further reactions with alkenes or epoxides that lead to the formation of polymer-grafted fullerene [54, 55]. Based on this, a few papers discussed chemically functionalizing nanotubes via the introduction of radicals or anions that can serve as initiators for the *in-situ* polymerization process, which can further lead to polymer-grafted CNT composites [4, 10, 56, 57]. Each of these functionalization methods has their own merit. In synthesizing different polymers, different initiators are used following different mechanisms. For example, polyethylene (PE) is prepared by free radical addition polymerization as described

Scheme 1

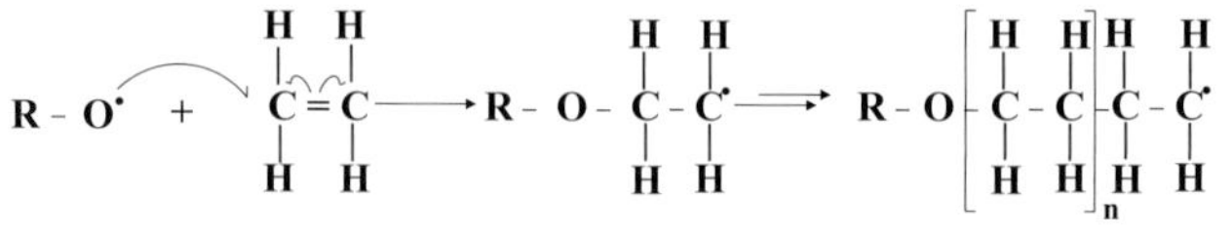

$$R\text{-}O^{\bullet} + \underset{H\ H}{\overset{H\ H}{C=C}} \longrightarrow R\text{-}O\text{-}\underset{H\ H}{\overset{H\ H}{C\text{-}C^{\bullet}}} \longrightarrow R\text{-}O\left[\underset{H\ H}{\overset{H\ H}{C\text{-}C}}\right]_n\underset{H\ H}{\overset{H\ H}{C\text{-}C^{\bullet}}}$$

Scheme 2

$$Bu^-\ Li^+\ \underset{H}{\overset{H}{C=C}}\underset{R_4}{\overset{O\ R_3}{}} \longrightarrow Bu\text{-}\underset{H\ R_4}{\overset{H\ R_3}{C\text{-}C\text{-}O^-}}Li^+ \longrightarrow Bu\text{-}\underset{H\ R_4}{\overset{H\ R_3}{C\text{-}C}}\left[O\text{-}\underset{H\ R_4}{\overset{H\ R_3}{C\text{-}C}}\right]_n O^-\ Li^+$$

$$\left(C_2H_5\text{-}\underset{CH_3}{\overset{H}{C}}\text{-}\right) \equiv Bu$$

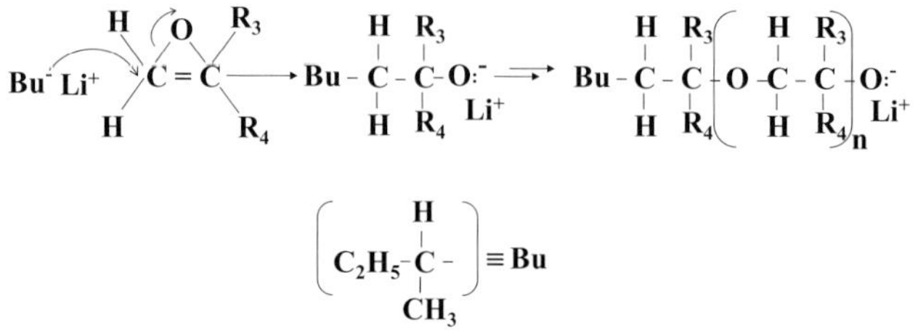

Figure 5.10 Free radical addition (Scheme 1) and anionic (Scheme 2) polymerization schemes.

in Scheme 1 of Figure 5.10, whereas polystyrene and polyepoxide are prepared by anionic polymerization as described in Scheme 2 of Figure 5.10.

5.5.1
Grafting Polyethylene

5.5.1.1 "Grafting from" Technique

Mylvaganam and Zhang [43] studied the reaction of C_2H_4 monomer with CNT–methoxy radical formed via the introduction of $^{\bullet}OCH_3$ (i.e., free-radical-functionalized CNT) by fully optimizing the geometry of the reactants and products using the DFT. This led to the formation of a new radical ($C_{60}H_{10}$–OCH_3–C_2H_4 radical, Figure 5.11a) that is 131 kJ/mol lower in energy than the reactants. According to the Mulliken spin density analysis [59], the free electron is mainly localized on the carbon at the free end of C_2H_4, indicating that the resulting radical is ready to undergo a propagation reaction with another C_2H_4 monomer. Addition of a second C_2H_4 molecule formed a bond with the C atom where the free electron was localized showing the typical propagation step in polymerization. The optimized geometry is shown in Figure 5.11b. The new radical is 107.8 kJ/mol lower in energy compared to the reactants. Thus the radical generated on the nanotube acts as an initiator so that a PE chain could be grafted onto the nanotube via *in-situ* polymerization.

5.5.1.2 "Grafting to" Technique

Fullerene can be free radically grafted onto saturated hydrocarbon polymers such as ethylene–propylene copolymer by heating them to 150 °C in the presence of *tert*-butyl peroxide [55]. CNTs are fullerene-related structures. Moreover, in general peroxides are capable of abstracting hydrogen atoms from polymer chains. On the

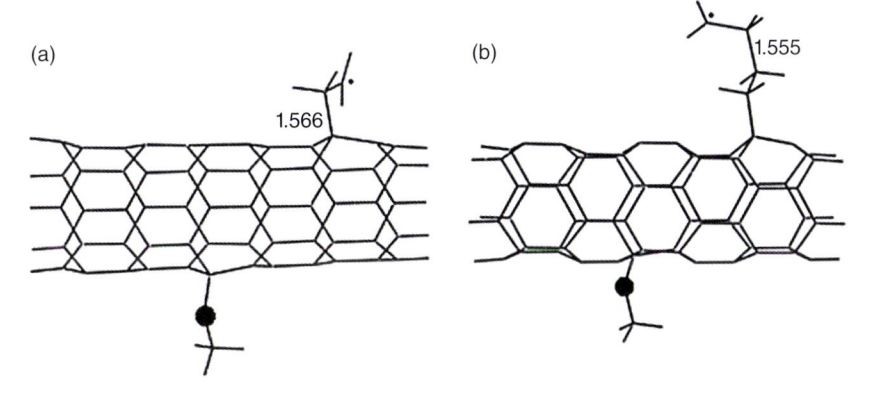

Figure 5.11 DFT/B3LYP optimized structure of (a) $C_{60}H_{10}$–OCH_3–C_2H_4 radical, (b) $C_{60}H_{10}$–OCH_3–CH_2–CH_2–CH_2–CH_2 radical. Reprinted from [58] with permission.

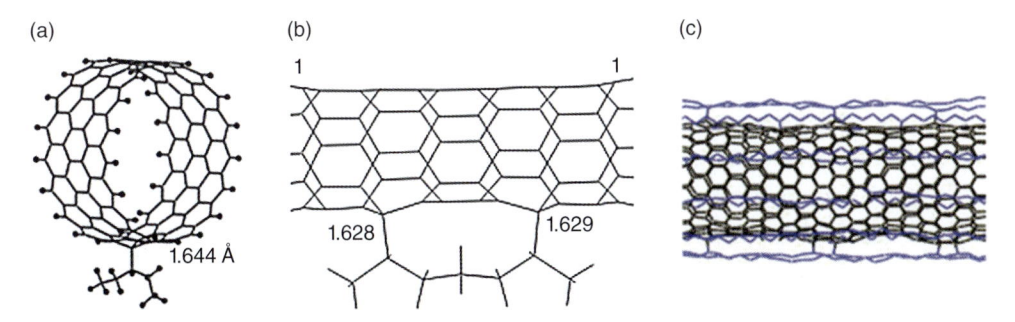

Figure 5.12 (a) DFT/B3LYP(3-21G) optimized structure of SWCNT–C_5H_{11}, (b) DFT/B3LYP(3-21G) optimized structure of SWCNT–C_7H_{14} biradical, (c) a section of molecular dynamics energy minimized structure of SWCNT–PE chains (PE backbone chains are shown in blue) [58, 60].

basis of the above idea, the possible chemical bond formation between the CNT and PE chain radicals was explored using model compounds [58]. Nanotube–polyethylene chain models were generated and investigated by quantum mechanics and molecular dynamics methods. The quantum mechanics calculations used a segment of a CNT with hydrogen atoms added to the dangling bonds of the perimeter carbons and alkyl radicals ($C_5H_{11}^{\bullet}$ and C_7H_{14} biradical with unpaired electrons on the 2nd and 6th carbon atoms) to represent the PE chains. The geometrical parameters were fully optimized using DFT with a hybrid functional B3LYP and a 3-21G basis set. A new C–C bond formed between the section of the nanotube sidewall and the alkyl radical ($C_5H_{11}^{\bullet}$) with a bond length of 1.644 Å demonstrating a covalent bond formation and the bonding at the attachment carbon become tetrahedral with elongated C–C bonds from the attachment site to neighboring carbon atoms (Figure 5.12a). Two new covalent bonds formed with bond lengths of 1.629 and 1.628 Å between CNT sidewall and C_7H_{14} biradical (Figure 5.12b).

In the molecular dynamics method, 12 PE chains, with each having 56 methylene groups, were placed around a (17,0) CNT of about 75 Å long. Five hydrogen atoms at equal distances along each PE chain were removed to generate radical centers (typically peroxides are capable of doing this). On minimizing the energy by conjugate gradient method, several new C–C bonds between the PE chains and CNTs were formed (Figure 5.12c).

Thus both the calculations at the molecular dynamics and quantum mechanics scales demonstrate that it is possible to form chemical bonds between the PE chains and the CNT by introducing radicals on the PE chains. In principle, radicals can be generated on PE chains by chemical or radiation attack. For example, an oxy radical formed by the pyrolysis of peroxide is capable of abstracting a hydrogen atom on the polymer chain. This method involves the attachment of premade polymer chains to the surface of CNTs (i.e., "grafting to" method).

5.5.2
Grafting Polypropylene Oxide

DFT studies of the reaction of propylene oxide with CNT–secondary butyl anion formed via the introduction of secondary butyl anion (i.e., the anion-functionalized CNT) resulted in the new $C_{60}H_{10}$–C_4H_9–propylene oxide anion as shown in Figure 5.13a. The new C–C bond length is 1.563 Å and the CNT anion attacks the methylene carbon of the propylene oxide. Here the reactants and the products have almost the same energy. The Mulliken charge density analysis showed that a portion of the negative charge is on the oxygen atom, indicating that the resulting anion can react with another propylene oxide monomer so that the reaction can propagate.

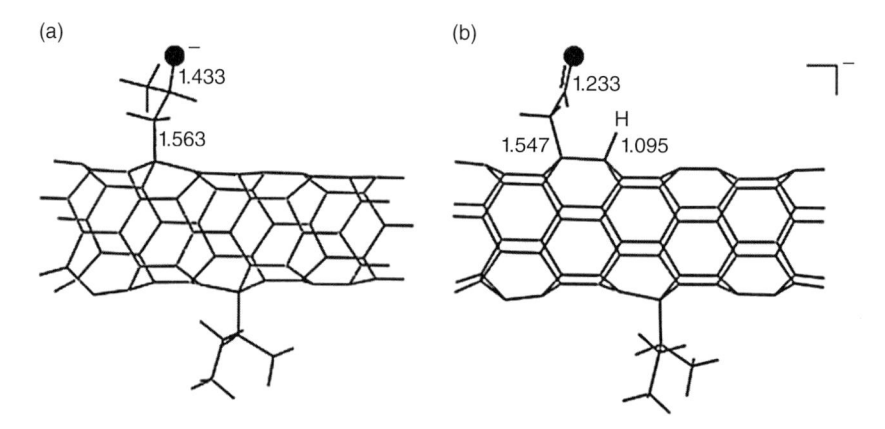

Figure 5.13 DFT/B3LYP optimized geometry of (a) $C_{60}H_{10}$–C_4H_9–propylene oxide anion and (b) $C_{60}H_{10}$–C_4H_9–ethylene oxide anion. Reprinted from [43] with permission.

Thus the anion generated on the CNT act as an initiator so that a poly propylene oxide chain could be grafted onto the nanotube via *in-situ* polymerization.

On the other hand, the reaction of anion-functionalized CNT with ethylene oxide resulted in a $C_{60}H_{10}$–C_4H_9–ethylene oxide anion as shown in Figure 5.13b. Unlike the reaction with propylene oxide, in this case although the epoxide ring opened up as in Scheme 2 of Figure 5.10, but the C–O bond shortened to –C=O and lost one hydrogen to the nanotube. As a result, the oxygen atom no longer possesses a negative charge for the reaction to propagate. This means that grafting a PE oxide chain onto nanotube via anionic polymerization may not be feasible. Thus the *ab-initio* DFT calculations are capable of predicting whether a reaction will take place or not.

5.5.3
Grafting Oligo-N-Vinyl Carbazole

Poly vinyl carbazole has interesting photoconductive properties [61, 62]. As such, very recently, Zaidi *et al.* [63] explored grafting of short-oligo-N-vinylcarbazole (OVK) onto SWCNTs by performing different experimental analyses (optical absorption, photoluminescence, infrared, and Raman) and theoretical studies based on *ab-initio* and semiempirical AM1 [64] methods. The structures of OVK having four VK units and a section of SWCNT having a diameter of 1.3 nm had been fully optimized at the DFT(B3LYP)/3-21G* level. The structure and the infrared vibrational frequencies of OVK-SWCNTs composite were calculated at the *ab-initio* Hartree–Fock and semiempirical AM1 level, which is an effective tool for qualitative study of functionalized nanotubes [65]. Zaidi *et al.* compared the force constants between neighboring atoms in both neutral and oxidized states to rationalize the reactive sites and find that the grafting process takes place on the nanotube sidewall. The appearance of a new aliphatic C–C band, disappearance of peaks characteristic to vinylidene groups, and the apparition of two –CH_2 features in the infrared vibrational frequencies further supported the covalent bonding of OVK vinylidene groups with the SWCNT.

5.6
Summary

Practical applications of CNTs critically depend on their properties. Since it is not practically feasible to grow the nanotube with specific physical property at the present, engineering the properties of as-grown CNTs through functionalization has been an important process. Theoretical investigations, as introduced in this chapter, have indicated that it is possible to attach a wide range of functional groups such as fluorine, carboxyl, amine, amide, etc., through covalent bonds and chloroform, metalloporphyrin complexes, amino acids, etc., through noncovalent bonds on the sidewall of CNTs. Functionalization of CNTs through encapsulation of organic molecules can lead to the realization of p- and n-type doping of CNTs that are essential for molecular electronic applications. In addition, the functionalization of CNTs via the introduction of radicals or anions serves as initiators for the subsequent grafting of polymers onto CNTs.

References

1 Geng, H., Rosen, R., Zheng, B., Shimoda, H., Fleming, L., Liu, J., and Zhou, O. (2002) *Adv. Mater.*, **14**, 1387.
2 Gong, X., Liu, J., Baskaran, S., Voise, R.D., and Young, J.S. (2000) *Chem. Mater.*, **12**, 1049.
3 Schadler, L.S., Giannaris, S.C., and Ajayan, P.M. (1998) *Appl. Phys. Lett.*, **73**, 3842.
4 Viswanathan, G., Chakrapani, N., Yang, H., Wei, B., Chung, H., Cho, K., Ryu, C.Y., and Ajayan, P.M. (2003) *J. Am. Chem. Soc.*, **125**, 9258.
5 Dieckmann, G.R., Dalton, A.B., Johnson, P.A., Razal, J., Chen, J., Giordano, G.M., Munoz, E., Musselman, I.H., Baughman, R.H., and Draper, R.K. (2003) *J. Am. Chem. Soc.*, **125**, 1770.
6 Zhu, W.H., Minami, N., Kazaoui, S., and Kim, Y. (2004) *J. Mater. Chem.*, **14**, 1924.
7 Xiao, K.Q. and Zhang, L.C. (2005) *J. Mater. Sci.*, **40**, 6513.
8 Meuer, S., Braun, L., and Zentel, R. (2009) *Macromol. Chem. Phys.*, **210**, 1528.
9 Homenick, C.M., Lawson, G., and Adronov, A. (2007) *Polym. Rev.*, **47**, 265.
10 Qin, S.H., Qin, D.Q., Ford, W.T., Resasco, D.E., and Herrera, J.E. (2004) *Macromolecules*, **37**, 752.
11 Hohenberg, P., and Kohn, W. (1964) *Phys. Rev. B*, **136**, B864.
12 Kohn, W., and Sham, L.J. (1965) *Phys. Rev.*, **140**, 1133.
13 Perdew, J.P. and Zunger, A. (1981) *Phys. Rev. B*, **23**, 5048.
14 Grimme, S., Antony, J., Schwabe, T., and Muck-Lichtenfeld, C. (2007) *Org. Biomol. Chem.*, **5**, 741.
15 Becke, A.D. (1988) *Phys. Rev. A*, **38**, 3098.
16 Becke, A.D. (1993) *J. Chem. Phys.*, **98**, 5648.
17 Lee, C.T., Yang, W.T., and Parr, R.G. (1988) *Phys. Rev. B*, **37**, 785.
18 Vosko, S.H., Wilk, L., and Nusair, M. (1980) *Can. J. Phys.*, **58**, 1200.
19 D'Souza, F., Zandler, M.E., Smith, P.M., Deviprasad, G.R., Arkady, K., Fujitsuka, M., and Ito, O. (2002) *J. Phys. Chem. A*, **106**, 649.
20 Tersoff, J. (1986) *Phys. Rev. Lett.*, **56**, 632.
21 Tersoff, J. (1989) *Phys. Rev. B*, **39**, 5566.
22 Brenner, D.W. (1990) *Phys. Rev. B*, **42**, 9458.
23 Brenner, D.W., Shenderova, O.A., Harrison, J., Stuart, S.J., Ni, B., and Sinnott, S. (2002) *J. Phys. Condens. Matter*, **14**, 783.
24 Mylvaganam, K. and Zhang, L. (2004) *Carbon*, **42**, 2025.
25 Mylvaganam, K. and Zhang, L.C. (2006) *Nanotechnology*, **17**, 410.
26 Binkley, J.S., Pople, J.A., and Hehre, W.J. (1980) *J. Am. Chem. Soc.*, **102**, 939.
27 Srivastava, D., Brenner, D.W., Schall, J.D., Ausman, K.D., Yu, M., and Ruoff, R. (1999) *J. Phys. Chem. B*, **103**, 4330.
28 Nakajima, T., Kasamatsu, S., and Matsuo, Y. (1996) *Eur. J. Solid State Inorg. Chem.*, **33**, 831.
29 Jaffe, R.L. (2003) *J. Phys. Chem. B*, **107**, 10378.
30 Bauschlicher, C.W. (2000) *Chem. Phys. Lett.*, **322**, 237.
31 Bauschlicher, C.W. (2001) *Nano Lett.*, **1**, 223.
32 Maseras, F. and Morokuma, K. (1995) *J. Comput. Chem.*, **16**, 1170.
33 Taylor, R., Holloway, J.H., Hope, E.G., Avent, A.G., Langley, G.J., Dennis, T.J., Hare, J.P., Kroto, H.W., and Walton, D.R.M. (1992) *J. Chem. Soc. Chem. Commun.*, 665.
34 Boul, P.J., Liu, J., Mickelson, E.T., Huffman, C.B., Ericson, L.M., Chiang, I.W., Smith, K.A., Colbert, D.T., Hauge, R.H., Margrave, J.L., and Smalley, R.E. (1999) *Chem. Phys. Lett.*, **310**, 367.
35 Mickelson, E.T., Chiang, I.W., Zimmerman, J.L., Boul, P.J., Lozano, J., Liu, J., Smalley, R.E., Hauge, R.H., and

Margrave, J.L. (1999) *J. Phys. Chem. B*, **103**, 4318.

36 Veloso, M.V., Souza, A.G., Mendes, J., Fagan, S.B., and Mota, R. (2006) *Chem. Phys. Lett.*, **430**, 71.

37 Mercuri, F. and Sgamellotti, A. (2007) *Inorg. Chim. Acta*, **360**, 785.

38 Mercuri, F. and Sgamellotti, A. (2006) *J. Phys. Chem. B*, **110**, 15291.

39 Vaska, L. (1963) *Science*, **140**, 809.

40 Clark, M., Cramer, R.D., and Vanopdenbosch, N. (1989) *J. Comput. Chem.*, **10**, 982.

41 Perdew, J.P. (1986) *Phys. Rev. B*, **33**, 8822.

42 Banerjee, S. and Wong, S. (2002) *Nano Lett.*, **2**, 49.

43 Mylvaganam, K. and Zhang, L.C. (2004) *J. Phys. Chem. B*, **108**, 15009.

44 Girao, E.C., Liebold-Ribeiro, Y., Batista, J.A., Barros, E.B., Fagan, S.B., Mendes, J., Dresselhaus, M.S., and Souza, A.G. (2010) *Phys. Chem. Chem. Phys.*, **12**, 1518.

45 Bonnett, R. (1995) *Chem. Soc. Rev.*, **24**, 19.

46 Zhao, J.X. and Ding, Y.H. (2008) *J. Phys. Chem. C*, **112**, 11130.

47 Perdew, J.P., Burke, K., and Ernzerhof, M. (1996) *Phys. Rev. Lett.*, **77**, 3865.

48 Wang, Y.B. and Lin, Z.Y. (2003) *J. Am. Chem. Soc.*, **125**, 6072.

49 Chen, R.J., Bangsaruntip, S., Drouvalakis, K.A., Kam, N.W.S., Shim, M., Li, Y.M., Kim, W., Utz, P.J., and Dai, H.J. (2003) *Proc. Natl. Acad. Sci. USA*, **100**, 4984.

50 Zorbas, V., Smith, A.L., Xie, H., Ortiz-Acevedo, A., Dalton, A.B., Dieckmann, G.R., Draper, R.K., Baughman, R.H., and Musselman, I.H. (2005) *J. Am. Chem. Soc.*, **127**, 12323.

51 Rajesh, C., Majumder, C., Mizuseki, H., and Kawazoe, Y. (2009) *J. Chem. Phys.*, **130**, 124911.

52 Lu, J., Nagase, S., Yu, D.P., Ye, H.Q., Han, R.S., Gao, Z.X., Zhang, S., and Peng, L.M. (2004) *Phys. Rev. Lett.*, **93**, 116804.

53 Sa, N., Wang, G., Yin, B., and Huang, Y. (2008) *Physica E*, **40**, 2396.

54 Ederle, Y. and Mathis, C. (1997) *Macromolecules*, **30**, 4262.

55 Patil, A.O. and Brois, S.J. (1997) *Polymer*, **38**, 3423.

56 Kong, H., Gao, C., and Yan, D.Y. (2004) *J. Am. Chem. Soc.*, **126**, 412.

57 Shaffer, M.S.P. and Koziol, K. (2002) *Chem. Commun.*, 2074.

58 Mylvaganam, K. and Zhang, L.C. (2004) *J. Phys. Chem. B*, **108**, 5217.

59 Mulliken, R.S. (1955) *J. Chem. Phys.*, **23**, 1833.

60 Mylvaganam, K. and Zhang, L.C. (2006) *Key Eng. Mater.*, **312**, 217.

61 Napo, K., Chand, S., Bernede, J.C., Safoula, G., and Alimi, K. (1992) *J. Mater. Sci.*, **27**, 6219.

62 Safoula, G., Touihri, S., Bernede, J.C., Jamali, M., Rabiller, C., Molinie, P., and Napo, K. (1999) *Polymer*, **40**, 531.

63 Zaidi, B., Bouzayen, N., Wery, J., and Alimi, K. (2010) *J. Mol. Struct.*, **971**, 71.

64 Dewar, M.J.S., Zoebisch, E.G., Healy, E.F., and Stewart, J.J.P. (1985) *J. Am. Chem. Soc.*, **107**, 3902.

65 Wongchoosuk, C., Udomvech, A., and Kerdcharoen, T. (2009) *Curr. Appl. Phys.*, **9**, 352.

6
Covalent Binding of Nanoparticles on Carbon Nanotubes
Daxiang Cui

6.1
Introduction

Since Iijima discovered carbon nanotubes (CNTs) in 1991 [1], many novel properties of CNTs such as unique mechanical, physical, and chemical properties have been gradually discovered [2–4], and CNTs including single-walled carbon nanotubes (CWCNTs) and multiwalled carbon nanotubes (MWNTs) have become ideal building blocks for nanoengineering [5–8]. Up to date, how to fully use unique properties of CNTs to develop new types of nanoelectronic devices or ultrasensitive detection method for disease diagnosis is a novel hotspot.

Nanoparticles (NP) such as magnetic nanoparticles [9, 10], quantum dots (QDs) [11, 12], gold NPs [13, 14], etc., own four basic nanoscale effects such as small size effects, quantum effects, surface effects, tunnel effects, and Raman-enhanced effects, exhibit great potential in applications such as microelectronic devices, biomolecules-absorption and separation, fluorescent detection, molecular imaging, drug delivery, etc. How to fully employ advantages of CNTs and NPs to develop new types of nanodevices or nanocomposites with special function has become a novel interdisciplinary frontier in materials science and life science.

NPs are conjugated with CNTs; the resultant NPs-conjugated CNTs composites own much more advantages than pure CNTs or only NPs. Such nanostructures possess unique physical and chemical properties and could serve as a basis for novel nanodevices. Integration of NPs and CNTs provides a novel opportunity in the construction of nanodevices with special properties. These NPs-conjugated CNTs composites own great potential in applications such as gene- or drug-delivery system, tumor-targeting therapeutic reagent, special nanodevices, etc. Here we review some of the main advances in this field over the past few years, explore the application prospects, with the aim of stimulating a broader interest in developing NPs-conjugated CNTs composites with special function.

Surface Modification of Nanotube Fillers, First Edition. Edited by Vikas Mittal.
© 2011 Wiley-VCH Verlag GmbH & Co. KGaA. Published 2011 by Wiley-VCH Verlag GmbH & Co. KGaA.

6.2
Covalent Binding of Quantum Dots on CNTs

CNTs have been fully investigated in recent years due to their interesting physical and chemical properties [15, 16]. Currently, a large amount of research in CNT chemistry is directed to the synthesis of diverse organic derivatives with the aim to promote surface functionalization and simultaneously provide colloidal stability to the resulting modified CNT in organic or aqueous media [17]. Such surface modification of CNT through covalent bonding of functional organic molecules opens the way to structural materials of high technological importance [18–20]. Recently, a new class of CNT products has been derived from the combination of CNT with metal ions [21], metals [22, 23], and semiconducting NPs [24]. This class of CNT derivatives shows promising features for nanoelectronics and other applications.

QDs have been subject to intensive investigations due to their unique properties and potential application prospect [25]. So far, several methods have been developed to synthesize water-soluble QDs for use in biological relevant studies [26, 27]. For example, QDs have been used successfully in cellular imaging [28], immunoassays [29], DNA hybridization [30], optical barcoding [31], and gene ultrasensitive detection [32]. QDs provide a new functional platform for bioanalytical sciences and biomedical engineering.

QDs also provide a new functional platform for novel materials and devices that benefit from the unique physical properties that arise from their quantum-confined nature, giving rise to properties between those of the molecular and the bulk materials [33, 34]. For semiconductor QDs such as CdSe, variation in particle size provides continuous and predictable changes in fluorescence emission [35]. Recently, several methods have been developed to disperse QDs in aqueous media for use in biologically relevant studies [28–32]. Functionalization of the water-soluble QD surface allows the formation of QD bioconjugates that can bind specifically to target molecules and form stable conjugate complexes.

Herein we introduced a strategy to derivatize CNTs with QD. Mercaptoacetic acid capped CdSe NPs were attached to CNTs by means of an interlinker molecule–ethylenediamine. This method provides a direct route to synthesize QD–CNT composites. This offers the possibility to predefine the size, shape, and amount of the NPs attached on CNTs, depending on the specific application. Immobilization of biomolecules on QD and CNT is an area of great interest. Such hybrid materials have numerous applications in the areas of biotechnology, immunosensing, and biomedical applications. The schematic is as shown in Scheme 6.1.

The concrete synthesis steps are as follows:

1) **Mercaptoacetic acid coating of CdSe nanoparticles**
   As-synthesized CdSe QDs are capped with trioctylphosphine oxide (TOPO) ligands, and thus are hydrophobic. To obtain QDs that can bind to the amine-functionalized MWNTs, the TOPO must be replaced with a suitable hydrophilic ligand. The resulting CdSe QDs were rendered water-soluble by replacing the native TOP/TOPO organic capping shell with mercaptoacetic acid

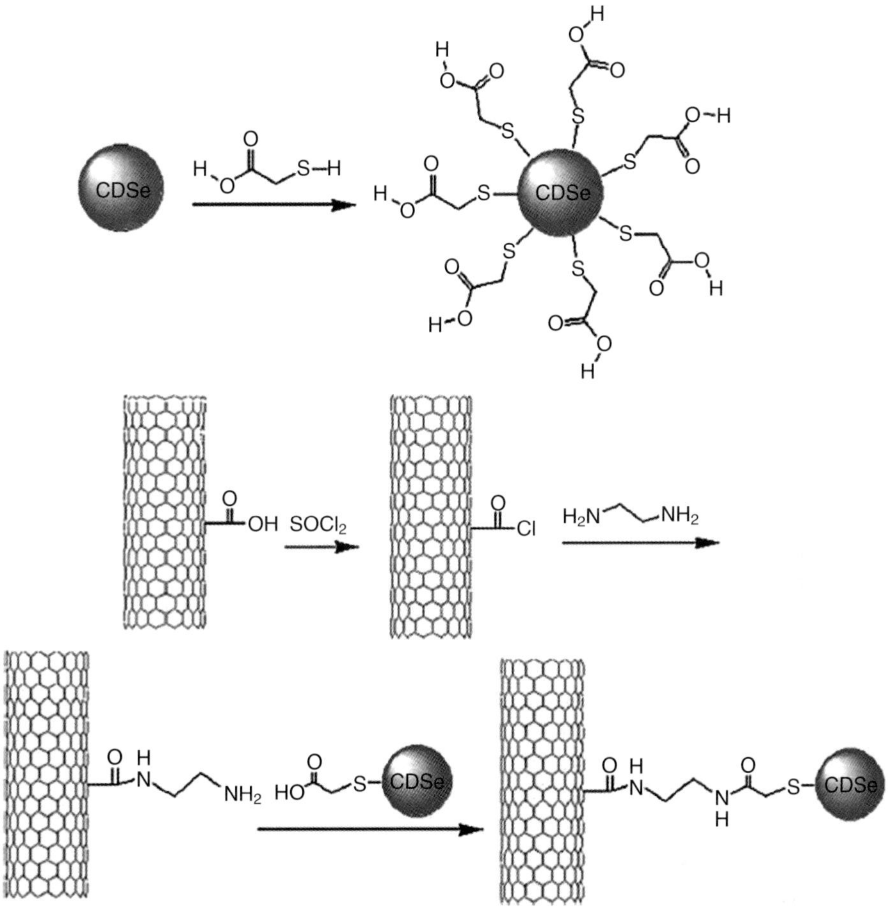

Scheme 6.1 Schematic of CdSe QD functionalized with mercaptoacetic acid, amine functionalization of MWNT, and covalently binding of MWNT–NH$_2$ with CdSe–COOH. Reprinted with permission from [24], B. Pan, et al. Covalent attachment of quantum dot on carbon nanotubes. *Chemical Physics Letters* 2006; 417: 419–424. © 2005 Elsevier B.V.

(SH–CH$_2$–COOH) ligands (Scheme 6.1). The colloidal QDs were dissolved in chloroform and were reacted with glacial mercaptoacetic acid (1.0 M) for 2 h. Water was added to this reaction mixture at a 1:1 volume ratio. After vigorous shaking and mixing, the chloroform and water layers separated spontaneously. The aqueous layer, which contained mercaptoacetic acid-coated QDs, was extracted. Excess mercaptoacetic acid was removed by four or more rounds of centrifugation. This provided aggregate-free water-soluble QD preparations with good quantum yields. Mercaptoacetic-capped CdSe QD samples stable over a period of 1 year have been routinely prepared in our laboratory.

The mercaptoacetic acid-capped nanocrystals have a homogeneous density of charge due to deprotonation of the carboxylic acid end groups.

2) **Preparation of MWNT–COOH from MWNT**
Crude MWNTs (0.5233 g) were added to H_2SO_4/ HNO_3 = 3:1 (v/v). The mixture was placed in an ultrasonic bath (40 kHz) for 30 min and then stirred for 24 h while being boiled under reflux. The mixture was then vacuum-filtered through a 0.22 μm Millipore polycarbonate membrane and subsequently washed with distilled water until the pH of the filtrate was ca. 7. The filtered solid was dried under vacuum for 12 h at 60 °C, yielding MWNT–COOH (0.5036 g).

3) **Generation of MWNT–COCl from MWNT–COOH**
MWNT–COOH (0.5036 g) was suspended in $SOCl_2$ (30 ml) and stirred for 24 h at 65 °C. The solution was filtered, washed three times with anhydrous THF, and dried under vacuum at room temperature for 24 h, generating MWNT–COCl (0.4926 g).

4) **Synthesis of MWNT–NH2 Initiators from MWNT–COCl**
MWNT–COCl (0.4926 g) was mixed with ethylene diamine (20.0 ml) and placed in an ultrasonic bath (40 kHz) for 5 h at 60 °C. The mixture was stirred for another 24 h at 60 °C. The resulting solid was separated by vacuum-filtration using 0.22 μm Millipore polycarbonate membrane filter and subsequently washed with water. After repeated washing and filtration, the resulting solid was dried overnight in a vacuum, generating MWNT–NH_2 (0.5042 g).

5) **Attachment of CdSe QDs to MWNT**
The MWNT–NH_2 aqueous solution was centrifugated and redispersed in water. The mercaptoacetic acid coated QDs aqueous solution was added to the MWNTs and the mixture was sonicated at 50 °C under nitrogen for 4 h. The suspension was then cooled to room temperature, centrifuged at 8000 rpm for 30 min, and rinsed thrice with water using several cycles of centrifugation and decantation. The coupled QD–MWNTs were filtered through a 3.0 lm Teflon membrane filter and resuspended in water. The ratio of QDs to MWNTs (QD/MWNT) was varied from 10:1 to 1:10 (w/w).

The prepared QD–MCNT composites were as shown in Figure 6.1, which showed that the immobilization of QD onto MWNT surfaces has been achieved. This presents a simple method of coupling individual MWNTs to single CdSe QDs. Amine groups present on the nanotube sidewalls, as well as the ends, lead to amide bond coupling with mercaptoacetic acid coated CdSe NPs. QDs attached to the MWNT sidewalls and ends by HRTEM characterization.

Shi et al. reported another kind of method to prepare QD-conjugated CNTs [36, 37]. The concrete steps are as follows: commercial-grade MWCNTs were obtained from Applied Science Inc. The CNTs were chosen for their relatively larger inner-wall diameters (50–80 nm), which are more suitable for cargo storage. An ultrathin (~1–2 nm) acrylic acid polymer film was deposited on CNTs by plasma coating to

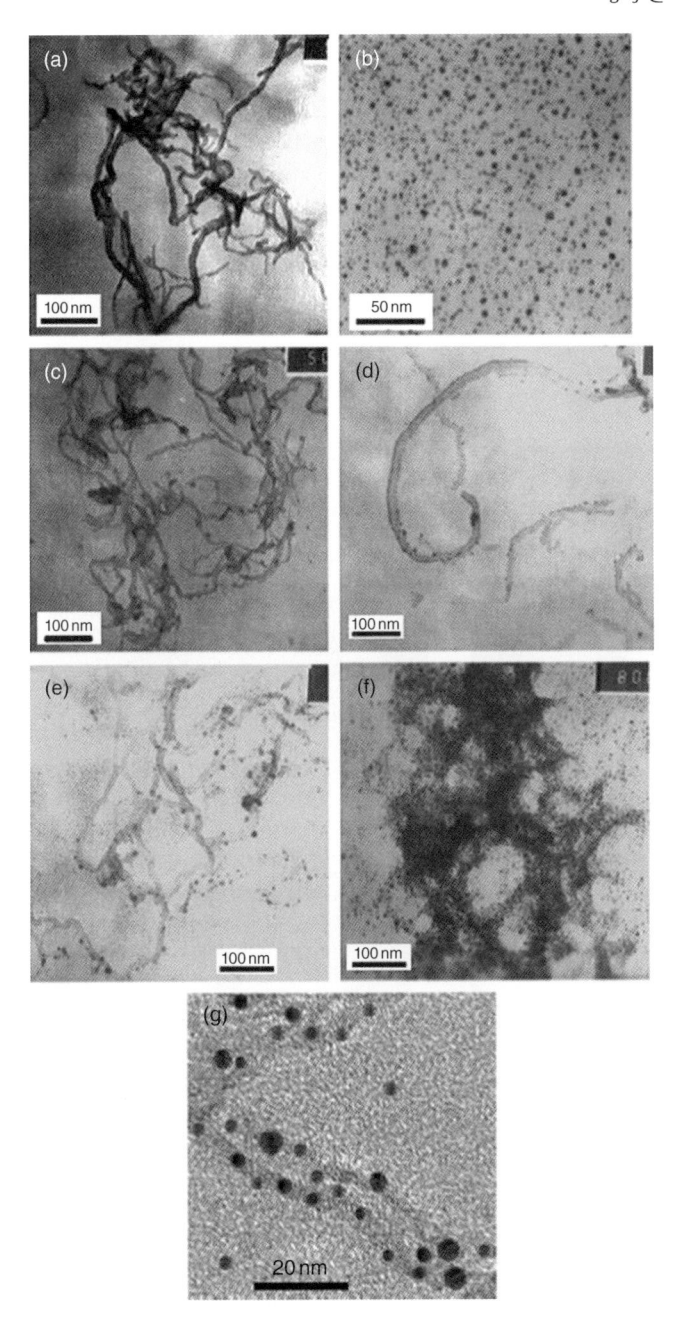

Figure 6.1 TEM images of (a) MWNT–NH$_2$; (b) 2.8–4.6 nm CdSe QDs; (c) mixture of MWNT–COCl and QD–COOH; (d) QD/MWNT = 1:10; (e) QD/MWNT = 1:1; (f) QD/WNT = 10:1; (g) HRTEM image of QD–MWNT (QD/MWNT = 1:1). Reprinted with permission from [24], Pan, et al. Covalent attachment of quantum dot on carbon nanotubes. *Chemical Physics Letters* 2006; 417: 419–424. © 2005 Elsevier B.V.

provide carboxyl functional groups for covalent coupling of QDs. CdSe/ZnS QDs (emission wavelength of 600 nm) were purchased from Evident Technologies, NY. They were functionalized with amine and packed in water at a concentration of 2.6 nmol ml^{-1} core QD. To generate the QD–CNT heterostructure, a previously published carboxyl/amine coupling protocol was applied using 1-ethyl-3-(3-dimethylaminopropyl) carbodiimide HCl (EDC) and N-hydroxysuccinimide (sulfo-NHS) in PBS, pH 7.4. Fabricated QD–CNTs were characterized by fluorescent spectroscopy and high resolution transmission electronic microscope (HR-TEM) (Figure 6.2).

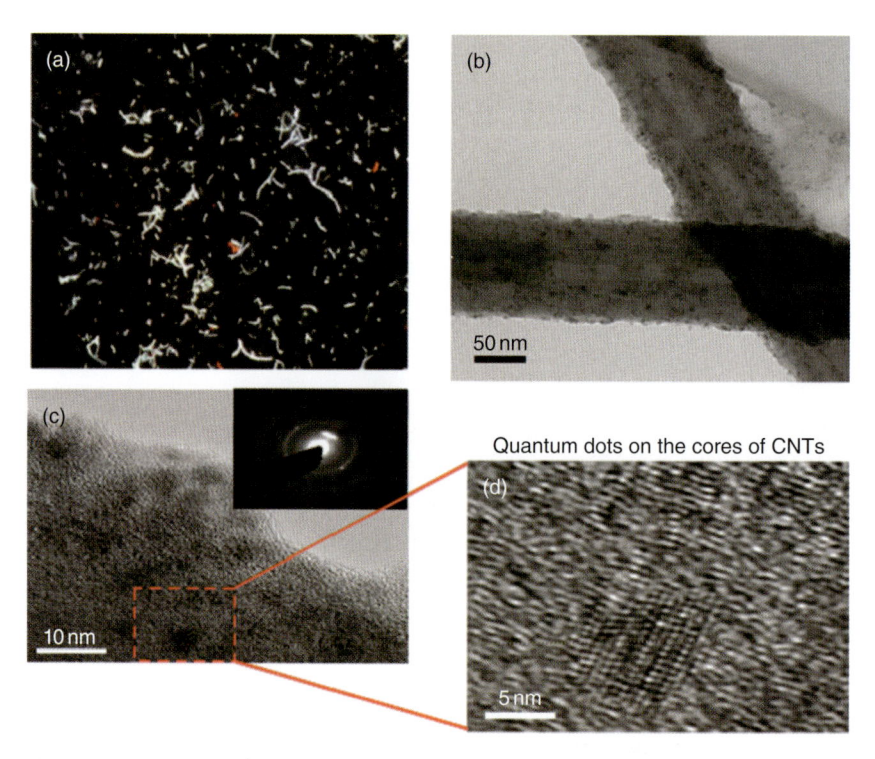

Figure 6.2 Microscopic characterization of QD–CNTs. (a) Representative fluorescent microscopy image (Olympus 1 × 51, EX = 350 nm) of fabricated fluorescent CNTs individually labeled with surface-conjugated QDs that exhibit strong emission against the dark background. (b) TEM image of CNTs with surface-coupled QDs. (c and d) Representative HRTEM images of the crystalline CdSe/ZnS QDs deposited on the CNTs. The inset in (c) visualizes the selected area electron diffraction pattern acquired from surface-functionalized CNTs. Reprinted with permission from [36, 37], Y. Guo, et al. In vivo imaging and drug storage by quantum-dot-conjugated carbon nanotubes. Adv. Funct. Mater. 2008; 18, 1–9; DL Shi, et al. Qd-activated luminescent carbon nanotubes via a uano scale surface functionalization for in vivo imaging. Adv Mater, 2007; 19: 4033–4037. © 2007 WILEY-VCH.

Attachment and control of the chemistry at the nanotube surface are important since this configuration could result in a bottom-up approach toward the assembly of MWNTs necessary for future nanoelectronic devices. The successful modification of MWNTs with semiconductor QDs has significant implications for materials research, and the door is now open for more extensive inquiries into the luminescent, electronic, and chemical properties of these unique building blocks as they are incorporated into new and functional multicomponent nanostructured materials.

6.3
Covalent Binding of Magnetic Nanoparticles on CNTs

One-dimensional (1D) nanostructures have considerable potential as building blocks in future electronics. The unique electrical, mechanical, and chemical properties of CNTs have made them intensively studied materials in the field of nanotechnology [1, 2]. So far these properties have essentially enabled the CNTs very useful for practical applications such as development of nanodevices [37], quantum wires [38], ultrahigh-strength engineering fibers [39], and sensors and support of electrocatalysts [40]. Furthermore, many studies have been focused on depositing metal or oxide metal NPs on the nanotubes surface.

Magnetite particles (microspheres, nanospheres, and ferrofluids) are widely studied for their applications in biology and medicine such as enzyme and protein immobilization [41], magnetic resonance imaging (MRI) [42], RNA and DNA purification [43], magnetic cell separation and purification [44], and magnetically controlled transport of anticancer drugs [45], as well as hyperthermia generation [46]. The challenge to take meaningful advantage of the extreme properties of magnetite nanoparticles (MNPs) and the tendency toward fabrication of 1D or quasi-1D nanostructures have stimulated recent investigations on the properties foreseen for magnetite in the form of tubular nanosystems, such as nanowires or nanorods.

Our group reported a facile strategy for fabricating MNP/CNT nanohybrid through covalent interaction as shown in Scheme 6.2 [48]. We described a simple method to covalently prepare nanohybrid materials from CNTs and NPs. The strategy demonstrated here through a covalent approach may be superior to those with noncovalent ways due to its ability to maintain the electrical and structural properties of the CNTs, offering a new route to prepare CNT nanohybrids, both of which are believed to be useful for the fundamental investigations of the CNTs and their possible applications.

The concrete synthesis steps are as follows:

1) **Preparation of MWNT–COOH from MWNT**
   Crude MWNTs (0.523 g) were added to aqueous HNO_3 (20.0 ml, 60%). The mixture was placed in an ultrasonic bath (40 kHz) for 40 min and then stirred for 48 h while being boiled under reflux. The mixture was then vacuum-filtered through a 0.22 µm Millipore polycarbonate membrane and subsequently

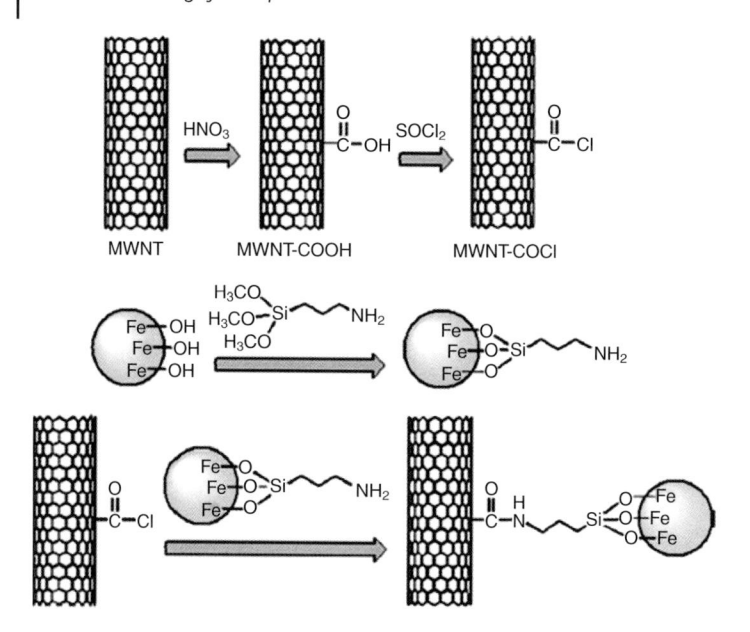

Scheme 6.2 Covalent attachment of MNPs on the CNT surface to form MNP/CNT nanohybrids. Reprinted with permission from [47], P. Xu, et al. A facile strategy for covalent binding of nanoparticles onto carbon nanotubes. *Applied Surface Science* 2008; 254: 5236–5240. © 2008 Elsevier B.V.

washed with distilled water until the pH of the filtrate was ca. 7. The filtered solid was dried under vacuum for 24 h at 70 °C, yielding MWNT–COOH (0.524 g).

2) **Generation of MWNT–COCl from MWNT–COOH**
   Dried MWNT–COOH (0.524 g) was suspended in $SOCl_2$ (30 ml) and stirred for 24 h at 70 °C. The solution was filtered, washed with anhydrous THF, and dried under vacuum at room temperature for 24 h, generating MWNT–COCl (0.549 g).

3) **Synthesis of magnetite nanoparticles**
   For the preparation of MNPs, 1.0 g ferrous chloride was dissolved in 3 ml 2 M HCl solution and 1.6 g ferric chloride in 10 ml 2 M HCl solution. Ferrous chloride and ferric chloride solution were mixed in a beaker with vigorously stirring. At the same time, 25 ml ammonium hydroxide solution was added slowly into the beaker by a pipette and magnetite in the form of a black precipitate was formed immediately. Stirring was continued for 15 min after adding ammonia hydroxide. The resulting precipitate was washed with water (200 ml) until the pH of the wash supernatant was reduced to 7. The final total volume for the suspension was adjusted to 300 ml using water.

4) **Magnetite nanoparticles coated by aminosilane**

 The 25-ml magnetite colloid ethanol solution prepared above was diluted to 150 ml by ethanol. The solution was then treated by ultrasonic waves for 30 min. A 10-ml amount of 3-aminopropyltrimethoxysilane ($NH_2(CH_2)_3$–Si–$(OCH_3)_3$, APTS) was added into it with rapid stirring for 7 h. The resulting solution was washed with methanol five times by magnetic separation. APTS-coated MNPs were dried under vacuum at room temperature for 24 h, generating amine-terminated MNP (MNP–NH_2).

5) **Fabrication of CNT/MNP nanohybrid**

 Ten milligram MNP–NH_2 and 50 mg CNT–COCl were added to 25 ml THF solution and the mixture was placed in an ultrasonic bath (40 kHz) for 60 min and then stirred for 24 h at room temperature. The resulting solution was washed magnetically with water for five times, and the precipitate was dissolved into 20 ml of water, obtaining a black solution.

As shown in Figure 6.3, the magnetic NPs were successfully attached on the CNTs by covalent bond.

Attachment and control of the chemistry at the CNT surface is important since this configuration could result in a bottom-up approach toward assembly of CNTs necessary for future nanoelectronic devices. CNTs have been well studied in recent years due to their interesting physical properties. Currently, a large amount of research in CNT chemistry is directed to the synthesis of diverse organic derivatives with the aim to promote surface functionalization and simultaneously provide colloidal stability to the resulting modified CNTs in organic or aqueous media. Nowadays, more and more people prefer to functionalize CNT by using the simple reaction between activated –COCl terminal with –OH/NH_2 groups. Gao and coworkers [47] reported an *in-situ* ring-opening polymerization strategy to grow multihydroxyl dendritic macromolecules on the convex surfaces of MWCNTs, affording novel 1D molecular nanocomposites. The crude MWNTs were oxidized using 60% HNO_3 and then reacted with thionyl chloride, resulting in MWNTs functionalized with chlorocarbonyl groups (MWNT–COCl). MWNT–COCl, when reacted with an excess of glycol [49], produced hydroxy-functionalized MWN-supported initiators (MWNT–OH). Using the MWNT–OH as the growth supporter and $BF_3 \cdot Et_2O$ as catalyst, multihydroxy hyperbranched polyethers-treelike macromolecules were covalently grafted on the sidewalls and ends of MWNTs via *in-situ* ring-opening polymerization of 3-ethyl-3-(hydroxymethyl)-oxetane (EHOX). By using the similar method, an *in-situ* repetitive divergent polymerization strategy was employed to grow multiamine poly(amidoamine) dendritic macromolecules on the surfaces of MWCNTs, affording novel 3D molecular nanocomposites. The crude MWNTs were oxidized using $H_2SO_4/HNO_3 = 3:1$ (v/v) and then reacted with thionyl chloride, resulting in MWNTs functionalized with chlorocarbonyl groups (MWNT–COCl). MWNT–COCl, when reacted with an excess of ethylenediamine, produced amine-functionalized MWNT-supported initiators (MWNT–NH_2). Using the MWNT–NH_2 as the growth supporter and methylacrylate/ethylenediamine as building blocks, multiamine dendritic poly(amidoamine) macromolecules were

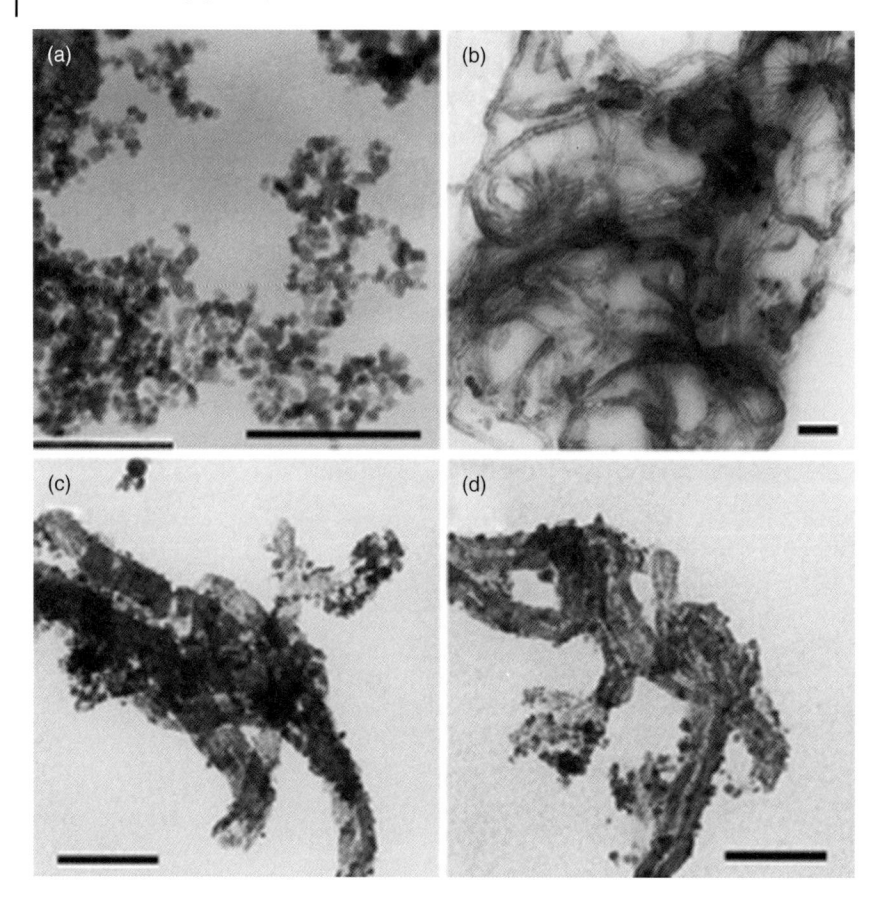

Figure 6.3 TEM photos of (a) MNP, (b) MWNT, (c and d) MNP/CNT nanohybrids. Scale bar: 100 nm. Reprinted with permission from [47], P. Xu, et al. A facile strategy for covalent binding of nanoparticles onto carbon nanotubes. *Applied Surface Science* 2008; 254: 5236–5240. © 2008 Elsevier B.V.

covalently grafted onto the sidewalls and ends of MWNTs via Michael addition reaction and amidation. Li and coworkers [50] reported SWNTs covalently functionalized with biocompatible poly-L-lysine, which is useful in promoting cell adhesion. For the potential application of functionalized CNT in biology, Dai and coworkers [51, 52] reported a simple nonspecific binding scheme that can be used to afford noncovalent protein–nanotube conjugates. A suspension of the oxidized and cut SWNTs was mixed with fluorescently labeled proteins. After this simple mixing step, proteins were found to adsorb nonspecifically onto nanotube sidewalls. The proteins are found to be readily transported inside various mammalian cells with nanotubes acting as the transporter via the endocytosis pathway. Because of the simpleness and cheapness of $-COCl + NH_2/-OH$ method, we start to assemble NPs on the surface of CNTs. This class of CNT derivatives shows promising features for nanoelectronics and other applications.

6.4
Covalent Binding of Gold Nanoparticles on CNTs

Bottom-up strategies for the fabrication of nanomaterials are integral components in nanoscience and nanotechnology [53, 54]. Recently, considerable interest has been focused on the possibility of construction of new assemblies from NPs and other components to yield superstructured nanomaterials with unique and useful optical, electronic, and magnetic properties, which go beyond those of single elements [55, 56]. Materials, such as CNTs, with outstanding mechanical and electrical properties represent ideal building blocks for nanoengineering through their integration into working devices [57, 58]. Because of their excellent physical properties and unique tubiform structures, which can be functionalized at one's discretion, CNTs in particular represent ideal units for the construction of nanostructured materials. NPs, such as metal and semiconductor colloids with diameters of one to tens of nanometers, are particularly attractive candidates. Metal and semiconductor NPs have already been used as the functional components of nanoscale electronic devices. Thus, the ability to organize, interconnect, and address NPs on surfaces is an important hurdle to overcome in order to fully realize nanoscale electronics. Many natural materials that have specific binding properties in molecular recognition and self-assembly can be used as a template to organize and interconnect assemblies of NPs on surfaces.

Both SWNT and MWNT have been proposed as substrates for biological devices [59, 60]. Examples of nanotube-based devices include single-electron transistors [61], molecular diodes [62], memory elements [21], and logic gates [63]. There is a great deal of interest in devising strategies to individually address each molecular unit, and interconnect them, without adversely affecting the local electronic structure [64]. One approach to enable this is to attach metal- or semiconductor nanoclusters to nanotubes. This approach is also attractive for creating molecular level hybrid units and will allow the exploration of new properties and effects that arise from electronic-structure-level interactions between the constituent molecular units and applications such as active nanodevices and heterogeneous nanocatalysts. Since CNTs are chemically inert, activating their surfaces is an essential prerequisite for linking nanoclusters to them. Chemical treatments such as wet oxidation in HNO_3 can functionalize nanotube surfaces with anchor groups such as hydroxyl (–OH), carboxyl (–COOH), and carbonyl (>C=O) that are necessary to tether metal NPs to the tube [22].

Herein we introduced a unique approach to connect gold NPs to CNTs by covalent interactions. This is a simple method for highly ordered assemblies of gold NPs (Au-NPs) along CNTs [65]. To securely attach Au-NPs to CNTs, preparations of mercaptoacetate-functionalized Au-NPs have been reported; such aqueous stable Au-NPs were prepared by thiolate ligand (methyl mercaptoacetate in this paper) exchange. Since the structure and chemistry of self-assembly hybrid materials are affiliated to many biomolecules such as proteins, cluster attachment will enable the design and creation of new bioinspired hybrid nanodevices. Creating metal–nanotube units will open up possibilities for replicating quantum effects such as single-electron hopping and coulomb blockade in nanotubes and NPs functionalized with proteins. Moreover, it is conceivable to modulate such effects by adjusting hybrid materials structure, enabling molecular level design of nanodevices for switching, sensing, and information storage. The schematic is as shown in Scheme 6.3.

The concrete synthesis steps are discussed next.

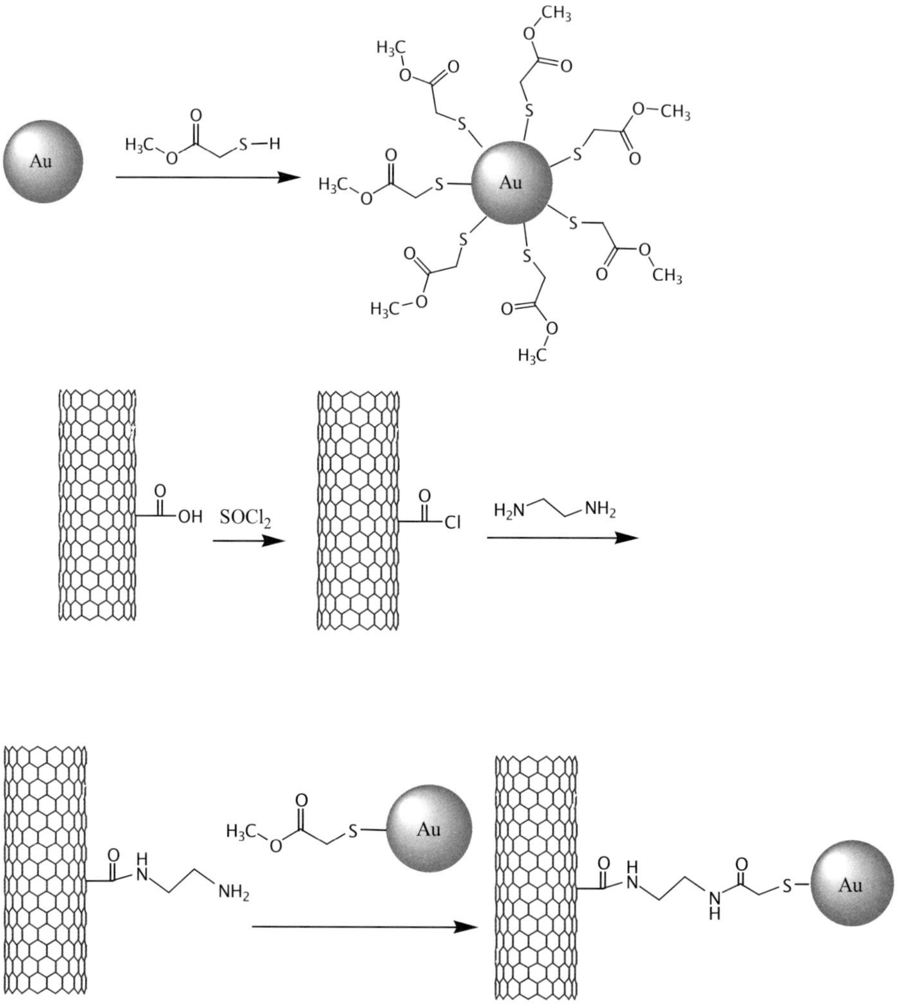

Scheme 6.3 Schematic illustration of methyl mercaptoacetate modification of gold NPs, amine functionalization of MWNT, and covalent binding of gold NPs onto MWNTs. Reprinted with permission from [66], B. Pan, et al. Attachment of gold nanoparticles on multi-walled carbon nanotubes. *Nanoscience* 2006; 11: 95–101. © 2006 American Association of Nanoscience and Technology, National Key Laboratory of Nanobiological Technology, Ministry of Health, P.R. China.

6.4.1
Preparation of MWNT–COOH

MWNTs of 0.1 g were suspended in 40 ml concentrated sulfuric acid/nitric acid mixture (3:1 v/v) and sonicated in a sonic bath for 2 h and then stirred for 24 h while being boiled under reflux. The mixture was then vacuum-filtered through a 0.22 μm Millipore polycarbonate membrane and subsequently washed with distilled water until the pH of the filtrate was ca. 7. The filtered solid was dried under vacuum for 12 h at 60 °C, yielding MWNT–COOH (0.1035 g). The carboxyl functional groups are hydrophilic and the nanotubes were dispersed easily in water.

Generation of MWNT–COCl from MWNT–COOH: 0.1035 g MWNT–COOH was suspended in 20 ml $SOCl_2$ and stirred for 24 h at 65 °C. The solution was filtered, washed three times with anhydrous DMF, and dried under vacuum at room temperature for 2 h, generating MWNT–COCl (0.1125 g).

Synthesis of MWNT–NH2 from MWNT–COCl: 0.1125 g MWNT–COCl was mixed with 30 ml ethylenediamine and placed in an ultrasonic bath (40 kHz) for 5 h at 60 °C. The mixture was stirred for another 24 h at 60 °C. The resulting solid was separated by vacuum-filtration using 0.22 μm Millipore polycarbonate membrane filter and subsequently washed with water. After repeated washing and filtration, the resulting solid was dried overnight in a vacuum, generating MWNT–NH_2 (0.1201 g).

Preparation of gold nanoparticle: A 100-ml sample of a 1.25×10^{-4} M concentrated aqueous solution of chloroauric acid ($HAuCl_4$) was reduced by 20 ml of 0.1 wt.% sodium citrate at 100 °C for 15 min to yield a ruby-red (surface plasmon absorption maximum at ca. 520 nm) solution containing 13-nm-diameter gold NPs.

6.4.2
Methyl Mercaptoacetate Modification of Gold Nanoparticles

As-synthesized 13-nm Au-NPs are capped with citrate ligands, and thus are negatively charged. To obtain Au-NPs that can bind to the amine-functionalized MWNTs, the colloidal gold NPs were dissolved in water and were reacted with methyl mercaptoacetate (~1.0 M) for 2 h. Excess methyl mercaptoacetate was removed by four or more rounds of centrifugation. This provided aggregate-free water-soluble gold NPs. Mercaptoacetate-capped gold NPs samples stable over a period of 1 year have been routinely prepared in our laboratory.

6.4.2.1 Formation of Au–MWNT Conjugates
MWNT–NH_2 powder of 10 mg was dispersed in 10 ml of water. To this solution, 10 ml of 1.0×10^{-4} M colloid mercaptoacetate-Au aqueous solution was added under vigorous stirring. After 5 h of sonicating at 60 °C, the Au–MWNT hybrid material was separated by centrifugation and was rinsed several times with water, resuspended in water, and stored at 4 °C prior to further characterization.

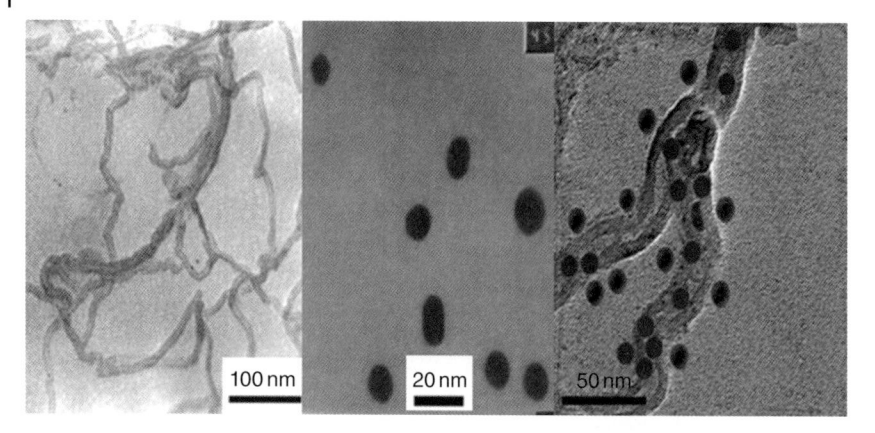

Figure 6.4 TEM images of (a) MWNT–NH$_2$, (b) mercaptoacetate-capped gold nanoparticles, (c) Au–MWNT composite with Au/MWNT = 1:2. Reprinted with permission from [66], B. Pan, et al. Attachment of gold nanoparticles on multiwalled carbon nanotubes. *Nanoscience* 2006; 11: 95–101. © 2006 American Association of Nanoscience and Technology, National Key Laboratory of Nanobiological Technology, Ministry of Health, P. R. China.

The Au-NPs-conjugated CNTs composites were shown in Figure 6.4. This is a simple covalent method for highly ordered assemblies of metal–nanotube hybrid materials, which enabled us to easily organize Au-NPs on CNTs surface in water. The covalent interaction between mercaptoacetate on gold NPs and amine groups on CNTs is the main driving force for the formation of the nanoassemblies. Along with the development of novel architectures derived from nanotube derivatives, the methodology of assembly described here is likely to afford a wide variety of supernanostructures that show extraordinary optoelectronic properties. Such molecular interactions are compatible with many biological systems and will be important for exploring and creating a rich variety of molecular nanostructures for device applications.

6.5
Growth of Poly(amidoamine) Dendrimers on Carbon Nanotubes

CNTs, because of their unique mechanical, physical, and chemical properties, have been actively investigated for their applications in various fields including molecular electronics, medical chemistry, and biomedical engineering [1–3]. CNTs can be functionalized to achieve improved biological properties and functions [66, 67]. Recently, some reports [68, 69] show that CNTs can take various cargos such as small peptides, streptavidin, and nucleic acids and penetrate mammalian cell membranes into the cytoplasm via sidewall functionalization. However, the amount of genes or peptides conjugated by each CNT is still limited. More important, these genes or peptides taken by CNTs have to be released out and then take effect in the cancer cells. As is known, a CNT is a foreign body to cells, the mechanism of the effects of CNTs on cells is not still well clarified [70, 71], and therefore the strategy of using less CNTs to take more genes or drugs into cells is an excellent choice to protect target cells.

Dendrimers are one class of novel special organic molecules: they can take different functional groups through a series of chemical modifications, and their interior cavities can serve as storage areas for a lot of genes or drugs because of the proton sponge effect [72–74]. Dendrimers are one kind of good nonviral gene-delivery systems because of their advantages of simplicity of use and ease of mass production compared to viral vectors with inherent risk [75, 76]. Our previous studies confirmed that the polyamidoamine (PAMAM) dendrimer-modified magnetic NPs can markedly enhance the efficiency of gene-delivery systems [77], dendrimer-modified SWCNTs can reduce the cytotoxicity of CNTs and enhance the cellular uptake of the CNTs [9], CNTs can be filled with biomolecules such as DNA and peptides [78, 79], and different generations of dendrimers can be grown on the surface of CNTs [80, 81]. The schematic is shown in Scheme 6.4.

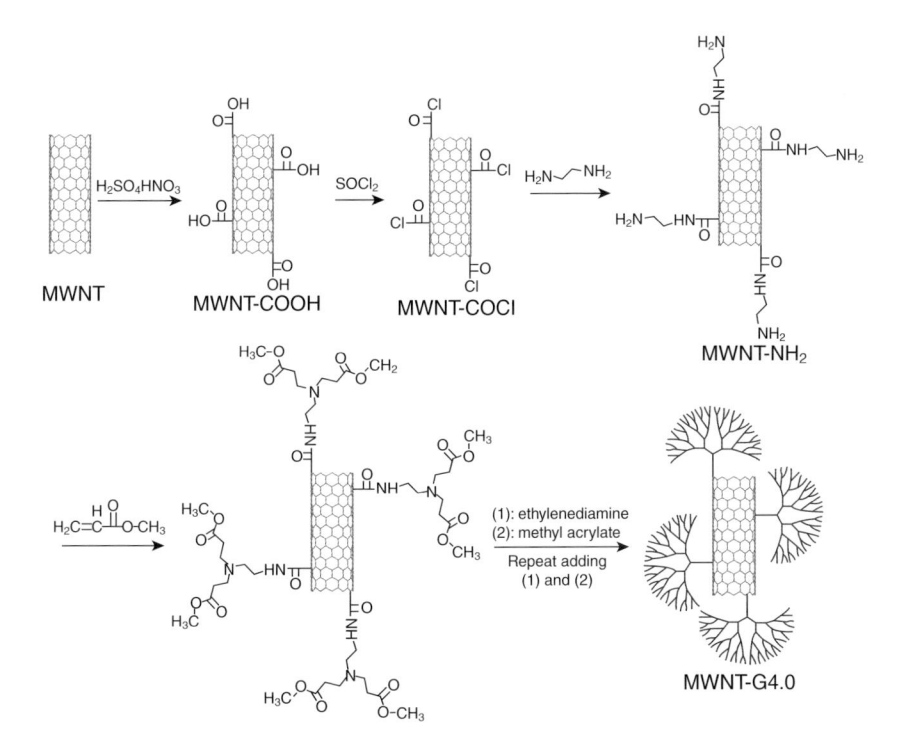

Scheme 6.4 Growth of poly(amidoamine) dendrimer on the surface of carbon nanotubes. Reprinted with permission from [81], B. Pan, et al. Growth of multi-amine terminated poly(amidoamine) dendrimers on the surface of carbon nanotubes. *Nanotechnology* 2006; 17: 2483–2489. © 2006 IOP Publisher.

The concrete synthesis steps are as follows:

1) **Preparation of MWNT–COOH from MWNT**
   Crude MWNTs (0.5233 g) were added to $H_2SO_4/HNO_3 = 3:1$ (v/v). The mixture was placed in an ultrasonic bath (40 kHz) for 30 min and then stirred for 24 h while being boiled under reflux. The mixture was then vacuum-filtered through a 0.22 μm Millipore polycarbonate membrane and subsequently washed with distilled water until the pH of the filtrate was about 7. The filtered solid was dried under vacuum for 12 h at 60 °C, yielding MWNT–COOH (0.5036 g).

2) **Generation of MWNT–COCl from MWNT–COOH**
   MWNT–COOH (0.5036 g) was suspended in $SOCl_2$ (30 ml) and stirred for 24 h at 65 °C. The solution was filtered, washed with anhydrous THF, and dried under vacuum at room temperature for 24 h, generating MWNT–COCl (0.4926 g).

3) **Synthesis of MWNT–NH2 initiators from MWNT–COCl**
   MWNT–COCl (0.4926 g) was mixed with ethylenediamine (20 ml) and placed in an ultrasonic bath (40 kHz) for 5 h at 60 °C. The mixture was stirred for another 24 h at 60 °C. The resulting solid was separated by vacuum-filtration using 0.22 μm Millipore polycarbonate membrane filter and subsequently washed with anhydrous methanol. After repeated washing and filtration, the resulting solid was dried overnight in a vacuum, generating MWNT–NH_2 (0.5042 g).

4) **Growth of poly(amidoamine) dendrimers on the MWNT surface initiated by MWNT–NH_2**
   Methylacrylate of 20 ml in 50 ml methanol was added to a 500-ml three-necked round-bottom flask. As-prepared MWNT–NH_2 initiator (0.1 g) in 20 ml methanol was carefully dropped into methylacrylate solution within 20 min, with continuously stirring. The solution was placed in an ultrasonic bath (40 kHz) for 7 h at 50 °C, and the mixture was stirred for another 24 h. To ensure that no ungrafted polymer or free reagents were present in the product, the filtered solid was dispersed in methanol, filtered, and washed three times with methanol. The product was dried overnight to give the dendritic polymer-grafted MWNTs (generation 0.5, MWNT-G0.5, 0.1023 g).

After rinsing, MWNT–G0.5 in 20 ml methanol was dropped into 40 ml of a 1:1 methanol/ethylenediamine solution. The solution was placed in an ultrasonic bath (40 kHz) for 5 h at 50 °C, and the mixture was stirred for another 24 h at 50 °C. The solid was filtered and washed. After rinsing, MWNT-G0.5 in 20 ml methanol was dropped into 40 ml of a 1:1 methanol/ethylenediamine solution. The solution was placed in an ultrasonic bath (40 kHz) for 5 h at 50 °C, and the mixture was stirred for another 24 h at 50 °C. The solid was filtered and washed three times with methanol to give a generation 1.0 dendrimer-modified MWNT (MWNT-G1.0). Stepwise growth using methylacrylate and ethylenediamine was repeated until the desired number of generations (up to generation 5.0) was achieved (Scheme 6.4).

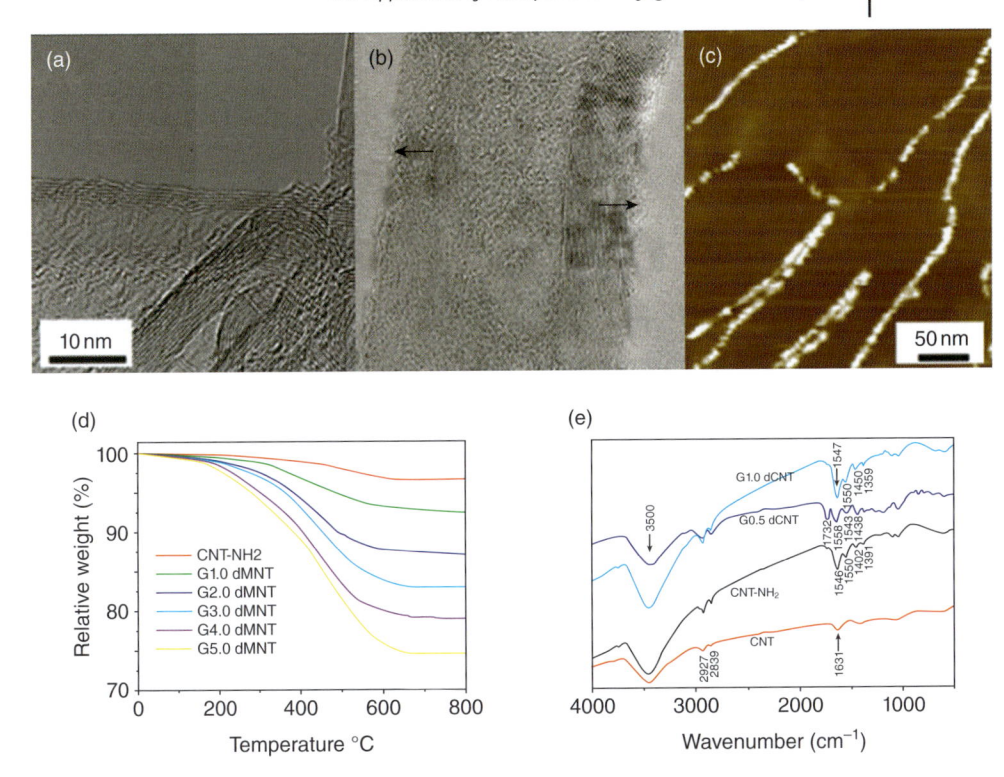

Figure 6.5 Characterization results of MNTs and dMNTs. (a) HR-TEM image of MNTs, (b) HR-TEM image of G5.0 dMNTs, arrow indicates the dendrimer layer, (c) AFM image of G5.0 dMNTs, (d) TGA curves of MNT–NH_2 and G1.0 to G5.0 dMNTs, (e) FT-IR spectra of dMNTs. CNT = pristine MNTs, CNT–NH_2 = amine-terminated MNTs (MNT–NH_2), G0.5 dCNT = G0.5 PAMAM dendrimer-modified MNTs and G1.0 dMNT = G1.0 PAMAM dendrimer-modified MNTs. Reprinted with permission from [82], B Pan, et al. Synthesis and characterization of polyamidoamine dendrimer-coated multi-walled carbon nanotubes and their application in gene delivery systems. *Nanotechnology* 2009; 20: 125101. © 2009 IOP Publisher.

The dendrimer-modified MWNT was then washed three times with 25 ml methanol, and three times with 25 ml water. The dendrimer-modified MWNT was then dissolved in water with a concentration of 0.2 mg ml^{-1}. The prepared dendrimer-modified CNTs composites are shown in Figure 6.5.

6.6
Application of Nanoparticles-Conjugated CNTs Composites

Herein we introduced several simple strategies for covalently attaching NPs onto the CNTs to fabricate hybrid nanostructure. The strategies introduced herein are quite generic and applicable to a variety of NPs, including metal, QD, and oxide as well as polymer NPs. These composite nanostructures should open up new possibilities in areas such as nanoelectronics, chemical sensing, field-emission displays, nanotribology, and cell adhesion/biorecognition investigations, and exhibit great potential applications in different fields such as biosensor, ultrasensitive detection, Scaffolds in tissue engineering, the polymeric solar cells, etc. [42, 83, 84–92].

For example, incorporation of QD and SWNTs into a poly(3-octylthiophene) (P3OT) composite has been shown to facilitate exciton dissociation and carrier transport in a properly structured device. The assembly of QD on the surface of MWNTs in aqueous solution results in forming the QD–MWNT hybrid structures. The assembly of QDs on the MWNTs surface occurs through interaction of the amine groups on the sidewall of MWNTs with the mercaptoacetic acid modified QDs. Furthermore, QD–MWNT hybrid material shows an excellent solubility in aqueous solution. Based on the QD–SWNT composites, an ideal electron acceptor for polymeric solar cells should exhibit high electron affinity and high electrical conductivity. The polymer composites including the P3OT complexes have been used to fabricate solar cells, showing limited efficiency due to recombination and surface effects. However, the optical absorption spectra for these nanomaterial–polymeric composites showed a marked enhancement in capturing the available irradiance of the air mass zero (AM0) spectrum. QD–CNTs composites were also used for gene or drug delivery and gene or drug tracking.

Self-assembly of CNTs and QDs also has been used for ultrasensitive detection of DNA and antigen [93]. As shown in Scheme 6.5, a highly selective, ultrasensitive, fluorescence detection method for DNA and antigen based on self-assembly of MWCNTs and CdSe QDs via oligonucleotide hybridization is showed. Mercaptoalkyl oligonucleotide molecules bind to the QDs, while aminoalkyl oligonucleotides bind to CNTs with –COCl surface groups. QDs and CNTs further assemble into nanohybrids through DNA hybridization in the presence of target complementary oligonucleotides. The method is achieved with good repeatability with the detection limit of 0.2 pM DNA molecules and 0.01 nM antigen molecules. This novel detection system can also be used for multicomponent detection and antigen–antibody immunoreaction. The novel system has great potential in the applications such as ultrasensitive pathogen DNA or antigen or antibody detection, molecular imaging, and photoelectrical biosensors. In general, the NPs-conjugated CNTs composites own great potential in applications such as nanoelectronics, biomedical engineering, etc. (Scheme 6.5).

6.7
Concluding Remarks

Covalent binding of NPs on the surface of CNTs provides many novel opportunities for developing the CNT-based nanoelectronics, nanodevices, and exploring potential application for nanobiomedical engineering. CNTs–NPs composites can

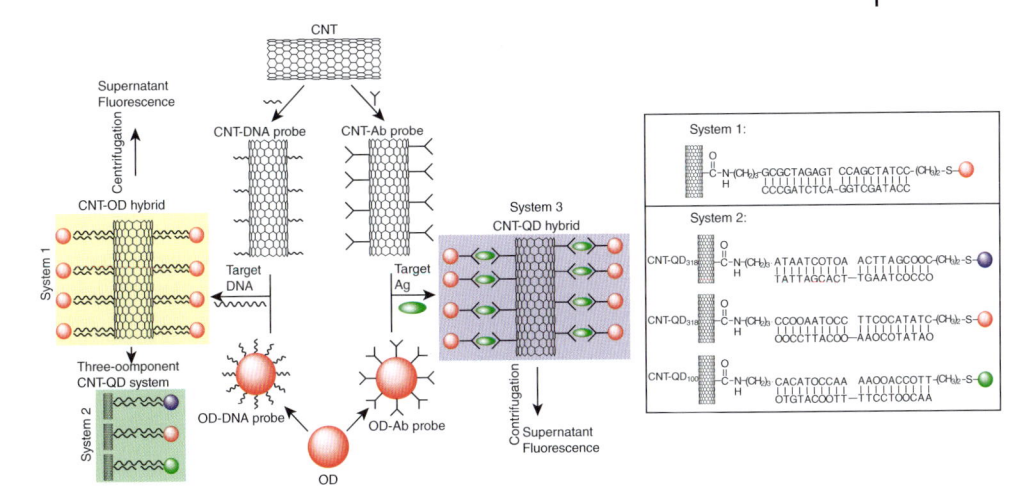

Scheme 6.5 Surface functionalization of CNT (or QD) with oligonucleotide/antibody (Ab), forming a CNT-DNA (or -Ab) probe and QD-DNA (or -Ab) probe, and subsequent addition of target oligonucleotide (or antigen) to form a CNT–QD assembly. The unbound QD probe was obtained by simple centrifugation separation, and the supernatant fluorescence intensity of QDs was monitored by spectrofluorometer. (System 1) Formation of CNT–QD hybrid in the presence of complementary DNA target. (System 2) Three-component CNT–QD system with the purpose to detect three different DNA targets simultaneously. (System 3) CNT–QD protein detection system based on antigen–antibody immunoreaction. Reprinted with permission from [93], D. Cui, et al. Self-assembly of quantum dots and carbon nanotubes for ultrasensitive DNA and antigen detection. *Anal. Chem.* 2008; 80: 7996–8001. © 2008 American Chemical Society.

improve mechanical, electric, thermal, electrochemical, optical, and superhydrophobic properties, and own potential application in industry, defense, diseases diagnosis, and therapy. Although CNTs' biocompatibility is still disputed, however, CNTs–NPs composites have been explored, their potential application for biomedical engineering and nanoelectronics, how to fully use advantages of CNTs and NPs to develop novel nanodevices, or nanohybridization materials with special function is still a great challengeable task.

Acknowledgments

This work was supported by the National Natural Science Foundation of China (No.20803040 and No.20471599), Chinese 973 Project (2010CB933901), 863 Key Project (2007AA022004), New Century Excellent Talent of Ministry of Education of China (NCET-08-0350), Special Infection Diseases Key Project of China (2009ZX10004-311), Shanghai Science and Technology Fund (10XD1406100).

References

1 Iijima, S. (1991) *Nature*, **354**, 56.
2 Dresselhaus, M.S., Dresslhous, G., and Avouris, P. (2000) *Carbon Nanotubes: Synthesis, Structure, Properties and Applications*, Springer, Berlin, Germany.
3 Dresselhaus, M.S., Dresslhous, G., and Eklund, P.C. (1996) *Science of Fullerenes and Carbon Nanotubes*, Academic Press, San Diego, USA.
4 Ajayan, P.M., Schadler, L.S., and Braun, P.V. (2003) *Nanocomposite Science and Technology*, Wiley-VCH Verlag GmbH & Co. KGaA, Weinheim, Germany.
5 Cui, D. (2007) *J. Nanosci. Nanotechnol.*, **7**, 1298.
6 Schnitzler, G.R., Cheung, C.L., Hafner, J.H., Saurin, A.J., Kingston, R.E., and Lieber, C.M. (2001) *Mol. Cell Biol.*, **21**, 8504.
7 Keren, K., Berman, R.S., Buchstab, E., Sivan, U., and Braun, E. (2003) *Science*, **302**, 1380.
8 Shim, M., Kam, N.M.S., Chen, R.J., Li, R., and Dai, H. (2002) *Nano. Lett.*, **2**, 285.
9 Pan, B., Cui, D., Sheng, Y., Ozkan, C.S., Gao, F., He, R., Li, Q., Xu, P., and Huang, T. (2007) *Cancer Res.*, **67**, 8156.
10 Yang, H., Chen, L., Lei, C., Zhang, J., Li, D., Zhou, Z., Bao, C., Hu, H., Chen, X., Cui, F., Zhang, S., Zhou, Y., and Cui, D. (2010) *Appl. Phys. Lett.*, **97**, 043702.
11 Lu, W., Ji, Z.Q., Pfeiffer, L., West, K.W., and Rimberg, A.J. (2003) *Nature*, **423**, 422.
12 Cui, D., Li, Q., Huang, P., Wang, K., Kong, Y., Zhang, H., You, X., He, R., Song, H., Wang, J., Bao, C., Asahi, T., Gao, F., and Osaka, T. (2010) *Nano. Biomed. Eng.*, **2**, 45.
13 Salem, A.K., Searson, P.C., and Leong, K.W. (2003) *Nat. Mater.*, **2**, 668.
14 Pan, B., Cui, D., Xu, P., Li, Q., Huang, T., He, R., and Gao, F. (2007) *Colloid Surf. A*, **295**, 217.
15 Martin, C.R. and Kohli, P. (2002) *Nature Rev. Drug Discovery*, **2**, 29.
16 Odom, T.W., Huang, J.L., Kim, P., and Lieber, C.M. (2000) *J. Phys. Chem. B*, **104**, 2794.
17 Nalwa, H.S. (2000) *Handbook of Nanostructured Materials and Nanotechnology*, vol. 5, Academic Press, New York, USA.
18 Zheng, M., Jagota, A., Semke, E.D., Diner, B.A., Mclean, R.S., Lustig, S.R., Richardson, R.E., and Tassi, N.G. (2003) *Nature Mater.*, **2**, 338.
19 Cui, D., Ozkan, C.S., Ravindran, S., Yong, K., and Gao, H. (2004) *MCB*, **1**, 113.
20 Bahr, J.L. and Tour, J.M. (2002) *J. Mater. Chem.*, **12**, 1952.
21 Quinn, B.M., Dekker, C., and Lemay, S.G. (2005) *J. Am. Chem. Soc.*, **127**, 6146.
22 Wang, X., Wang, H., Coombs, N., Winnik, M.A., and Manners, I. (2005) *J. Am. Chem. Soc.*, **127**, 8924.
23 Banerjee, S. and Wong, S.S. (2002) *J. Am. Chem. Soc.*, **124**, 8940.
24 Pan, B., Cui, D., He, R., Gao, F., and Zhang, Y. (2006) *Chem. Phys. Lett.*, **417**, 419.
25 Li, X.Q., Wu, Y.W., Steel, D., Gammon, D., Stievater, T.H., Katzer, D.S., Park, D., Piermarocchi, C., and Sham, L.J. (2003) *Science*, **301**, 809.
26 Lu, W., Ji, Z.Q., Pfeiffer, L., West, K.W., and Rimberg, A.J. (2003) *Nature*, **423**, 422.
27 Parak, W.J., Pellegrino, T., and Plank, C. (2005) *Nanotechnology*, **16**, R9.
28 Kaul, Z., Yaguchi, T., Kaul, S.C., Hirano, T., Wadhwa, R., and Taira, K. (2003) *Cell Res.*, **13**, 503.
29 Medintz, I.L., Uyeda, H.T., Goldman, E.R., and Mattoussi, H. (2005) *Nature Mater.*, **4**, 435.
30 Hoshino, A., Fujioka, K., Manabe, N., Yamaya, S., Goto, Y., Yasuhara, M., and Yamamoto, K. (2005) *Microbiol. Immunol.*, **49**, 461.
31 (a) Edgar, R., McKinstry, M., Hwang, J., Oppenheim, A.B., Fekete, R.A., Giuliani, G., Merril, C., Nagashima, K., and Adhya, S. (2006) *Proc. Natl. Acad. Sci. USA*, **103**, 4841.(b) Han, M., Gao, X., Su, J.Z., and Nie, S.M. (2001) *Nature Biotechnol.*, **19**, 631.
32 Sepahvand, R., Adeli, M., Astinchap, B., and Kabiri, R. (2008) *J. Nanoparticle. Res.*, **6**, 1309.

33 Huang, X.Y., Li, L., Qian, H.F., Dong, C.Q., and Ren, C.J. (2006) *Angew. Chem. Int. Ed.*, **45**, 5140.
34 Kirchner, C., Liedl, T., Kudera, S., Pellegrino, T., Javier, A.M., Gaub, H.E., Stolzle, S., Fertig, N., and Parak, W.J. (2005) *Nano. Lett.*, **5**, 331.
35 Dong, C.Q., Qian, H.F., Fang, N.H., and Ren, J.C. (2006) *J. Phys. Chem. B*, **110**, 11069.
36 Guo, Y., Shi, D., Cho, H., Dong, Z.Y., Kulkarni, A., Pauletti, G.M., Wang, W., Lian, J., Liu, W., Ren, L., Zhang, Q., Liu, G.K., Huth, C., Wang, L.M., and Ewing, R.C. (2008) *Adv. Funct. Mater.*, **18**, 1.
37 Shi, D.L., Guo, Y., Dong, Z.Y., et al. (2007) *Adv. Mater.*, **19**, 4033.
38 Guo, X., Small, J.P., Klare, J.E., Wang, Y., Purewal, M.S., Tam, I.W., Hong, B.H., Caldwell, R., Huang, L., O'Brien, S., Yan, J., Breslow, R., Wind, S.J., Hone, J., Kim, P., and Nuckolls, C. (2006) *Science*, **311**, 356.
39 Summerscales, O.T., Cloke, G.N., Hitchcock, P.B., Green, J.C., and Hazari, N. (2006) *Science*, **311**, 829.
40 Veedu, V.P., Cao, A., Li, X., Ma, K., Soldano, C., Kar, S., Ajayan, P.M., and Ghasemi-Nejhad, M.N. (2006) *Nat. Mater.*, **5**, 457.
41 He, J., Chen, B., Flatt, A.K., Stephenson, J.J., Doyle, C.D., and Tour, J.M. (2006) *Nat. Mater.*, **5**, 63.
42 You, X., He, R., Gao, F., Shao, J., Pan, B., and Cui, D. (2007) *Nanotechnology*, **18**, 035701.
43 Lee, H., Lee, E., Kim, D.K., Jang, N.K., Jeong, Y.Y., and Jon, S. (2006) *J. Am. Chem. Soc.*, **128**, 7383.
44 Ito, A., Ino, K., Kobayashi, T., and Honda, H. (2005) *Biomaterials*, **26**, 6185.
45 Sincai, M., Ganga, D., Ganga, M., Argherie, D., and Bica, D. (2005) *J. Magn. Magn. Mater.*, **293**, 438.
46 Cui, D., Han, Y., Li, Z., Song, H., Wang, K., He, R., Liu, B., Liu, H., Bao, C., Huang, P., Ruan, J., Gao, F., Yang, H., Cho, H.S., Ren, Q., and Shi, D. (2009) *Nano. Biomed. Eng.*, **1**, 61.
47 Xu, P., Cui, D., Pan, B., Gao, F., He, R., Li, Q., Huang, T., Bao, C., and Yang, H. (2008) *Appl. Surface Sci.*, **254**, 5236.
48 Kim, D.H., Lee, S.H., Kim, K.N., Kim, K.M., Shim, I.B., and Lee, Y.K. (2005) *J. Magn. Magn. Mater.*, **293**, 287.
49 Xu, Y., Gao, C., Kong, H., Yan, D., Jin, Y.Z., and Watts, P.C.P. (2004) *Macromolecules*, **37**, 8846.
50 Yan, X., Tay, B., Yang, Y., and Po, W.Y.K. (2007) *J. Phys. Chem. C*, **111**, 17254.
51 Zhang, Y., Li, J., Shen, Y., Wang, M., and Li, J. (2004) *J. Phys. Chem. B*, **108**, 15343.
52 Kam, N.W.S., Jessop, T.C., Wender, P.A., and Dai, H. (2004) *J. Am. Chem. Soc.*, **126**, 6850.
53 Kam, N.W.S. and Dai, H. (2005) *J. Am. Chem. Soc.*, **127**, 6021.
54 Yan, H., Park, S.H., Finkelstein, G., Reif, J.H., and LaBean, T.H. (2003) *Science*, **301**, 1882.
55 Seeman, N.C. (1999) *Trends Biotechnol.*, **17**, 437.
56 Bashir, R. (2001) *Superlattice Microstruct.*, **29**, 1.
57 Bonard, J.M., Weiss, N., Kind, H., Stockli, T., Forro, L., Kern, K., and Chatelain, A. (2001) *Adv. Mater.*, **13**, 184.
58 Batalia, M., Protozanova, E., Macgregor, R., and Erie, D. (2002) *Nano. Lett.*, **2**, 269.
59 Chen, D., Wu, X., Wang, J., Han, B., Zhu, P., and Peng, C. (2010) *Nano. Biomed. Eng.*, **2**, 61.
60 Styers-Barnett, D.J., Ellison, S.P., Park, C., Wise, K.E., and Papanikolas, J.M. (2005) *J. Phys. Chem. A*, **109**, 289.
61 Auvray, S., Derycke, V., Goffman, M., Filoramo, A., Jost, O., and Bourgoin, J.P. (2005) *Nano. Lett.*, **5**, 451.
62 He, P. and Bayachou, M. (2005) *Langmuir*, **21**, 6086.
63 Stewart, D.A. and Léonard, F. (2005) *Nano. Lett.*, **5**, 219.
64 Baskaran, D., Mays, J.W., Zhang, X.P., and Bratcher, M.S. (2005) *J. Am. Chem. Soc.*, **127**, 6916.
65 Hu, H., Ni, Y., Mandal, S.K., Montana, V., Zhao, B., Haddon, R.C., and Parpura, V. (2005) *J. Phys. Chem. B*, **109**, 4285.
66 Pan, B., Cui, D., Gao, F., and He, R. (2006) *Nanoscience*, **11**, 90.
67 Pan, B., Cui, D., Xu, P., Huang, T., Li, Q., He, R., and Gao, F. (2006) *Intl Conf on Biomedical and Pharmaceutical Engineering*, pp. 541–546.

68 Tsang, S.C., Davis, J.J., Green, M.L.H., Hill, H., Leung, Y.C., and Sadler, P.J. (1995) *Chem. Commun.*, **24**, 1803.
69 Bhattacharyya, S., Sinturel, C., Salvetat, J.P., and Saboungi, M.-L. (2005) *Appl. Phys. Lett.*, **86**, 113104.
70 Pantarotto, D., Singh, R., McCarthy, D., Erhardt, M., Briand, J.P., Prato, M., Kostarelos, K., and Bianco, A. (2004) *Angew. Chem. Int. Ed.*, **43**, 5242.
71 Cui, D., Tian, F., Ozkan, C.S., Wang, M., and Gao, H. (2005) *Toxicol. Lett.*, **155**, 73.
72 Cui, D., Wang, H.Z., Asahi, T., and Osaka, T. (2008) *ECS Transact.*, **13**, 111.
73 Lee, J.W., Kim, B.K., Kim, H.J., Han, S.C., Shin, W.S., and Jin, S.H. (2006) *Macromolecules*, **39**, 2418.
74 Hong, S., Leroueil, P.R., Janus, E.K., *et al.* (2006) *Bioconjug. Chem.*, **17**, 728.
75 Majoros, I.J., Myc, A., Thomas, T., Mehta, C.B., and Baker, J.R. (2006) *Biomacromolecules*, **7**, 572.
76 Radu, D.R., Lai, C.Y., Jeftinija, K., Rowe, E.W., Jeftinija, S., and Lin, V.S.Y. (2004) *J. Am. Chem. Soc.*, **126**, 13216.
77 Majoros, I.J., Thomas, T.P., Mehta, C.B., and Baker, J.R. (2005) *J. Med. Chem.*, **48**, 5892.
78 Pan, B., Gao, F., and Ao, L. (2005) *J. Magn. Magn. Mater.*, **293**, 252.
79 Gao, H., Kong, Y., Cui, D., and Ozkan, C.S. (2003) *Nano. Lett.*, **3**, 471.
80 Kong, Y., Cui, D., Ozkan, C.S., and Gao, H. (2003) *Materials Research Society Symposium Proceedings – Biomicroelectrome Chanical Systems (BIOMEMS)*, vol. 773, Materials Research Society, USA, pp. N8.5.1–6.
81 Pan, B., Cui, D., Gao, F., and He, R. (2006) *Nanotechnology*, **17**, 2483.
82 Pan, B., Cui, D., Xu, P., Ozkan, C.S., Gao, F., Ozkan, M., Huang, T., Chu, B., Li, Q., He, R., and Hu, G. (2009) *Nanotechnology*, **20**, 125101.
83 Cui, D., Tian, F., Coyer, S.R., Wang, J., Pan, B., Gao, F., He, R., and Zhang, Y. (2007) *J. Nanosci. Nanotechnol.*, **7**, 1639.
84 Li, Z., Huang, P., Lin, J., He, R., Liu, B., Zhang, X., Yang, S., Xi, P., Zhang, X., Ren, Q., and Cui, D. (2010) *J. Nanosci. Nanotechnol.*, **10**, 4859.
85 Yang, H., Guo, Q., He, R., Li, D., Zhang, X., Bao, C., Hu, H., and Cui, D. (2009) *Nanoscale Res. Lett.*, **4**, 1469.
86 Pan, B., Cui, D., Ozkan, C.S., Ozkan, M., Xug, P., Huang, T., Liu, F., Chen, H., Li, Q., He, R., and Gao, F. (2008) *J. Phys. Chem. C*, **112**, 939.
87 Pan, B., Cui, D., Xu, P., Chen, H., Liu, F., Li, Q., Huang, T., You, X., Shao, J., Bao, C., Gao, F., He, R., Shu, M., and Ma, Y. (2007) *Chinese J. Cancer Res.*, **19**, 1.
88 Valentini, L. and Kenny, J.M. (2005) *Polymer*, **46**, 6715.
89 Kumar, S. (2004) *Polym. Compos.*, **25**, 630.
90 Jin, W.J., Sun, X.F., and Wang, Y. (2004) *New Carbon Mater.*, **19**, 312.
91 Li, H.J., Wang, X.B., Song, Y.L., Liu, Y.Q., Li, Q.S., Jiang, L., and Zhu, B.D. (2001) *Angew. Chem. Int. Ed.*, **40**, 1743.
92 Wang, S.Q., Humphreys, E.S., Chung, S.Y., Delduco, D.F., Lustig, S.R., Wang, H., Parker, K.N., Rizzo, N.W., Subramoney, S., Chiang, Y.M., and Jaqota, A. (2003) *Nature Mater.*, **2**, 196.
93 Cui, D., Pan, B., Zhang, H., Gao, F., Wu, R., Wang, J., He, R., and Asahi, T. (2008) *Anal. Chem.*, **80**, 7996.

7
Amine-Functionalized Carbon Nanotubes
Ramaiyan Kannan and Vijayamohanan K Pillai

7.1
Introduction

Over the past few years, increasing attention by several reputed groups has been devoted on functionalization of carbon nanotubes (CNTs) mainly because CNTs are considered as promising materials for many next generation devices due to a combination of properties related to its unique electronic structure, electrical conductivity, chemical inertness, mechanical stability, and thermal conductivity [1]. For example, a defect-free CNT could carry 1000 times higher current density than that of wires made up of copper [2], and due to many of these unique properties CNTs can be used in a number of applications such as nanoelectronics, molecular switches, nanotransistors, sensing, gas storage, field emission devices, catalyst supports, probes for scanning probe microscopy, and energy applications such as batteries, fuel cells, and supercapacitors. However, many of these require application-specific CNTs with tailor-made properties that necessitate the chemical modification of CNTs. Chemical processing of pristine nanotubes is also a demanding task, necessitating several prerequisites of purification and separation due to the possibility of different types of nanotubes during the preparation stage, and some of these steps obviously cause inadvertent functionalization often in an uncontrollable manner. Thus, a proper understanding of CNT surface and tip chemistry is important to tune their properties in a systematic manner.

One of the revealing features of CNT is its unique composition of rolling up of graphitic sheets in to cylindrical bundles enabling the arrangement of hexagonal rings along the tubular axis suitable for specific functionalization purpose. Consequentially, CNTs with defects or surface functional groups can be used in many applications depending on the kind of modification and atoms or molecules that are attached to it. For example, Au, TiO_2, and SiO_2 nanoparticles covalently bound to CNTs are expected to find interesting applications in semiconductors and solar cells [3]. Similarly, after functionalization CNTs have given more than 100 000 turnover numbers for the H_2 evolution reaction in aqueous sulfuric acid with very low overvoltage [4]. Sulfonated CNTs are shown to improve the

---

Surface Modification of Nanotube Fillers, First Edition. Edited by Vikas Mittal.
© 2011 Wiley-VCH Verlag GmbH & Co. KGaA. Published 2011 by Wiley-VCH Verlag GmbH & Co. KGaA.

proton conductivity and mechanical stability of perfluorosulfonic acid based electrolytes dramatically in polymer electrolyte membrane fuel cells by providing better channel network for proton transport [5]. Different functional groups such as carboxylic acid, sulfonic acid, phosphonic acid, and basic groups like hydroxy moieties, amines, polymers, DNA, and proteins can be attached on the CNT surface through careful chemical strategies, each having their own advantages for specific applications.

In this chapter, we explore various contemporary aspects of amine-functionalized CNTs. We first discuss all benefits associated with amine group attachment on different types of CNTs offering selected examples of the preparation of amine functionalization followed by their characterization. Although single-walled CNTs are mainly considered for all the chemical transformations, double- and multi-walled CNTs could as well be used for similar amine-functionalization approaches except for some important changes in band structure [6]. We further focus on the control of degree of functionalization, which highlights some contentious issues related to the quantitative control of functionalization, especially since many independent methods offer different values. A majority of the topics covered in this chapter deal with sidewall amine functionalization with their promising application potential, despite the difficulty in estimating quantitative yields to select more efficient and less hazardous amine-functionalization procedures. In the end, we identify the current research directions and future needs since CNT functionalization itself is an emerging area, as explained in several other excellent chapters of this book. Before going into the details of amine functionalization and its influence in different applications, let us briefly have a look at the regular functionalization strategies that are prevalent in the surface chemistry of CNTs.

7.2
Functionalization Strategies for CNTs

CNT functionalization can broadly be classified into three categories: covalent functionalization, noncovalent functionalization, and innercavity filling [1d]. All these classes have some common features, although noncovalent functionalization could be further divided to other types based on the type of interaction like electrostatic, H-bonding, van der Waals interactions, π–π stacking, etc. CNTs after functionalization by innercavity filling, however, show new properties that the pristine nanotubes do not show and hence these types of CNTs are called some times as metananotubes [7]. Among the different modes of functionalization, noncovalent functionalization normally involves the wrapping of high-molecular-weight species such as polymers, DNA, and proteins on the CNT surface, thereby disrupting the hydrophobic interface with water. Further it also interrupts the intertubular interactions in a tubular aggregate, thus individualizing the nanotube. For example, SWNT has been attached with protein molecules by noncovalent attachment of pyrenebutanoic acid. One of the main advantages associated with noncovalent functionalization is the preservation of sp^2 in CNTs, which is the key for its unique electronic and electrical properties. Noncovalent attachment mainly happens through the π–π stacking of the aromatic rings of CNTs with incoming aromatic systems such as pyrene, naphthalene, etc., to which further chemical modifications are required to generate desired functional groups. However, noncovalent functionalization suffers from limitations such as the difficulty in controlling the degree of functionalization due to its weak nature and hence the attached aromatic groups to the CNT are not robust enough as required for some applications. When electrostatic interactions are used for functionalization especially to generate amino and carboxy groups, the resultant CNTs show pH-dependent behaviors, which could pose difficulties for certain types of applications.

Covalent functionalization, on the other hand, provides a better methodology to attach application-specific functional groups onto CNT surface, thereby improving the nanotube processability and dispersing ability in different solvents. Further the possibility to vary the level of covalent attachment provides a unique means to tune the CNT properties, which is not possible with noncovalent attachment. Covalent modification of CNT surface normally is carried out through two methods. First one involves the CNT surface oxidation through strong oxidizing agents such as HNO_3 for prolonged refluxing to create carboxylic acid moieties. Further, the precise locations of functionalization can be controlled by varying the oxidation conditions. For instance, milder oxidizing conditions specifically attack the CNT tips and defect sites to leave the sidewall unaffected, thus important for molecular wires and transistors-based applications while strong oxidizing reagents attack both sidewalls and tips, thereby making the functionalized CNTs suitable for composite making. This carboxylic acid groups could be further modified to incorporate specific functionalities ranging in a wide variety of biomolecules and semiconductor quantum dots (QDs) by appropriate chemical conversions.

Direct surface functionalization of CNTs without the need to create carboxylic acid groups are also well documented and a variety of chemical reactions such as 1,3-dipolar cycloaddition, diazotization, etc., are reported in this category [8]. Further, fluorination of CNT tends to open a new route for subsequent functionalization procedures. Moreover, fluorinated CNTs can be reduced back to the starting material using reagents like hydrazine, which provides many simpler ways to control the level of functionalization [9].

Electrochemical attachment of functional groups on the sidewalls is proven to be another attracting way for the chemical modification of CNTs. An array of compounds has been attached on the sidewalls of CNTs by electrochemical reduction of aryl diazonium salts [10]. The electrochemical approach is highly desired for chemical alteration because the extent of reaction can be directly adjusted by an applied potential.

Recently, the use of microwave-assisted methods of functionalization has also been reported as a viable technique due to the reduced time of reaction and increased efficiency of functionalization [11]. However, since a large amount of localized energy is used in a short time scale, microwave methods can have undesirable side effects such as chopping of CNTs, uncontrolled defect generation, and contamination due to the presence of a dipolar liquid. Ability to scale up for industrial application is also a serious limitation of microwave methodology. Overfunctionalization could also possibly cause the transformation of CNTs (especially multiwalled case) to graphene layers, which has been recently named as "unzipping of CNTs" as reported recently by strong chemical reagents under aggressive conditions [12].

With this general background of chemical functionalization of CNTs, we will now explain the different ways to attach amine groups specifically on various types of CNTs, along with their relative merits and demerits for applications in different fields.

7.3
Importance of Amine Functionalization

There are several reasons why an amine-functionalized CNT would be interesting. Some of these are as follows:

- Controlled solubilization of CNTs is critical for many applications and ability to aminate different sites in the CNTs by means of varying the method of functionalization allows the fine-tuning of CNT solubility toward various solvents and manipulates their property. For example, CNTs could be made soluble in several organic solvents by functionalization using long-chain amines through an amide coupling, while short-chain amines and diamines and preponderance of amino groups on the sidewall itself make these interestingly water soluble. For instance, the highest solubility of 2.6 mg/mL in water has been reported for oxalyl-modified CNTs [13].

- Due to the ability of amine groups to replace fluorine atom through nucleophilic substitution reactions and to form amide linkage with acyl chloride groups, they can be attached covalently to either on the sidewalls or on tips. Further, amine groups can also be attached to the CNTs noncovalently by controlling electrostatic charges or even as a side-chain moiety in aromatic rings by means of π–π stacking.

- The presence of amine groups in many biomolecules makes this amine-functionalization strategy important as these biomolecules can be attached to CNTs through condensation (amidation) or substitution (fluorinated CNTs) reaction. For example, aminated CNTs have been tested for the delivery of drug, antigen, and gene by Bianco *et al.* [14].

- It can help to separate the metallic and semiconducting CNTs as amine groups are shown to attach well with semiconducting nanotubes and hence are very important in nanoelectronic applications. As a typical illustration, octadecylamine (ODA) is shown to separate the semiconducting CNTs from the metallic ones by retaining the semiconducting nanotubes on the solution due to the higher affinity of amine groups, while the metallic nanotubes get precipitated. This method can be used to separate the CNTs even in the bulk scale [15].

- CNTs attached with bioactive molecules can be used as molecular transporters (MTs) to take molecules that normally cannot go inside the cell by endocytosis (endocytosis is a process by which cells absorb molecules such as proteins from outside the cell by engulfing them with their cell membrane). For example, alexa fluor streptavidin (SA) could not transverse the cell membrane alone; however, when attached on CNT surface these molecules easily entered the cell membranes [16].

- A proper understanding of interaction between CNTs and amine groups would help in forming aligned network that are crucial for fabricating CNT-based devices. Many first-principle calculations and molecular dynamics simulations have helped to provide important insights about transformation of many types of CNTs to graphene layers during chemical functionalization.

The higher surface area of CNTs enables higher catalytic activity in many cases, especially if one aims to use them for biological or electrochemical applications after amination for attaching either using complexes or nanoparticles for their unique catalytic abilities. More significantly CNTs provide higher stability for electrocatalysts due to their chemical inertness, while amine groups act as a starting point to connect many inorganic, organic, and biomolecules on the surface of CNTs, thereby creating a synergy in desired properties. Further, the high electrical conductivity of CNTs help in electrochemical applications such as fuel cells, batteries, and super capacitors as demonstrated by several illuminating recent demonstrations [17].

7.4
Methods of Amine Functionalization

Amine functionalization of CNTs can also be classified mainly into three categories based on the mode of attachment such as covalent, electrostatic, and noncovalent bonding. Covalent bonding includes the surface oxidation of CNTs to which the functional groups are attached to the sidewalls and tips either by condensation reaction (amidation) or by nucleophilic substitution reaction (mainly on fluorinated CNTs). Further, many electrochemical methods are also being developed by several illustrious groups to attach amine directly onto the sidewalls of CNTs [10]. Microwave-generated N_2/NH_3 plasma is another method where amine functionalities can be introduced, although not selectively, on the CNT surface [18].However, it lacks specificity in terms of attaching only amine or amide group or N atom doped on CNT. On the other hand, noncovalent attachment mainly happens through the π–π stacking of the aromatic rings of CNTs with incoming aromatic systems such as pyrene, naphthalene, etc., with amine at the chain ends. The covalent functionalization of CNT can be further classified depending on the location into many types as that occur on tips and on the sidewalls and innerwalls,

```
                          Carbon nanotubes
                                 |
         ----------------------------------------------
         |                       |                    |
  Noncovalent           Endohedral filling of      Covalent
  functionalization     inner empty cavity         functionalization
         |                                            |
    -----------                              -------------------
    |         |                              |      |          |
  Polymer   π-π                             On     On         On
  wrapping  stacking                        tip    sidewall   inner wall
```

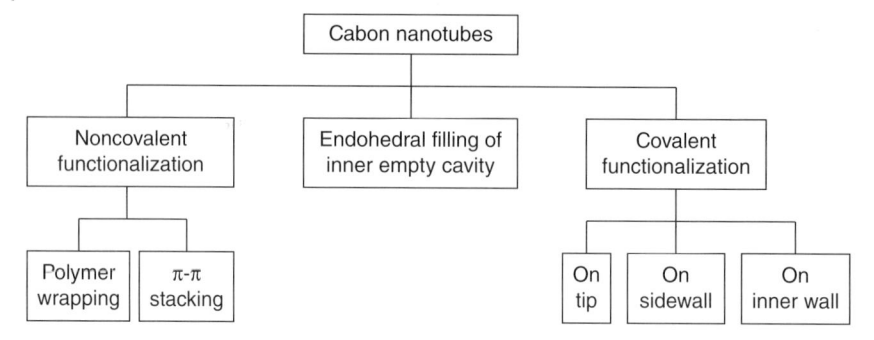

Figure 7.1 Different modes of amine functionalization of CNTs.

although there is a difficulty for a clear distinction especially for multiwalled CNTs (Figure 7.1).

7.4.1
Sidewall Functionalization

Sidewall functionalization of CNTs can be carried out through many methods, out of which, mainly two are more popular to adopt due to their ease of preparation. First, amine groups with long-chain organic molecules to be attached onto fluorinated CNTs, which are prepared by treating CNTs with F_2/Ar mixture at temperatures ranging from 150 °C to 350 °C [19]. These are subsequently treated with amino alkyl compounds in pyridine medium at 120 °C to obtain amine groups attached directly without the amide linkage on the sidewalls of CNTs [20]. The general procedure for amine functionalization is given in Figure 7.2, where various modes for sidewall functionalization are especially presented to reveal their significance with respect to the scheme displayed in Figure 7.1. Another important sidewall functionalization comprises of electrochemical reduction of amines in the presence of CNTs resulting in sidewall functionalization through a radical induced mechanism for attacking the double bonds in the CNTs. Further, electrochemical attachment can also be used to form a monolayer of functional groups by varying the duration of reaction.

7.4.2
Tip and Defect Site Functionalization

Tips and defect functionalization of CNTs is achieved mainly by acid treatment, which opens up the tips by creating carboxylic acid groups at the tube openings along with creating some defects on the sidewalls. The level of carboxylic acid creation on the surface of CNTs can be varied by varying the reagents (of diverse oxidizing strength), temperature, and time used for the functionalization. After this, there are three major modes for attaching the amine groups on the carboxylic sites:

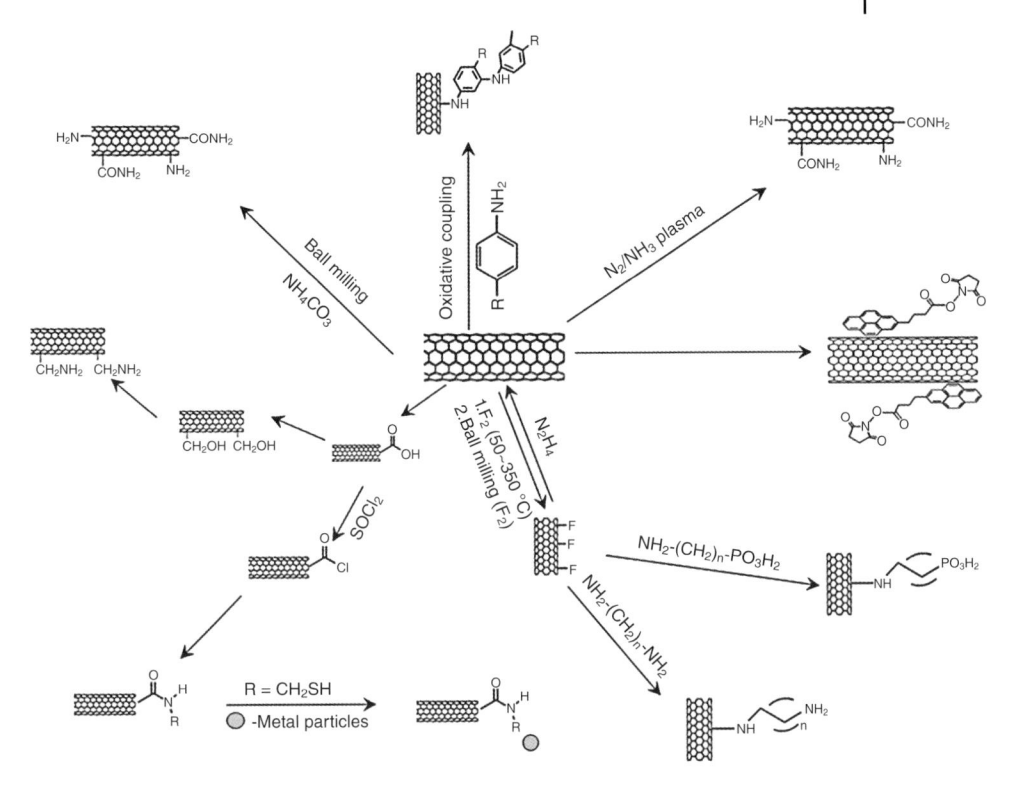

Figure 7.2 Pictorial representation of different functionalization strategies to prepare amino-functionalized CNTs for various applications.

1) Reaction with thionyl chloride creating acyl chloride which can be readily converted to amides when reacted with amines [21]

2) Reaction with oxalyl chloride is also known to create acyl chloride moieties that are converted into amide groups for attaching long-chain organic molecules with high solubility in organic solvents such as $CHCl_3$ and CS_2 [13a]

3) Reaction with carbodiimides creates a temporary intermediate which upon exposure to amines forms the amide linkage

However, the use of thionyl chloride route produces more individual nanotubes in comparison with carbodiimide route especially in the case of SWCNT bundles [22]. After attaching the amine groups onto CNTs by varying the chain-end functional groups, we can prepare application-oriented CNTs. For example, use of ethylendiamine gives amino end groups, which in turn can be used to bind proteins, amino acids, and nucleic acids while aminoethyl phosphonic acid helps in binding TiO_2 and in forming composites with polybenzimidazole membranes for fuel-cell electrolyte applications.

Both tip and sidewall functionalization have their own merits as tip functionalization ensures that the electronic properties of CNT remain more or less intact and thus helpful in many applications like molecular electronics where conductivity is important, while the sidewall functionalization often reduces the electronic conductivity and hence could be suitable for composites with polymers especially where more strength is the key issue. It is also further proved that simultaneously both tip and sidewall can be functionalized in a controlled manner by choosing proper functionalization strategies. For example, an array of CNTs have been attached with amine-functionalized DNA oligonucleotides on the tips through amide linkage while pyrene-based attachment strategy allows attaching them on the sidewalls [23].

7.4.3
Noncovalent Functionalization

Noncovalent attachment of functional groups onto CNTs normally involves the initial adsorption of bifunctional aromatic molecule irreversibly attached onto the graphitic basal planes of CNTs through $\pi-\pi$ stacking. For example, pyrene groups with side-chain provisions to bind amine groups have been attached on CNT sidewalls by $\pi-\pi$ stacking after which proteins were immobilized on the CNT surface [24]. The different functionalization strategies used and their main applications are briefly given in Table 7.1.

7.5
Characterization Techniques

Amino-functionalized CNTs can be characterized by a number of techniques including FT-IR, Raman, TGA, TEM, SEM, and AFM although some of these techniques like fluorescent measurement studies are mainly useful in the case of biomedical applications. Information from different complementary techniques are often necessary to arrive at a conclusion regarding the nature of interactions, and each technique in this regard has its own limitations. It is very important to determine the nature of bonding between CNTs and functional groups since depending upon the mode of attachment, their properties also would change dramatically. We will specifically discuss the use of FT-IR, Raman, and AFM techniques for the characterization of amino-functionalized CNTs here since several excellent chapters are available in this volume itself on the applications of various characterization techniques for functionalized CNTs.

7.5.1
FT-IR

Among the different techniques, FT-IR plays a vital role in determining the mode of attachment as it can directly reveal the creation of new bonds formed along with

Table 7.1 Different amine functionalization strategy widely used and their main applications.

Functionalization strategy	Mode of attachment/ the type of CNT	Applications	Reference
Electrochemical oxidation	Covalent sidewall	Biocompatible CNTs for biomedical applications	[10b]
Amidation from thionyl chloride route	Covalent tip and defect sides	Binding gold nanoparticles on CNTs	[41d]
	Covalent tip and defect sides	Enzyme-attached CNT expecting biosensor activity	[11]
	Covalent attachment defect sites	Electrocatalysts methanol oxidation	[20a]
	Covalent tip and defect sides	Biocompatible CNTs for biomedical applications	[20b]
Amidation by carbondiimide route	Covalent tip and defect sides	Attaching QDs onto CNTs	[3a]
Amidation route	Covalent attachment defect sites	Molecular transporter	[16]
Chemical substitution reaction	Sidewall	Solid basic catalyst for transesterification of triglycerides	[38a]
Chemical substitution on fluorinated CNTs	Sidewall/multiwall carbon nanotubes	High-performance composite materials with superior mechanical properties	[20b]
π–π stacking	Noncovalent attachment	Protein immobilization	[24]

specific information on the disappearance of existing bonds upon functionalization. While noncovalent functionalized CNTs are not expected to show any dramatic change in the CNTs IR spectrum, covalently modified CNTs display many characteristic peaks for the functional groups that would disclose the bonding information in the make up of the modified CNTs. Accordingly, Table 7.2 describes some of the key features of CNT-IR spectra with amine functional groups attached onto it by various modes. For example, in the case of amide route, carboxylic acid functionalized CNTs show C=O and C–O stretching at around 1700–1720 and 1100–1300 cm^{-1}, respectively. However, the formation of amide linkage is normally associated with a shift in C=O stretching to lower values revolving around 1650 cm^{-1} while C–O stretching is often replaced with C–N stretching. Although C–O and C–N stretching vibration frequencies overlap each other causing difficulty to ploy

Table 7.2 FT-IR peak positions observed for the carbon nanotubes aminated through various routes.

CNT type	Stretching modes (cm^{-1})								
	C–C	C=O	C–O	C–N	C–F	N–H	C–H	C–H	O–H
Pristine CNT	1540–1580	–	–	–	–	–	–	–	–
Carboxylated CNT	1540–1580	1710–1730	1100–1300	–	–	–	–	–	3000–3600
Amidated CNT	1540–1580	1630–1670	Disappears	1075	–	–	2800–3000	1360–1460	
Fluorinated CNT	1540–1580	–	–	–	1140	1638 (I), 1570 (II)			
Aminated CNT	1540–1580	–	–	1120–1210	Disappears	1638 (I), 1570 (II)	2800–3000	1360–1460	

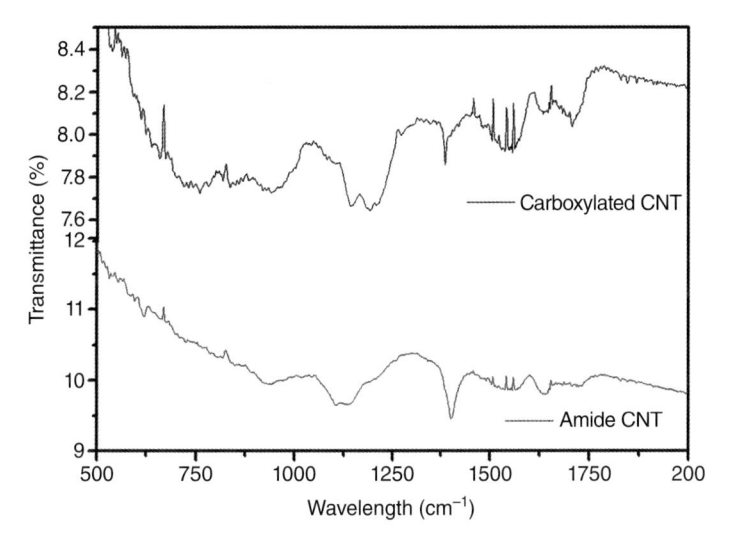

Figure 7.3 IR spectra of CNTs functionalized with carboxylate groups and aminoethyl phosphonate groups attached through amide bonds.

a definite range, there is always a change in the peak position after functionalization suggesting the attachment. While amine group attachment to carboxylated CNTs is confirmed by the change in C=O stretching frequency, amines replacing the fluorine in F-CNTs can be easily diagnosed looking at the disappearance of C–F bond stretching by C–N. However, we need to be careful in assigning all these peaks since despite pristine CNTs are expected to show only the C=C stretching bands, it is often found with many other peaks depending upon the mode of preparation of CNTs and the method of purification used. Figure 7.3 shows the FT-IR spectra of carboxylated CNT and aminoethyl phosphonate attached CNTs. The carboxylated CNTs show a peak at 1720 cm^{-1} corresponding to the C=O stretching of carboxylic acid groups while it is shifted to 1650 cm^{-1} in the amide CNT suggesting the successful formation of amide linkage.

7.5.2
Raman Spectroscopy

Raman spectroscopy is a very useful technique in determining the covalent and noncovalent bonding characteristics as covalent bonding is normally carried out by creating defects on the nanotube surface, while noncovalent attachment of molecules generally precludes such changes in the surface morphology [24]. Raman spectra of CNTs normally reveal three characteristic features: the G band arising from the tangential vibration of graphitic carbon atoms, the D band characteristic of defective graphitic structures, and the third one called radial breathing mode (RBM), which is extremely sensitive to the changes in the diameter of the SWCNTs. However, RBM is not well resolved in the case of DWNTs and MWCNTs due to the variation in diameter from inner rings toward the outer ones.

Covalent attachment through the carboxylic acid-amide route generally increases the defects in the CNTs thereby increasing the D band in the Raman spectra along with interesting changes in G/D ratio. Further, the defects produced significantly affect the RBM, displaying diminished peak values. For example, CNTs covalently attached with cis-Myrtanylamine show increase in intensity of the disorder mode peak at $1320\,cm^{-1}$ due to the defects created during the functionalization [25]. Similarly, covalently aminated CNTs prepared from F-CNTs also are expected to show an increase in the D band along with an alteration in the RBM mode as F-CNTs are known to reveal increase in D band along with the remarkable disappearance of RBM band [19]. Further the extent of functionalization can be qualitatively measured by the ratio of G/D band. Even though CNTs with covalent modification can be monitored by Raman spectroscopy, it has some limitations like inability to explicitly show the nature of bonding and thus can be considered as complete technique when coupled with IR spectroscopy.

7.5.3
AFM

AFM measurements are normally used to observe the structural and morphological changes on the CNTs after functionalization. AFM has been used to specially study the changes to single CNT before and after functionalization. Normally, the change in the height profile before and after modification is used to estimate the level of functionalization. Also AFM can be used to confirm the end group and sidewall functionalization, and attachment of long-chain biomolecules such as DNA has been detected by AFM measurements using the contrast between the CNT and the attached molecule. For example, AFM studies carried out on DNA-attached CNTs show a bright line for CNTs, while a pale line prove the presence of DNA on the sidewalls of CNTs [26].

7.5.4
XPS

X-ray photoelectron spectroscopy (XPS) is another potential tool for determining the kind of functionalization and interaction between the different functional groups due to its ability to detect the chemical state of the atoms present in the CNT. For example, Ramanathan *et al.* studied the amide- and amine-functionalized CNTs, prepared through amide link formation and chemical reduction in carboxylic acid to primary amine through XPS and were able to confirm the mode of attachment between the CNTs and amine groups [27]. For example Table 7.3 presents a comparison of binding energy values of C and N 1s signals in XPS for the amine- and amide-functionalized CNTs with different chemical environment. Further,

Table 7.3 Variations of binding energy with changing chemical environment for carbon and nitrogen 1s level in the XPS of amino-functionalized CNTs [27].

Type of atom	Carbon peak positions (C1s) (eV)	Nitrogen peak positions (N1s) (eV)	Type of functionalization
C–C	284.6	NA	Pristine CNT
C–C–O	286.1	NA	Carboxylated CNT
O=C–O	288.8	NA	Carboxylated CNT
O=C–N	287.9	399.5	Amidated CNT
C–C–N	285.8	400.5	Aminated CNT

XPS is a reliable method for estimating the degree of functionalization, especially if amino, carboxy, and thiol groups are involved as the area of appropriate peak could be estimated after deconvolution. Functionalization using nanoclusters of gold and Pt on CNT surface provides a revealing example, where extent of functionalization can be obtained from XPS analysis. However, sometimes this could give rise to problems due to limitations like multiple peaks and initial/final state effects.

7.5.5
Other Characterization Techniques

Thermogravimetric analysis (TGA) is normally used to find the amount of functionalization on the CNT and their thermal stability. In some instances, functionalized CNTs are reported to revert back to their pristine form after TGA, thus helping to find the structural integrity of CNTs after functionalization. This is mainly because most of the organic molecules tend to decompose before the CNT decomposition temperature. For example, PPEI–EI–MWCNT gets thermal defunctionaization at 500 °C leaving the pristine nanotubes as evidenced by SEM measurements [22, 28]. TEM is another commonly used technique to study the morphology change especially in the case of SWCNTs.

7.6
Degree of Amine Functionalization

The degree of amine functionalization, if estimated reliably using quantitative methods, can have a significant influence in tailoring the properties of CNTs for specific applications. For example, 10% covalent functionalization is found to be optimum in terms of elastic modulus devastation of sidewalls [29]. It could be defined as the ratio of number of functional groups per one unit area obtained by a given functionalization method to the maximum possible amount of functional groups which could be attached. This is also intimately related to the surface coverage of amino groups, which could be quantitatively estimated by any of the surface spectroscopy. A typical value could be 0.3, as for example, estimated by XPS in case of sulfonate functionalization and this could be also valid for N signal in case of amino functionalization. This would also, of course, depend on the method of characterization as well as preparation. Although there is no accepted way of defining this, it can be expressed as number density (i.e., number of amino groups in unit area or unit weight) which can be controlled by choosing the desired method of amine functionalization along with selected characterization techniques like Raman, IR, or XPS.

One of the most successful functionalization strategies with maximum degree of functionalization is fluorination. For example, fluorination of CNTs attaches fluorine atoms on the surface to the maximum extent of C_2F and by substitution reaction theoretically one would expect to reach the level of C_2NH_2 [19]. However, it is hard to reach as CNTs tend to loose their integrity by this high extent of functionalization. An N/C ratio of approximately 1 is achieved by varying the ball-milling time with CNT and ammonium bicarbonate even though not all the N atoms are covalently bound to the surface as determined by XPS [26]. A more tangible value of 0.3–0.5 is much more useful especially for SWCNTs. Electrochemical methods have provided a functionalization extent of 1 out of 20 carbon atoms resulting in significant improvement in solubility. The importance of controlled amination is further demonstrated by CNT–biotin–streptavidin conjugate where overdosage of SA resulted in cell death, which could be managed by varying the degree of amination [16].

7.7
Changes in the Band Structure

The modification of electronic band gap is an important step to make electronic devices for specific applications. Doping is the normal method to tune the electronic structure, and attachment of functional groups in small levels can be considered to create surface states. In this aspect, B and N are considered as very good choices for doping in CNTs due to their almost similar atomic radius as carbon atoms with one electron less or more than carbon, respectively. The lone pair of electrons on the nitrogen if gets donated to CNT could make semiconducting nanotubes metallic with concomitant changes in the electronic structure. To verify this, a theoretical study on the different modes of nitrogen incorporation onto the CNT, such as replacement of carbon atom by nitrogen atom with and without vacancy, amine attachment, parallel adsorption of nitrogen on the CNT, has been carried out [30]. Interestingly, this incorporation results in the formation of an impurity band near the Fermi level despite creating an sp^3 carbon in the CNT matrix. What is the critical value of degree of surface amination where surface-state treatment is no longer valid and the miniband description is more useful is not clear despite the availability of many studies using density functional theory. However most of the calculations suggest that depending upon the level

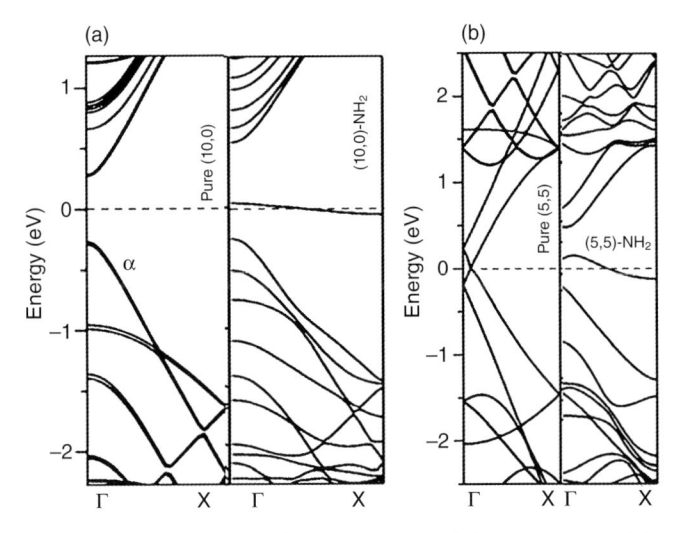

Figure 7.4 Band structures of (a) pure (10,0) and (b) (5,5) CNTs and that of CNTs attached with amine groups on the sidewalls. Fermi level is represented by the dotted horizontal lines. Reprinted with permission from Ref. [30]. Copyright 2007 American Physical Society.

of attachment, semiconducting nanotubes can be converted to metallic nanotubes, while in the case of metallic tubes the density of states near the Fermi level increases. Covalent attachment of other functional groups such as carboxylic acid and hydroxyl groups are also reported to create the impurity band near Fermi level [31]. However, their role in converting semiconducting nanotubes into metallic is not discussed possibly due to the lack of electron-donating mechanism (Figure 7.4).

7.8
Applications of Amine-Functionalized CNTs

7.8.1
Solubility Manipulation

One of the critical advantages of amine functionalization is the ability to tune the solubility of CNTs to variable degree in different solvents ranging from polar aqueous solvents to nonpolar organic solvents. For example, ODA-attached CNTs show substantial solubility in chloroform, dichloromethane, carbon disulfide, aromatic solvents such as benzene, toluene, chlorobenzene, and 1,2-dichlorobenzene with the solubility exceeding 1 mg/mL in some solvents [32]. The long alkyl chain helped in dissolving the CNTs in the above solvents. However, CNTs functionalized with small-chain organic molecules with polar end-group show solubility in polar solvents. For example, functionalization using poly(allylamine) chains using cation–π interactions on the surface of the CNTs makes them soluble in water [33]. CNTs attached with ethanolamine, propanolamine, and diethanolamine all showed improved solubility in water in comparison with pristine CNTs [20a]. Further, covalent attachment of ethylene diamine on CNTs resulted in reduced intertubular interactions due to steric hindrances and thus resulted in less compact and more soluble CNTs. This is very helpful in forming a composite with polyamide 6 (PA6), resulting in improved storage modulus, glass transition temperature, and yield strength of PA6 composites with EDA-functionaized CNTs [34]. Hence similar functionalization strategies are extremely important to manipulate solubility irrespective of our processing objective.

7.8.2
Selective Organic Reactions

CNTs, appropriately functionalized, can play an important role in many organic chemical reactions as catalysts and scavengers due to their ability to interact with chemical functionalities while at the same time easy to separate due to their size. Organic reactions involving base catalysts recently found ample advantages by the use of N-doped CNTs as solid-base catalysts. For example, N-CNTs are known to catalyze the Knoevenagel condensation of benzaldehyde and ethyl-α-cyanoacetate with significant reduction in waste production [35]. CNTs functionalized with amine groups on the sidewalls are reported to play a vital role in the separation of by-products, which is normally a time- and effort-consuming process that also restricts the final yield in many reactions. Polymer-based scavenging reagents are reported to remove the excess reagents and by-products from a crude reaction mixture by forming polymer-bound products that can be easily removed by simple filtration. This has significantly simplified the work-up procedure and has been widely used even though their biphasic nature hinders the access to functional groups there by taking several hours for the complete removal. However, amine-functionalized CNTs, in this particular instance CNT–triethylenetetramine or CNT–tetraethylenepentamine, when used for the removal of by-products and excess reagent in the amidation reaction of amines, such as butylamine, cyclohexylamine, etc., remove the by-products in less than 30 min. The scheme of the reaction as given in (Figure 7.5) thus illustrates an elegant example of a fast by-product removal kinetics as explained by the easy access of functional groups that are on the sidewalls of CNTs [36].

Transesterification of glyceryl tributyrate with methanol is considered as a test reaction for the production of biodiesel due to the similarities between them. This reaction is catalyzed by basic catalysts although with poor recyclability [37]. Amine-functionalized CNTs have shown remarkable catalytic activity for this conversion reaction. The deactivation in the performance after first cycle is mainly due to the adsorption of triglycerides on the CNTs as is duly rectified by including washing steps, which show a significant improvement in the recyclability of the catalysts. Again the mode of aminated CNT fabrication plays a crucial role as amide-link

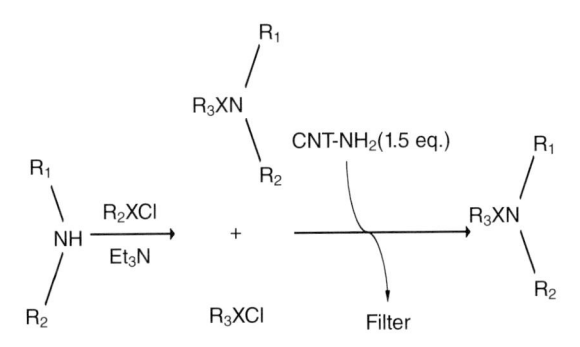

Figure 7.5 Role of aminated CNTs in removing the by-products and excess reagent in organic alkylation reactions. Reprinted with permission from Ref. [36].

CNTs show poor catalytic performance in comparison with amine directly attached on the CNT sidewalls [38].

7.8.3
Chemical and Biological Sensors

Carbon monoxide (CO) generated by the incomplete combustion of fossil fuels, oil, and coal is a significant atmospheric pollutant, which is harmful to human health. Among different sensors developed for detecting CO concentration in the air, amperometric CO sensors, which utilize semiconductors such as ZnO and TiO_2 are shown to have better resolution and high sensitivity although with some disadvantages like baseline drift upon interaction with poisoning species. Amine-functionalized CNTs tend to show better CO-sensing characteristics when coupled with conducting polymers. For example, amine-functionalized CNTs grafted with polydiphenylamine (PDPA) are shown to have enhanced sensitivity over Au-modified Pt electrode and Sn-modified Pt/Nafion electrodes. A comparison of cyclic voltammograms of $CNT-NH_2$–PDPA, $CNT-NH_2$, and PDPA systems in the presence and absence of CO reveals that $CNT-NH_2$–PDPA electrode has higher sensitivity than other electrodes, while $CNT-NH_2$ electrode shows moderate activity. However, PDPA alone dose not show any considerable sensitivity suggesting the importance of $CNT-NH_2$ for the sensing application. Further, variation in the loading of $CNT-NH_2$ alters the onset potential for the CO oxidation, while variation in PDPA does not alter the peak position. Thus the combined presence of $CNT-NH_2$ and PDPA is the key for the pronounced sensitivity for CO oxidation [39].

CNTs attached with specific biomolecules could act as an effective sensor for PNA and many protein molecules. CNTs attached with peptide nucleic acid through an amide bond by carbodiimide process have shown remarkable DNA recognition. By varying the PNA, one could prepare a specific biosensor that will bind to a particular DNA sequence [26].

7.8.4
Molecular Electronics

Designing strategies for the delivery of drugs and molecular probes inside the cells is plagued by problems like poor cellular penetration of many macromolecules including proteins and nucleic acids. Several classes of MTs have been tested despite with limited success. CNTs can be attached with many biomolecules thorough amide linkage and thus attempted as MT in human promyelocytic leukemia (HL60) cells. Fluorescein-functionalized CNTs were incubated with HL60 for 1 h and then allowed to enter the cell by endocytosis at 4 °C. Confocal images (Figure 7.6) of the cell after incubation revealed appreciable fluorescence on the surface and in the cell interior. This proved the ability of CNTs to carry molecules inside the cells through endocytosis. Proteins attached through amide linkage to CNTs also showed similar penetration inside the cells revealing the ability of CNTs to carry large consignments inside the cell membrane. Further, CNTs alone are shown not to have any toxicity and thus can provide the foundation for new classes of materials for drug, protein, and gene delivery applications [16].

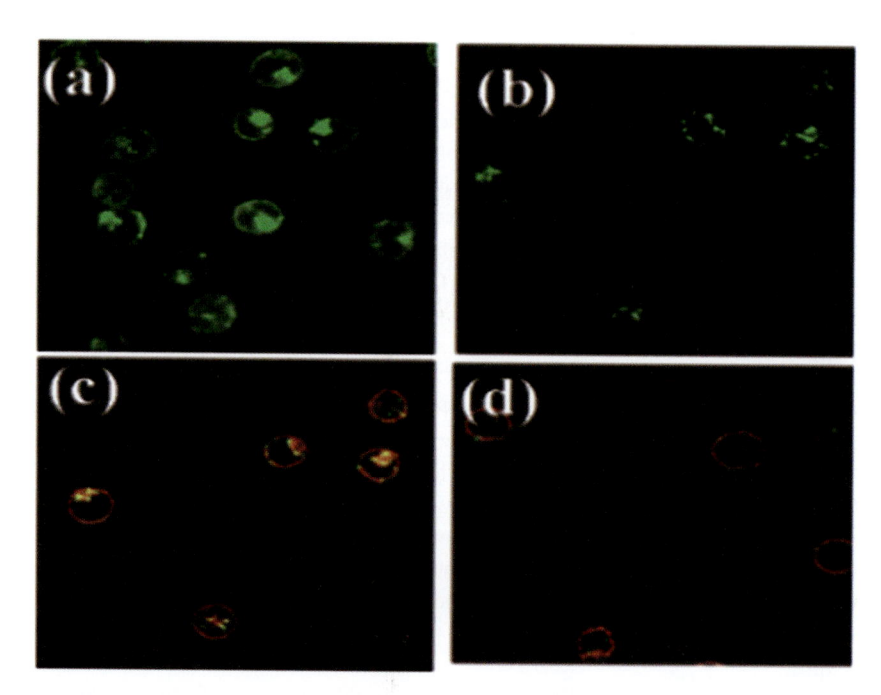

Figure 7.6 Confocal images of cells after incubation in solutions of CNT conjugates: (a) after incubation in CNT thioureidyl fluorescein, (b) after incubation in a mixture of biotin, SA, CNT composite, (c) same as (b) with additional red fluorescence shown due to FM-4-64 stained endosomes, (d) same as (b) after incubation at 4 °C. Reprinted with permission from Ref. [16]. Copyright American Chemical Society.

7.8.5
Semiconductor Chemistry

Due to their varying electronic properties that depend very much on the size and shape of the particles, QDs show great potential for future photonic applications such as light-emitting devices and solar cells. For instance, CdSe semiconductor QDs show significant potential for such applications [40] due to quantum confinement on the photoexcited excitons and tunable electronic properties. However, the application of QDs still in practice faces a number of obstacles largely due to the difficulty in effectively transferring the charge carriers to the substrate molecule. CNTs interacting through amine groups seem to offer better prospects as they can bind these QDs and transfer the charge either way depending upon the nature of QDs, that is, from nanotube to QD nanocrystal (TiO_2) or QD nanocrystal to CNTs (CdSe). Further, the possibility of selectively oxidizing the CNTs at the edges provides a means for capping the CNTs with these QDs and use as nanojunctions in nanoelectronics application [41].

7.8.6
Biocompatibility

Despite its huge promise for biomedical and bioelectronic applications, CNTs still face issues regarding the biocompatibility especially for *in-vivo* applications of implantable bioelectronic devises. This is particularly due to the inherent cytotoxicity of CNTs demanding many effective strategies for controlling the interaction between CNTs and cells for eliminating their toxicity, and amination provides a convenient method. Biocompatibility tests carried out on CNTs attached with bovine serum albumin (BSA) via amino groups by diimide-activated amidation suggest that over 90% of the BSA proteins in the CNT–BSA conjugate remain bioactive revealing no significant toxicity for CNTs and the compatibility for *in-vivo* experiments [42]. Similarly, CNTs with different functional groups including alexa fluor SA are incubated on the H60 cells separately to study any potential toxicity of CNTs. It was found that the CNTs with functional groups other than SA do not pose any toxicity. However, the observed cell death in the case of SA–CNT is attributed to the presence of SA as control experiment with lower SA contents shows no toxicity [16]. Further functionalized CNTs are shown to have lesser toxicity than pristine- and surfactant-stabilized CNTs [43]. Similarly, the development of amine-functionalized CNTs may pave the way for new types of biocompatible devices whose operation relay mainly on the physical properties of nanomaterials along with high recognition capabilities of biomolecules [44].

7.8.7
Composite Electrolytes

CNT connected with phosphonic acid groups through amide linkage with amino ethyl phosphonic acid could be used as an electrolyte in proton exchange membrane fuel cells where phosphoric acid plays a vital role in proton transport. Similar observation has been made with phosphonated fullerenes, which after forming a pellet showed proton conductivity in the order of 10^{-3} S cm^{-1} [45]. Phosphonated CNTs, when used for composite formation with phosphoric acid doped polybenzimidazole membranes, show improved proton conductivity, mechanical stability, and fuel-cell performance [46]. The improved proton conductivity is attributed to the increased functional group density, while the enhanced mechanical stability is accounted by the presence of CNTs. This fact further explains the better processability of CNTs in composite making when tailor-made groups are covalently attached onto it. Similarly poly(oxyalkylene)diamine functionalized CNTs were able to attach siloxan (3-glycidoxypropyl)trimethoxysilane by means of amine functionalization. This modified CNTs are able to form homogeneous composite membranes with Nafion, a perfluorosulfonic acid based membrane. The resulting composite membranes show improved proton conductivity at higher temperatures (>100 °C) due to stronger binding of water in the silica framework in the membrane [17].

7.8.8
Separation of Semiconducting Carbon Nanotubes

Amine groups have the tendency to attract semiconducting CNTs specifically over metallic CNTs. For example, a self-assembled monolayer of aminopropyltrimethoxysilane on silane substrate is shown to attach specifically semiconducting CNTs in desired alignment [47]. This could find important applications in field-effect nanotransistors that specifically require semiconducting nanotubes. Further, ODA is shown to attach specifically the semiconducting CNTs, thus allowing only the metallic nanotubes to get precipitate by centrifugation. It is also reported that attachment of amine groups onto CNTs results in the conversion of semiconducting nanotubes from p- to n-type and in some instance to metallic nanotubes [48].

7.9
Limitations of Amine-Functionalized CNTs

Although amine functionalization offers many advantages in various applications ranging from energy storage to biomedical devices, there are still certain challenges that need to be solved for a complete understanding so that aminated CNTs could be used more efficiently. The main limitations of amine-functionalized CNTs are as follows:

1) Functionalization often has a negative effect on elastic modulus as the introduction of sp^3 carbon in the sp^2 carbon network leaves kinks in the surface that ultimately reduces the mechanical strength. A surface functionalization level of 10% has been theoretically found to be optimum for the balance of efficient load-transfer and conspicuous devastation from the highly covalent functionalization [29].

2) Since amine functionalization produces sp^3-hybridized carbon in the CNT matrix, it is almost impossible to avoid the resulting change in the texture and morphology of the CNT. For example, many TEM micrographs clearly reveal increased corrugation and disorder on sidewalls after functionalization [49]. However, by keeping degree of functionalization under control, this type of a defect generation and their impact on morphology can be ameliorated, especially for MWCNTs.

3) Amine-functionalized CNTs have an important limitation of displaying pH-dependent properties especially when they are in contact with aqueous electrolytes. Some of them depending on the nature of electrostatic interactions will have different structures with varying pH, and both acidic and alkaline conditions can thwart their stability due to hydrolysis. However, this could be a boob where pH-sensitive reversible confirmational changes are sought for certain applications. This can be easily observed in the case of forming aligned networks of CNTs on aminated Si surfaces where depending upon the pH the alignment and coverage of CNTs varied drastically.

4) The chemical stability of the functional groups under the operating environment of many devices is not yet well studied, which makes the lifetime of the functionalized CNT based devices uncertain. Defunctionalization of aminated CNTs is also a major issue during certain applications dealing with prolonged heating and radiation exposure. For example, defunctionalization of CNTs has been observed under electron beam irradiation and also during heating in argon environment for different types on CNTs, and in some extreme cases these generate graphene layers due to unzipping of the CNT structure. The mechanism of this requires further studies.

7.10 Conclusions

Amine-functionalized CNTs provide many exciting challenges by altering or fine-tuning chemical, physical, and electronic properties of CNTs and hence are important for many applications in areas such as catalysis, energy conversion, nanoelectronics where important benefits could be accomplished in terms of enhanced efficiency and stability. Some of the procedures adopted for amine functionalization will undergo rapid changes in the foreseeable future due to the need for adopting green chemistry principles where more efficient and site-selective functionalization will be evolved. These greener methods would obviously provide more precise control of the functionality, illustrating the power of blending multiscale modeling and simulation with innovative synthetic methods that require use of atom-efficient and environment-friendly reagents with CNTs. The beneficial role of amination of CNTs mainly is that it increases the processability which is a prerequisite for most of the applications, especially for realizing the complete potential of bioelectronic devices. While a variety of amine-functionalization schemes and synthetic protocols currently exists, the need for site-specific and efficient techniques may bring more innovative approaches like generating multiple functionalities in one step, use of microwave, optical methods, use of patterned microreactor setups, etc. highlighting unique opportunities.

References

1 (a) Dresselhaus, M.A., Dresselhaus, G., and Avouris, P.H. (2001) *Carbon Nanotubes Synthesis, Structure, Properties and Applications*, Springer, Berlin. (b) Rao, C.N.R., and Govindaraj, A. (2005) *Nanotubes and Nanowires*, Royal Society of Chemistry, London. (c) Eder, D. (2010) *Chem. Rev.*, **110**, 1348. (d) Tasis, D., Tagmatarchis, N., Bianco, A., and Prato, M. (2006) *Chem. Rev.*, **106**, 1105. (e) Sun, Y.P., Fu, K., Lin, Y., and Huang, W. (2002) *Acc. Chem. Res.*, **35**, 1096. (f) Rao, C.N.R., Satishkumar, B.C., Govindaraj, A., and Nath, M. (2001) *ChemPhysChem*, **2**, 78.

2 Wei, B.Q. (2001) *Appl. Phys. Lett.*, **79**, 1172.

3 (a) Sainsbury, T. and Fitzmaurice, D. (2004) *Chem. Mater.*, **16**, 3780. (b) Ellis, A.V., Vijayamohanan, K., Goswami, R., Chakrapani, N., Ramanathan, L.S., Ajayan, P.M., and Ramanath, G. (2003) *Nano Lett.*, **3**, 279.

4 Goff, A.L., Artero, V., Jousselme, B., Tran, P.D., Guillet, N., Metaye, R., Fihri, A., Palacin, S., and Fontecave, M. (2009) *Science*, **326**, 1384.

5 (a) Kannan, R., Kakade, B.A., and Pillai, V.K. (2008) *Angew. Chem. Int. Ed.*, **120**, 2693. (b) Kannan, R., Meera, P., Maaraveedu, S.U., Kurungot, S., and Pillai, V.K. (2009) *Langmuir*, **25**, 8299.

6 (a) Liu, K., Wang, W., Xu, Z., Bai, X., Wang, E., Yao, Y., Zhang, J., and Liu, Z. (2009) *J. Am. Chem. Soc.*, **131**, 62. (b) Kim, Y.A., Muramatsu, H., Hayashi, T., Endo, M., Terrones, M., and Dresselhaus, M.S. (2004) *Chem. Phys. Lett.*, **398**, 87. (c) Peng, B., Locascio, M., Zapol, P., Li, S., Mielke, S.L., Schatz, G.C., and Espinosa, H.D. (2008) *Nature Nanotechnol.*, **3**, 626. (d) Pumera, M. (2009) *Chem. Eur. J.*, **15**, 4970.

7 Monthioux, M. and Flahaut, E. (2007) *Mater. Sci. Eng. C*, **27**, 1096.

8 Coleman, K.S., Bailey, S.R., Fogden, S., and Green, M.L.H. (2003) *J. Am. Chem. Soc.*, **125**, 8722.

9 (a) Banerjee, S., Benny, T.H., and Wong, S.S. (2005) *Adv. Mater.*, **17**, 18. (b) Osuna, S., Sucarrat, M.T., Sola, M., Geelings, P., Ewels, C.P., and Lier, G.V. (2010) *Phys. Chem. C*, **114**, 3340.

10 (a) Bahr, J.L., Yang, J., Kosynkin, D.V., Bronikowski, M.J., Smalley, R.E., and Tour, J.M. (2001) *J. Am. Chem. Soc.*, **123**, 6536. (b) Kooi, S.E., Schlecht, U., Burghard, M., and Kern, K. (2002) *Angew. Chem. Int. Ed.*, **41**, 1353.

11 Wang, Y., Iqbal, Z., and Mitra, S. (2005) *Carbon*, **43**, 1015.

12 (a) Zhang, Z., Sun, Z., Yao, J., Kosynkin, D.V., and Tour, J.M. (2009) *J. Am. Chem. Soc.*, **131**, 13460. (b) Paiva, M.C., Xu, W., Proenca, M.F., Novais, R.M., Laegsgaard, E., and Besenbacher, F. (2010) *Nano Lett.*, **10**, 1764.

13 (a) Kakade, B., Patil, S., Sathe, S., Gokhale, S., and Pillai, V. (2008) *J. Chem. Sci.*, **120**, 599. (b) Bianco, A. and Prato, M. (2003) *Adv. Mater.*, **15**, 1765.

14 Bianco, A., Kostarelos, K., Partidos, C.D., and Prato, M. (2005) *Chem. Commun.*, **5**, 571.

15 Chattopadhyay, D., Galeska, I., and Papadimitrakopouos, F. (2003) *J. Am. Chem. Soc.*, **125**, 3370.

16 Kam, N.W.S., Jessop, T.C., Wender, P.A., and Dai, H. (2004) *J. Am. Chem. Soc.*, **126**, 6850.

17 Chen, W.F., Wu, J.S., and Kuo, P.L. (2008) *Chem. Mater.*, **20**, 5756.

18 (a) Khare, B., Wilhite, P., Tran, B., Teixeira, E., Fresquez, K., Mvondo, D.N., Bauschlicher, C., and Meyyappan, M. (2005) *J. Phys. Chem. B*, **109**, 23466. (b) Khare, B.N., Wilhite, P., Quinn, R.C., Chen, B., Schingler, R.H., Tran, B., Imanaka, H., So, C.R., Bauschlicher, C.W., and Meyyappan, M. (2004) *J. Phys. Chem. B.*, **108**, 8166.

19 Mickeson, E.T., Huffman, C.B., Rinzler, A.G., Smalley, R.E., Hauge, R.H., and Margrave, J.L. (1998) *Chem. Phys. Lett.*, **296**, 188.

20 (a) Zhang, L., Kiny, V.U., Peng, H., Zhu, J.M., Lobo, R.F., Margrave, J.L., and Khabashesku, V.N. (2004) *Chem. Mater.*, **16**, 2055. (b) Oki, A., Adams, L., Khabashesku, V., Edigin, Y., Biney, P., and Luo, Z. (2008) *Mater. Lett.*, **62**, 918.

21 (a) Santhosh, P., Gopalan, A., and Lee, K.L. (2006) *J. Catal.*, **238**, 177. (b) Shen, M., Wang, S.H., Shi, X., Chen, X., Huang, Q., Petersen, E.J., Pinto, R.A., Baker, J.R., Jr, and Weber, W.J., Jr (2009) *J. Phys. Chem. C.*, **113**, 3150.

22 (a) Hill, D.E., Lin, Y., Allard, F., and Sun, Y.P. (2002) *Int. J. Nanosci.*, **1**, 213. (b) Sun, Y.P., Fu, K., In, Y., and Huang, W. (2002) *Acc. Chem. Res.*, **35**, 1096.

23 Taft, B.J., Lazareck, A.D., Withey, G.D., Yin, A., Xu, J.M., and Kelley, S.O. (2004) *J. Am. Chem. Soc.*, **126**, 12750.

24 Chen, R.J., Zhang, Y., Wang, D., and Dai, H. (2001) *J. Am. Chem. Soc.*, **123**, 3838.

25 Wang, Y., Iqbal, Z., and Malhotra, S. (2005) *Chem. Phys. Lett.*, **402**, 96.

26 (a) Williams, K.A., Veenhuizen, P.T.M., Torre, B.G.D.L., Eritja, R., and Dekker, C. (2002) *Nature*, **420**, 761. (b) Daniel, S., Rao, T.P., Rao, K.S., Rani, S.U., Naidu, G.R.K., Lee, H.Y., and Kawai, T. (2007) *Sens. Actuators B*, **122**, 672.

27 Ramanathan, T., Fisher, F.T., Ruoff, R.S., and Brinson, L.C. (2005) *Chem. Mater.*, **17**, 1290.

28 Lin, Y., Rao, A.M., Sadanadan, B., Kenik, E.A., and Sun, Y.P. (2002) *J. Phys. Chem. B*, **106**, 1294.

29 Wang, S. (2009) *Curr. Appl. Phys.*, **9**, 1146.

30 Lim, S.H., Li, R., Ji, W., and Lin, J. (2007) *Phys. Rev. B*, **76**, 195406.

31 Zhao, J., Park, H., Han, J., and Lu, J.P. (2004) *J. Phys. Chem. B*, **108**, 4227.

32 Chen, J., Hamon, M.A., Hu, H., Chen, Y., Rao, A.M., Eklund, P.C., and Haddon, R.C. (1999) *Science*, **282**, 95.

33 Kim, J.B., Premkumar, T., Lee, K., and Geckeler, K.E. (2007) *Macromol. Rapid Commun.*, **28**, 276.

34 Meng, H., Sui, G.X., Fang, P.F., and Yang, R. (2008) *Polymer*, **49**, 610.

35 Dommele, S.V., Jong, K.P.D., and Bitter, J.H. (2006) *Chem. Commun.*, 4859.

36 Li, Y., Zhao, Y., Zhang, Z., and Xu, Y. (2010) *Tetrahedron Lett.*, **51**, 1434.

37 (a) Kelly, G.J., King, F., and Kett, M. (2002) *Green Chem.*, **4**, 392. (b) Tanabe, K., and Holderich, W.F. (1999) *Appl. Catal. A*, **181**, 399.

38 (a) Villa, A., Tessonnier, J.P., Majoulet, O., Su, D.S., and Schlogl, R. (2009) *Chem. Commun.*, 4405. (b) Tessonnier, J.P., Villa, A., Majoulet, O., Su, D.S., and Schlogl, R. (2009) *Angew. Chem.*, **48**, 6543.

39 Santhosh, P., Manesh, K.M., Gopalan, A., and Lee, K.P. (2007) *Sens. Actuators B*, **125**, 92.

40 Huynh, W.U., Dittmer, J.J., and Alivisatos, A.P. (2002) *Science*, **295**, 2425.

41 Banerjee, S. and Wong, S.S. (2002) *Nano Lett.*, **2**, 195. (b) Haremza, J.M., Hahn, M.A., Krauss, T.D., Chen, S., and Calcines, J. (2002) *Nano Lett.*, **2**, 1253. (c) Ravindran, S., Chaudhary, S., Coburn, B., Ozkan, M., and Ozkan, C.S. (2003) *Nano Lett.*, **3**, 447. (d) Azamian, B.R., Coleman, K.S., Davis, J.J., Hanson, N., and Green, M.L.H. (2002) *Chem. Commun.*, 366.

42 Huang, W., Taylor, S., Fu, K., Zhang, D., Hanks, T.W., Rao, A.M., and Sun, Y.P. (2002) *Nano. Lett.*, **2**, 311.

43 Sayes, C.M., Liang, F., Hudson, J.L., Mendez, J., Guo, W., Beach, J.M., Moore, V.C., Doyle, C.D., West, J.L., Billups, W.E., Ausman, K.D., and Colvin, V.L. (2006) *Toxicol. Lett.*, **161**, 135.

44 Shim, M., Kam, N.W.S., Chen, R.J., Li, Y., and Dai, H. (2002) *Nano Lett.*, **2**, 285.

45 (a) Cheng, F., Yang, X., Zhu, H., and Song, Y. (2000) *Tetrahedron Lett.*, **41**, 3947. (b) Cheng, F., Yang, X., Zhu, H.,

and Song, Y. (2001) *Tetrahedron*, **57**, 7331. (c) Li, Y.M. (2006) US Patent 7, 128, 888.

46 Kannan, R., Aher, P.P., Palaniselvam, T., Sreekumar, K., Kharul, U.K., and Pillai, K.V. (2010) *J. Phys. Chem. Lett.*, **1**, 2109.

47 Opatkiewicz, J.P., LeMieux, M.C., and Bao, Z. (2010) *ACS Nano*, **4**, 1167.

48 Ma, P.C., Tang, B.Z., and Kim, J.K. (2008) *Chem. Phys. Lett.*, **458**, 166.

49 Gu, Z., Peng, H., Hauge, R.H., Smalley, R.E., and Margrave, J.L. (2002) *Nano Lett.*, **2**, 1009.

8
Functionalization of Nanotubes by Ring-Opening and Anionic Surface Initiated Polymerization

Georgios Sakellariou, Dimitrios Priftis, and Nikos Hadjichristidis

8.1
Introduction

Carbon nanotubes (CNTs), first reported by Iijima in 1991 [1], possess very attractive properties such as high flexibility [2], low mass density [3], large aspect ratio, and a unique combination of mechanical, electrical, and thermal properties [2, 3]. Since the first report by Iijima, much work has been performed on the production, purification, characterization, functionalization, and polymer grafting of CNTs, in order to fully understand and explore the immense potential of these materials.

The exceptionally high material properties of CNTs render them ideal functional reinforcing agents for polymers in a number of applications. The first polymer composite, using CNT as filler, was reported in 1994 by Ajayan [4]. The successful applications of CNT nanocomposite systems require well-dispersed nanotubes and consequently good adhesion with the host matrix. Unfortunately, due to the extended π-conjugated network (strong intramolecular interactions), CNTs form large aggregations (bundles), which are completely insoluble in organic and aqueous solvents. Moreover, nanotube bundles have limited chemical affinity and dispersivity into polymeric matrices.

Modification of CNT surfaces with polymers enhances their chemical affinity as well as their dispersivity into polymer matrices. Two strategies have been used to modify the CNT surfaces with polymers. The covalent modification involves either "grafting from" or "grafting to" methods, and the noncovalent fixation involves the "grafting to" strategy through supramolecular interactions. The "grafting from" (also called surface-initiated polymerization) is the most promising method, since it provides control over the functionality, density, thickness, composition, and architecture of the polymer brushes. This method involves first the covalent immobilization of initiator molecules on the surface of CNTs, followed by *in-situ* polymerization to generate polymer brushes. All known types of living and controlled/"living" polymerization techniques have been utilized for the polymer grafting on CNT molecules [5–7]. Herein, we will focus on recent developments in CNT surface-initiated ring-opening polymerization (ROP) of cyclic monomers and anionic polymerization of linear monomers.

Surface Modification of Nanotube Fillers, First Edition. Edited by Vikas Mittal.
© 2011 Wiley-VCH Verlag GmbH & Co. KGaA. Published 2011 by Wiley-VCH Verlag GmbH & Co. KGaA.

8.2
Surface-Initiated Polymerization

8.2.1
Surface-Initiated Ring-Opening Polymerization

Aliphatic polyesters, such as poly(p-dioxanone) (PPDX), poly(ε-caprolactone) (PCL), poly(L-lactide) (PLLA), as well as other polymers belonging to polyethers (e.g., polyethylenoxide) and polypeptides, are biodegradable and biocompatible materials. These materials have been proven to be very useful as implantable medical devices, such as surgical sutures and drug-delivery systems, because they eliminate the need for device removal after being used. Their synthesis can be achieved by ROP techniques, using a rich variety of catalytic/initiating systems [8]. Recently, interest has been shown in the grafting of such macromolecules onto the surface of CNTs (Table 8.1). In addition, interest has been also shown for grafting polymers like nylon-6 on CNTs, via ROP of caprolactame, in order to use them as nanocomposites in commodity plastics. This chapter will deal with the synthesis of such hybrid polymers by anionic, cationic, and coordination anionic ROP.

In an early work, Yoon et al. [9] used ROP to synthesize PPDX chains, from a single-walled carbon nanotube (SWNT) surface, via the "grafting from" method. 6-Amino-1-hexanol was covalently attached on the surface of SWNTs by immersing the acid chloride functionalized s-SWCNTs into a dimethyl formamide (DMF) solution of 6-amino-1-hexanol. The produced hydroxy-terminated SWNTs were then served as initiators for the polymerization of p-dioxanone, in the presence of tin(II) 2-ethylhexanoate/$Sn(Oct)_2$. The thermal stability of the PPDX increased after the strong covalent attachment on the CNTs, while no noticeable peaks (corresponding to T_g and T_m) were observed up to 125 °C, although pure PPDX showed T_g and T_m at −13.4 °C and 103 °C, respectively. These changes in PPDX properties were attributed to the effective interactions between SWNTs and PPDX and the consequent mobility decrease in PPDX polymer chains.

The growth of hyperbranched macromolecules on CNT surfaces provides a new route to highly functionalized CNTs. The numerous hydroxy groups on the periphery of hyperbranched chains can be further functionalized producing fascinating novel nanomaterials and nanodevices. Xu et al. [10] employed ROP to grow multihydroxy dendritic chains from the surface of multiwalled carbon nanotubes (MWNTs) (Figure 8.1). Hydroxy-functionalized MWNTs, in the presence of $BF_3 \cdot Et_2O$, were used as initiators for the in-situ surface cationic ROP of 3-ethyl-3-(hydroxymethyl)-oxetane (EHOX). The produced hyberbranched polyether CNTs (HP-CNTs) exhibited good dispersivity in polar solvents such as methanol, ethanol, DMF, and dimethyl sulfoxide (DMSO). The amount of polymer grafted on the surface of the MWNTs was controllable, allowing the formation of nanocomposite materials with up to 87 wt.% of polymer. The molecular weight of the HP-CNTs increased with the feed ratio of monomer to macroinitiator. Free HPs, resulted from the self-initiation of active hydroxy groups of EHOX with

Table 8.1 Monomers that have been polymerized by surface-initiated ROP on CNTs.

Substrate	Precursor initiator	Catalyst	Monomer	Polymer	Reference
SWNTs	$H_2N\sim\sim\sim OH$	$Sn(Oct)_2$	(1,4-dioxepan-2-one)	PPDX	[9]
MWNTs	$HO\sim OH$	$BF_3 \cdot OEt_2$	(oxetane with CH$_2$OH)	PEHOX_	[10]
SWNTs, MWNTs	caprolactam	Na	caprolactam	Nylon-6	[12, 13]
MWNTs SWNTs	$HO\sim OH$, CH_2CH_2OH-benzocyclobutene, NH_2-benzyl alcohol, $N_3\sim OH$	$Sn(Oct)_2$	ε-caprolactone	PCL	[14, 15, 18, 21, 22, 23]
MWNTs	–OH, Ar + N$_2$ microwave plasma	AlEt$_3$, AlMe$_3$	ε-caprolactone	PCL	[16, 17]
MWNTs SWNTs	$H_2N\sim NH_2$, $N_3\sim NH_2$	–	NCA (R)	PBLG	[24, 26]
MWNTs_	$H_2N\sim\sim\sim NH_2$	–	NCA (R)	PLL	[25]
MWNTs	Fe$_3$O$_4$, Fe$_3$O$_4$, $HO\sim OH$	$Sn(Oct)_2$	lactide	PLLA	[27–30]

Table 8.1 Continued.

Substrate	Precursor initiator	Catalyst	Monomer	Polymer	Reference
MWNTs	MeO$^-$K$^+$	–	(epoxide with CH$_2$OH)	PG	[11, 19]
MWNTs	SOCl$_2$	FeCl$_3$	(epoxide with CH$_2$Cl)	PETH	[20]

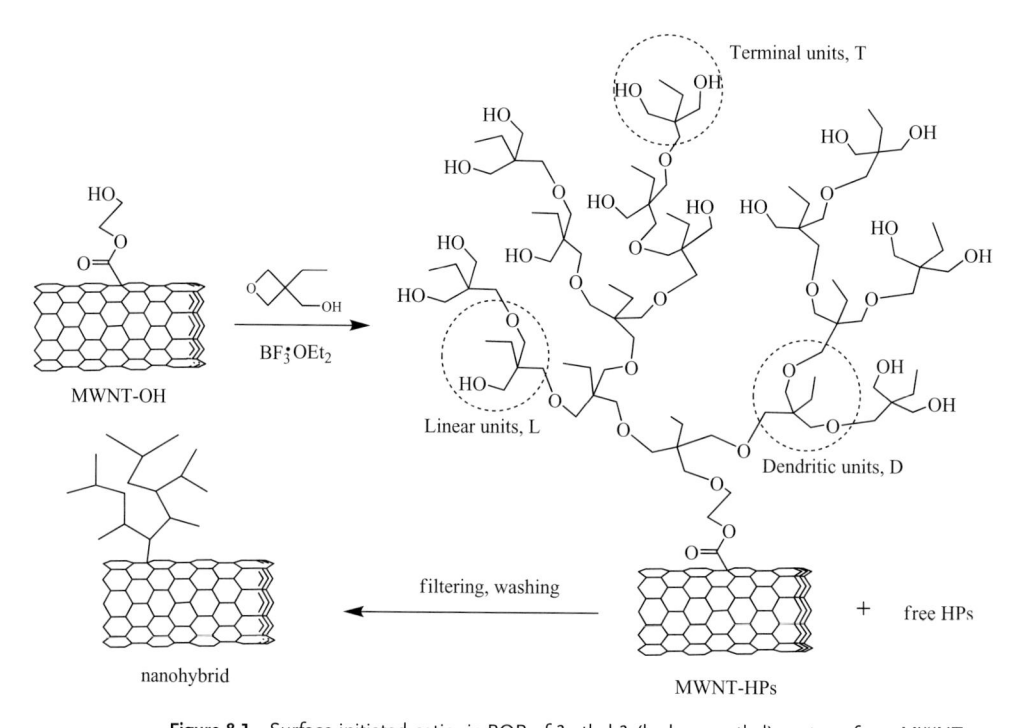

Figure 8.1 Surface-initiated cationic ROP of 3-ethyl-3-(hydroxymethyl)-oxetane from MWNTs [10].

BF$_3 \cdot$ Et$_2$O, were removed by successive filtration and washing. The degree of branching (DB parameter), which depends on the number of dendritic, linear, and terminal units of the hyperbranched polymer, was obtained from ^{13}C NMR spectra. It was found that an increase in the M_w of the grafted HPs results in a dramatic increase in the DB. Interestingly, the DB of the grafted HPs was lower than that of free HPs, although grafted HPs had larger M_n. Transmission electron microscopy (TEM) indicated that the polymer shell thickness around the CNTs was uniform due to evenly distributed initiation sites or to the defect density on the surface.

In another work, the same group [11] performed anionic ROP of glycidol (2,3-epoxy-1-propanol) monomer in dioxane and in the presence of potassium methoxide, to grow hyperbranched polyglycerol (HPG) from OH-MWNT surfaces. The macroinitiator, MWNT-OH, was prepared by one-pot nitrene chemistry in N-methyl-2-pyrrolidone. The amount of grafted HPG was adjusted, up to 90 wt.%, by changing the feed ratio of glycidol to MWNT-(OH). When the polymerization was conducted in bulk, the grafted polymer content was never higher than 45 wt.%, even when very high feed ratios of glycidol to MWNT-(OH) were used. The authors ascribed this phenomenon to the higher viscosity of the reaction mixture in bulk polymerization. By reaction of hydroxy groups of the outer shell with palmitoyl chloride afforded MWNTs-g-amphiphilic hyperbranched polymers. The resulting product showed good solubility in weak polar or nonpolar solvents in contrast to MWNTs-g-HPG, which were soluble only in strong polar solvents. Fluorescent MWNTs were also prepared by attaching rhodamine 6B molecules to the surface of MWNT-g-(HPG). The abundance of functional groups on the surface of CNTs hybrids can be used as a nanoplatform to conjugate functional molecules for further application in drug delivery, cell imaging, and bioprobing.

Anionic ROP was used by Qu et al. [12] to functionalize SWNTs with nylon-6. Nylon-6 is an important commodity polymer with a wide variety of applications and the preparation of nylon-functionalized CNTs is highly relevant and beneficial. ε-Caprolactam molecules were covalently attached to SWNT surfaces followed by the polymerization of the same monomer in bulk (Figure 8.2). The final nanocomposite material contained approximately 30 wt.% polymer and was soluble in some organic solvents, such as formic acid and m-cresol.

The covalent functionalization of MWNTs with nylon-6 was also reported by Adronov et al. [13]. Isocyanate-functionalized CNTs, prepared by reacting OH-MWNTs with excess of toluene 2,4-diisocyanate, were used as initiators in anionic ROP of ε-caprolactam, in the presence of a sodium caprolactamate catalyst. A series of polymerizations were performed in order to investigate the effect of the reaction time and of the feed ratio (monomer/MWNT initiator) on the grafting efficiency ($f_{wt.\%}$). The results revealed that the $f_{wt.\%}$ increased with polymerization time, reaching a limiting value after 6 h and was roughly controlled by adjusting the monomer to initiator feed ratio within the range between 150 and 800.

Buffa et al. [14] functionalized SWNTs, produced by the CoMoCAT process, with 4-hydroxymethylaniline (HMA) via the diazonium salt method. The OH-SWNTs, generated by the functionalization, were then used to PCL, in the presence of stannous octanoate. The suspendibility of SWNTs grafted by PCL was tested in chloroform and showed a dramatic increase compared to that of neat SWNTs, due to the high solubility of PCL. TG analysis indicated that the amount of PCL that grafted was 63 wt.%, while a combined temperature-programmed desorption

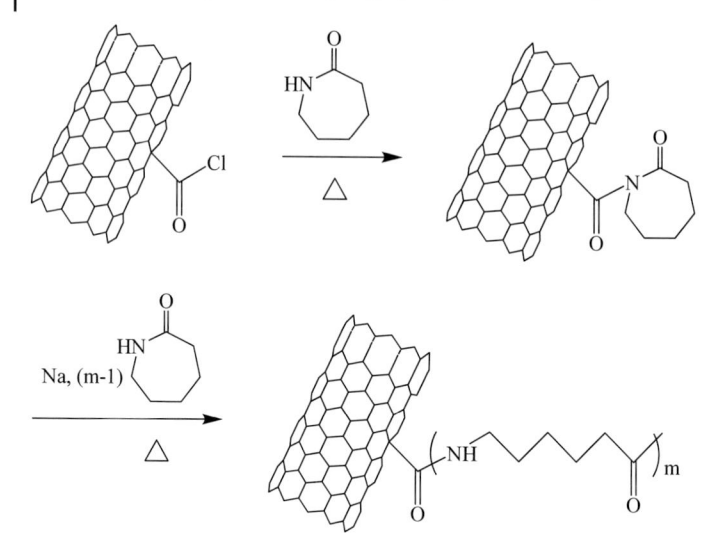

Figure 8.2 Surface-initiated anionic ROP of ε-caprolactam from caprolactam–SWNT [12].

(TPD)/temperature-programmed oxidation(TPO) technique revealed that the average polymer length was only 5 monomeric units. According to the authors, the two main factors for the limited growth of the grafted chains were the presence of nanotubes in the polymerization medium, which may act as terminators and the presence of adsorbed amount of water on the CNT surface that may act as initiator generating large amount of "free" PCL. In order to increase the length of the PCL attached to the nanotubes, one should eliminate the adsorbed water from the nanotubes.

In a similar work, Zeng et al. [15] introduced OH-initiating sites on the MWNT surfaces by reacting oxidized MWNTs with excess thionyl chloride and then glycol. The OH-functionalized MWNTs were used for surface-initiated ROP of ε-caprolactone, in butanol at 120 °C in the presence of stannous octanoate. Biodegradation of the filtered MWNT-g-(PCL) products was tested by using pseudomonas lipase (PS) as a bioactive enzyme catalyst. It was revealed that the grafted PCL chains retained their biodegradability and were completely degraded within 4 days. On the contrary, PS lipase and MWNTs had no strong effect on each other. These results open a new path for the application of CNT hybrids in bionanomaterials, biomedicine, and artificial organs and bones.

The influence of MWNT-g-PCL on vapor-sensing properties has been investigated by Castro et al. [16], for a series of conductive polymer composite (CPC) transducers. MWNT-g-PCL was prepared by first reacting OH-MWNTs with trimethylaluminum (catalyst) to produce an intermediate compound followed by the anionic coordination ROP of ε-caprolactone (Figure 8.3). Atomic force microscopy observations allowed an evaluation of MWNT coating and dispersion level. Sensors were prepared by sprayed layer-by-layer MWNT-g-PCL solutions onto interdigi-

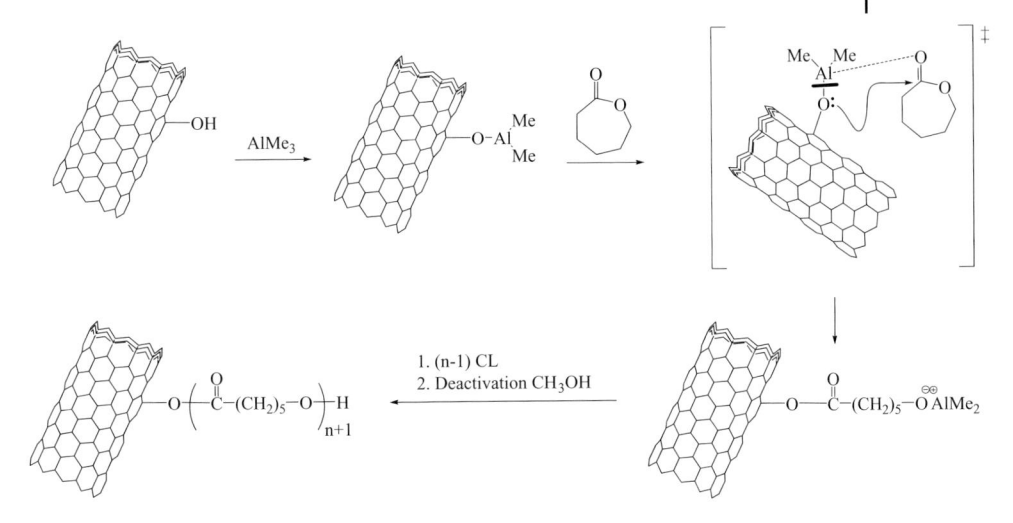

Figure 8.3 Steps followed for the synthesis of MWNT-g-PCL conductive polymer composite [16].

tated copper electrodes that had been etched by photolithography onto an epoxy substrate. Chemoelectrical properties of CPC sensors exposed to different vapors – water, methanol, toluene, tetrahydrofuran, and chloroform – have been analyzed in terms of signal sensitivity, selectivity, reproducibility, and stability. The CNT-g-PCL sensors displayed very good discrimination capability for all the vapors above (polar and nonpolar) which allow for the preparation of sensors with a large detection spectrum. The chemoelectrical properties of the sensors exposed to the different vapors were reproducible and the electrical signals displayed reversibility and fast recovery.

A different synthetic approach of grafting ε-caprolactone on MWNTs was presented by Ruelle et al. [17]. MWNTs were placed under atomic nitrogen flow formed through an Ar + N_2 microwave plasma, in order to covalently functionalize with primary and secondary amines. The amino-MWNTs were activated by triethylaluminum and then used as initiators for ROP of ε-caprolactone via the coordination–insertion mechanism shown in Figure 8.4. As revealed by TEM, functionalization was not homogeneous and the CNTs of the upper side of the sample had the highest amine surface functionalization.

Biodegradable supramolecular hybrids of MWNT-g-PCL and α-cyclodextrins (α-CDs) (inclusion complexes), with potential applications in medicine and biology, have been prepared by Yang et al. [18] (Figure 8.5). MWNTs were first functionalized via nitrene cycloaddition to give the OH-MWNTs, followed by the surface-initiated ROP of ε-caprolactone in the presence of catalyst stannous octanoate to afford MWNT-g-PCL. Thermogravimetric analysis showed that about 58 wt.% of PCL grafted onto the CNT surface and TEM experiments revealed polymer layer thicknesses in the range of 5–7 nm. The hybrids were

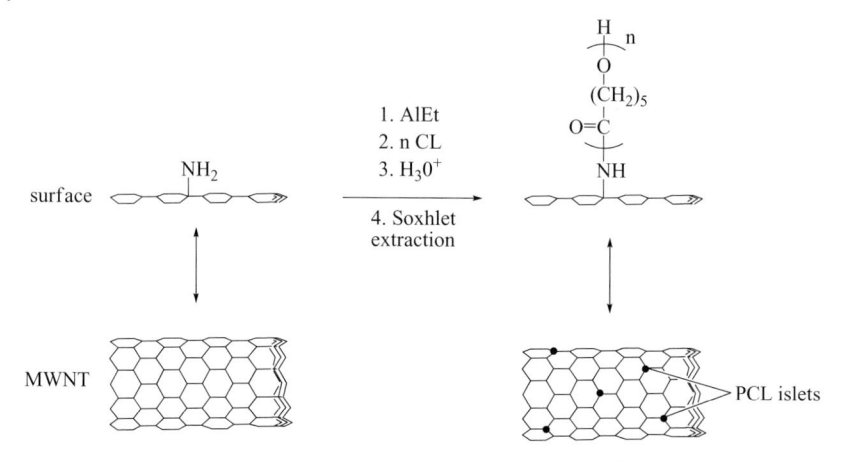

166 | *8 Functionalization of Nanotubes by Ring-Opening and Anionic Surface Initiated Polymerization*

Figure 8.4 ROP of ε-CL initiated by amines grafted on MWNTs [17].

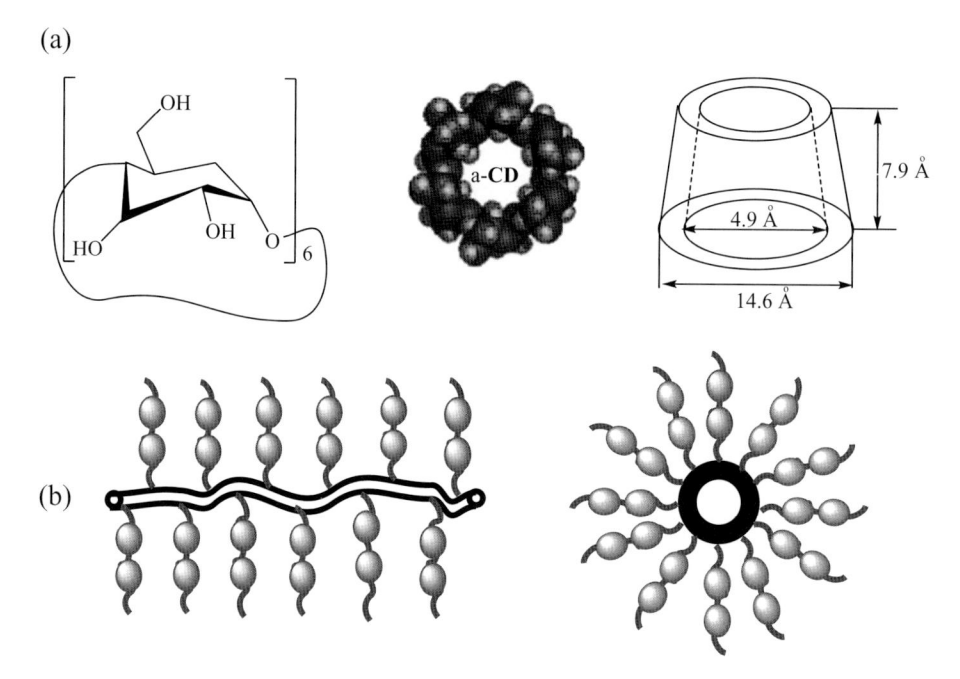

Figure 8.5 (a) Schematic representation for the chemical structure and size of α-CD and (b) the formation of inclusion complex from α-CD and grafted-PCL on the MWNT surface. Reproduced with permission from Elsevier [18].

formed by hydrogen bonds between the hydroxyl groups of α-CDs and the carbonyl groups of PCL, indicating that supramolecular polypseudorotaxanes were formed through α-CD channel threading onto the PCL chains of MWNT-g-PCL.

Another route for the synthesis of biocompatible, biodegradable CNTs hybrids was reported by Adeli et al. [19]. Hyperbranched molecular trees were grown from the surface of MWNTs via *in-situ* anionic ROP of glycidol (Figure 8.6). Short-term *in-vitro* cytotoxicity and hemocompatibility tests were conducted on HT 1080 cell line (human fibrosarcoma). The results showed no sign of toxicity for concentration up to 1 mg ml^{-1} after 24-h incubation.

In-situ cationic ROP along with quaternization reactions have been employed to confine ferricyanide onto MWNTs in order to prepare integrative nanostructured electrochemical biosensors. The strategy, demonstrated by Xiang et al. [20], was based on the creation of positively charged moieties onto CNTs by first grafting CNTs with polyether (via an *in-situ* cationic ROP of epoxy chloropropane) and then introducing positively charged methylimidazolium moieties through a quaternization reaction between methylimidazole and polyether grafted onto CNTs (Figure 8.7). The positively charged moieties were electrostatically interacted with ferricyanide anions [Fe(CN)$_6^{3-}$] to form efficient electronic transducers. The produced electrochemical biosensors displayed good stability and excellent electrochemical properties.

Very recently, our group [21] studied the kinetics of ROP of ε-CL from MWNT surfaces (Figure 8.8) and the crystallization behavior of the nanocomposite

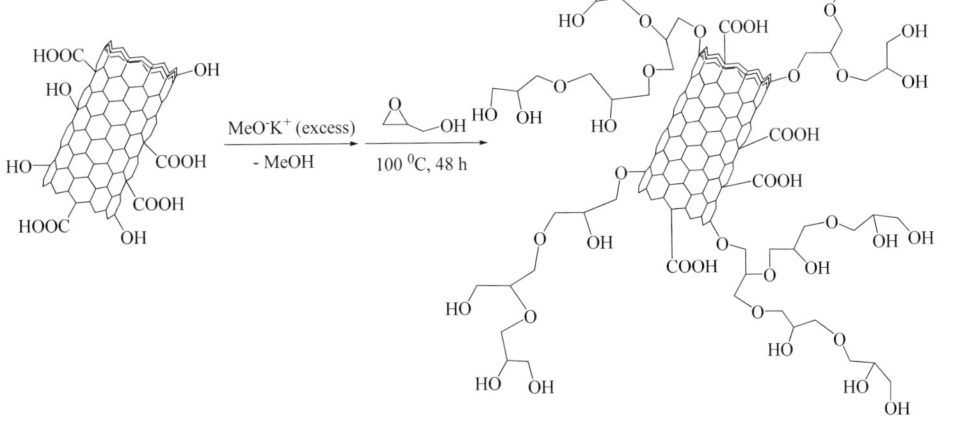

Figure 8.6 Synthetic route for MWNT-g-PG hybrid materials [19].

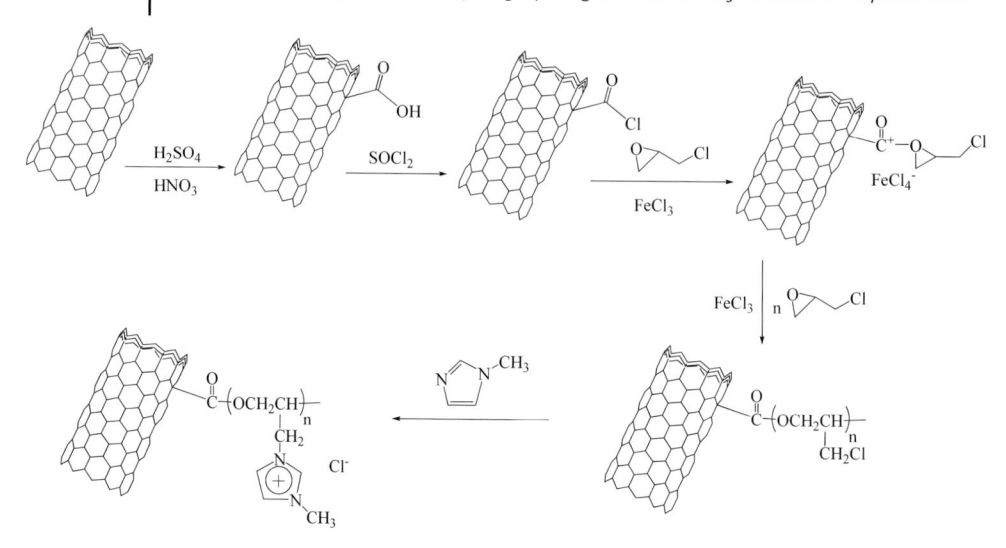

Figure 8.7 Synthetic route of the poly(epoxychloropropane)-grafted and methylimidazolium-tethered MWNTs [20].

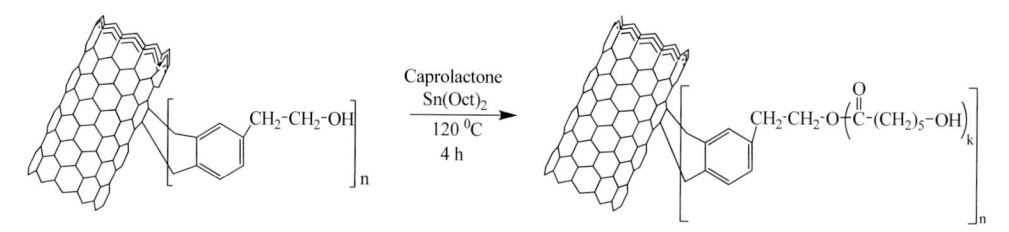

Figure 8.8 Surface-initiated ROP of ε-CL from MWNTs [21].

materials produced. The necessary initiating sites were covalently attached to the MWNT surface through a [4 + 2] Diels–Alder cycloaddition reaction. It was revealed that even though the polymerization proceeded very rapidly, it could be controlled with time. A remarkable nucleation effect was observed by the incorporation of MWNT, since the crystallization temperature increased by a relatively significant amount when compared to that of neat PCL. The isothermal crystallization kinetics of grafted PCL was substantially accelerated compared to the neat polymer due to the strong impact of the covalent MWNT–polymer bond on the nucleation and crystallization kinetics.

We also reported [22] a strategy for the synthesis of polymer-grafted Janus-MWNTs, where two different polymer chains are grown from MWNTs in one step. A [4 + 2] Diels–Alder cycloaddition reaction was used to functionalize MWCNTs with two different precursor initiators, one for ROP and one for atom transfer radical polymerization (ATRP). Those binary functionalized MWNTs were used for simultaneous surface-initiated polymerizations of various monomers through different polymerization mechanisms (Figure 8.9). The nanocomposite materials could form Janus-type structures when placed in two selective and immiscible solvents (Figure 8.10).

In another study in our laboratory [23], novel diblock copolymers were grafted onto MWNTs via a combination of polymerization techniques, including

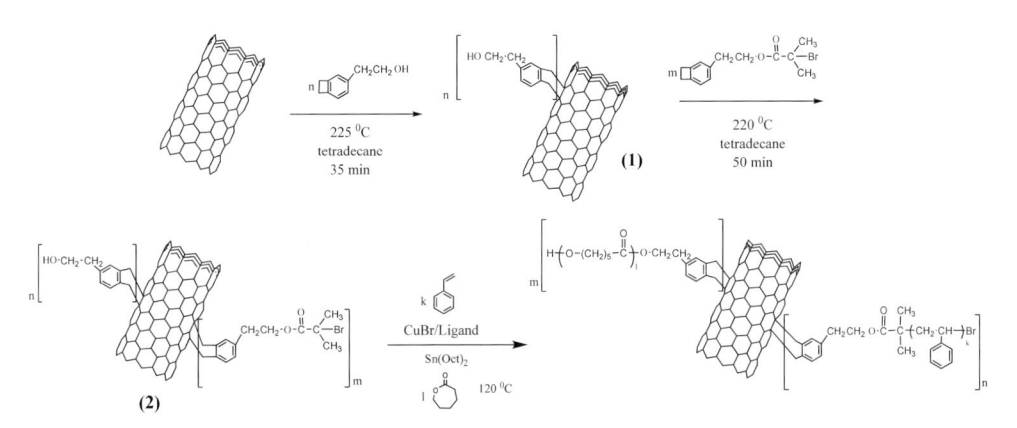

Figure 8.9 One-step synthesis of chemically different brushes grafted from a CNT surface [22].

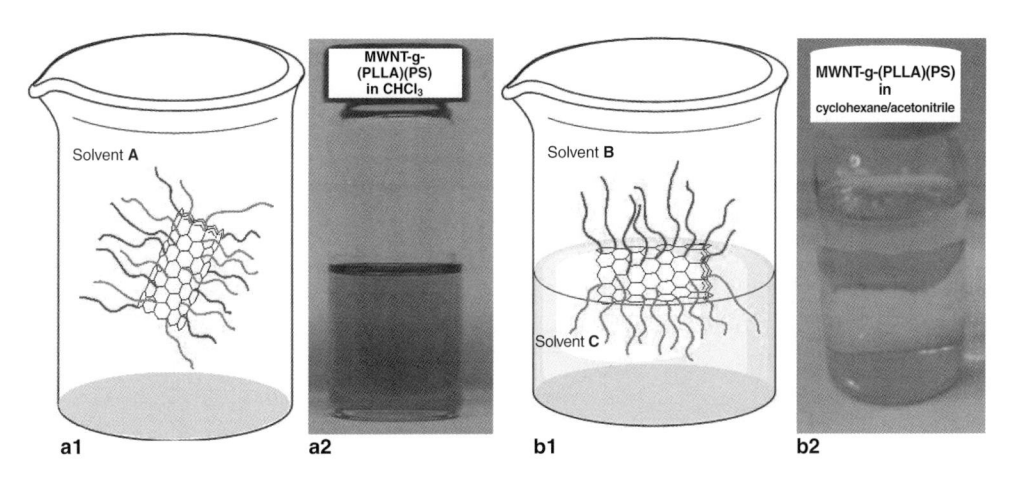

Figure 8.10 (a1) Schematic representation of chemically different brushes grafted on MWNTs in good solvent A. (a2) Experimental evidence: MWNT-g-[(PLLA)$_n$(PS)$_m$] in good solvent CHCl$_3$. (b1) Schematic representation of chemically different brushes grafted on MWNTs in a mixture of selective and immiscible solvents B and C. (b2) Experimental evidence: MWNT-g-[(PLLA)$_n$(PS)$_m$] in cyclohexane (upper layer: selective solvent for PS) and acetonitrile (bottom layer: selective solvent for PLLA). Reproduced from Royal Society of Chemistry [22].

ring-opening anionic polymerization, in a specially designed apparatus (Figure 8.11b).

Using the same method above, a substituted benzocyclobutene was attached on the surface of MWNTs and was used for the polymerization of EO in the presence of a phosphazane base (Figure 8.12). These nanocomposite materials bearing a hydroxyl end group were used as ROP macroinitiators for the polymerization of ε-CL resulting in biocompatible MWNT-g-(PEO-b-PCL) hybrid materials. By esteri-

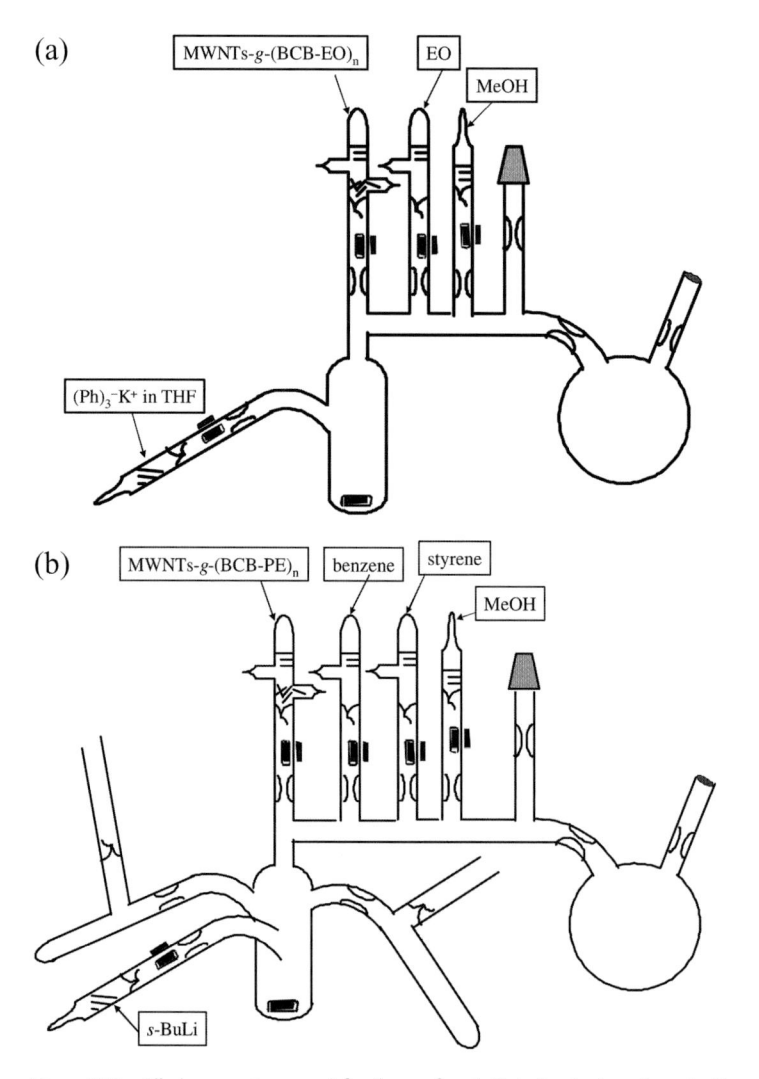

Figure 8.11 All-glass reactors used for the surface-initiated anionic polymerization of (a) ethylene oxide and (b) styrene from MWNTs-g-(BCB-EO)$_n$ and MWNT-g-(BCB-PE)$_n$, respectively. Reproduced with permission from Wiley [23].

Figure 8.12 Surface-initiated anionic ring-opening polymerization of EO from MWNT-g-(BCB-PE)$_n$. Synthesis of ATRP macroinitiator and polymerization of styrene [23].

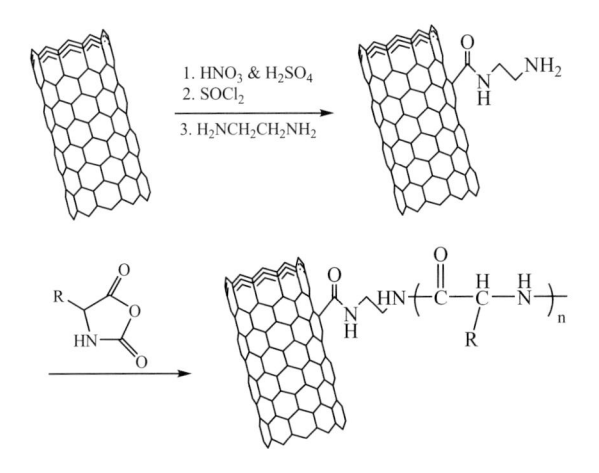

Figure 8.13 Surface-initiated ROP of γ-benzyl-L-glutamate from MWNTs [24].

fication of the hydroxyl PEO end-groups with excess 2-bromoisobutyryl bromide, an ATRP macroinitiator was obtained and used for the polymerization of St and 2-(dimethylamino)ethylmethacrylate (DMAEMA) resulting in diblock copolymer grafted-MWNTs.

In another approach by Yao et al. [24], ROP was used to graft a polypeptide from the surface of MWNTs. Amino groups were covalently bonded to the MWNTs (Figure 8.13), which were then used for ROP of -benzyl-L-glutamate N-carboxyanhydride (BLG-NCA). The resulting hybrid material exhibited core–shell morphology with thickness of the polypeptide layer ranging from 4 to 22 nm. The thickness of the polypeptide could not be controlled by the feed ratio of NCA to MWNT-NH$_2$. Solubility experiments revealed that MWNT-g-(PBLG) was soluble in strong polar solvents, such as DMF and DMSO, and insoluble in weak polar solvents, such as acetone and esters.

Poly(L-lysine) is a biocompatible and biodegradable polymer and its structure facilitates various modifications, including the conjugation with transferrin, epidermal growth factor, and fusogenic peptides, and thus has been widely used in the field of gene and drug delivery. Oxidized MWNTs were converted into amino-functionalized MWNTs by amidation of the carboxylic groups with excess of 1,6-diaminohexane. Surface-initiated ROP of ε-(benzyloxycarbonyl)-L-lysine N-carboxyanhydride using the created amino groups resulted in MWNT-g-PLys(Z) [25]. Acidolysis of the benzylcarbamate groups afforded water-soluble MWNT-g-PLL. The amount of PLys(Z) attached on CNTs could be altered by adjusting the feed ratio of monomer to MWNT-NH$_2$. Core–shell structures were determined by high-resolution TEM with the polymer shell thickness varying between 4 and 18 nm. A thicker polymer layer was observed at the bends and tips than at the straight sections.

A more detailed work involving the surface-initiated ROP of γ-benzyl-L-glutamate N-carboxyanhydrides from amine-functionalized SWNTs (Figure 8.14) revealed that chemically grafted PBLGs adopt random-coil conformations in contrast to the physically adsorbed PBLGs, which exhibit α-helical conformations [26]. Microfibers of SWNT-g-PBLG/PBLG were prepared by electrospinning of a SWNT-g-PBLG/PBLG solution. Wide-angle X-ray scattering diffractograms suggested that SWNTs were evenly distributed among PBLG rods in solution and in the solid state, where PBLGs formed a short-range nematic phase interspersed with amorphous domains.

PLLA is an aliphatic polyester with good thermal plasticity, shape memory property, biocompatibility, and biodegradability, and thus has been widely used in tissue engineering, drug-delivery systems, and implant materials. However, the mechanical properties of PLLA for high load bearing biomedical applications are insufficient. To improve the mechanical properties of PLLA, nanocomposite materials of PLLA with CNTs can be utilized. The first attempt to polymerize L-lactide from the surface of MWNTs was reported by Chen et al. [27] (Figure 8.15). The amount of grafted polymer increased with reaction time but in a different way for each solvent. When the polymerization took place in DMF, the amount of grafted polymer on MWNTs and its uniformity was higher than that obtained in toluene.

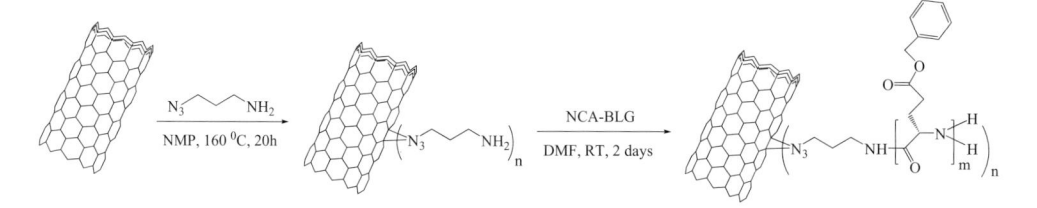

Figure 8.14 Synthetic route for PBLG-functionalized SWNTs [26].

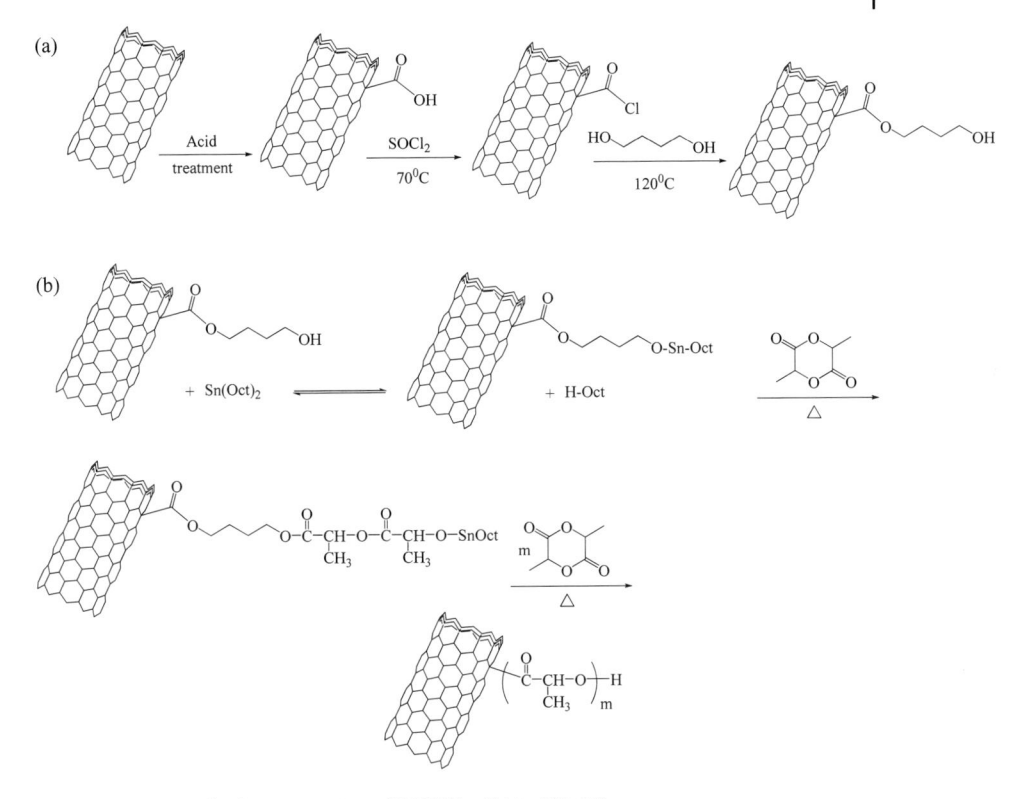

Figure 8.15 Route for the preparation of MWNT-g-PLLAs [28, 29].

The better dispersion of CNTs in DMF than in toluene was one possible reason for such behavior. Incorporation of 1 wt.% of MWNT-g-PLLA in PLLA matrix did not reveal any indication of aggregation and increased the tensile strength and modulus rendering PLLA more resistant to deformation.

Thermal stability, electrical conductivity, as well as mechanical properties of MWNT-g-PLLA nanocomposites were also studied [28, 29] (Figure 8.15). MWNTs were treated with strong acids to introduce carboxylic and hydroxyl functional groups on the surface of CNTs. Acid-treated MWNTs were reacted with $SOCl_2$ and then with glycol to yield hydroxyl-functionalized MWNTs. The resulting MWNTs-OH were then reacted with tin(II) 2-ethylhexanoate under sonication in a DMF solution, and L-lactide was subsequently introduced. The polymerization was allowed to proceed at 140 °C for 24 h.

A high degree of dispersion of MWNTs in the PLLA and MWNT-g-PLLA composites was obtained after PLLA grafting on MWNTs. As a result, the mechanical properties of PLLA/MWNT-g-PLLA were enhanced compared to those of the PLLA/MWNT composite. Good interfacial adhesion between MWNTs and the polymer matrix in PLLA/MWNT-g-PLLA composite also increased the activation

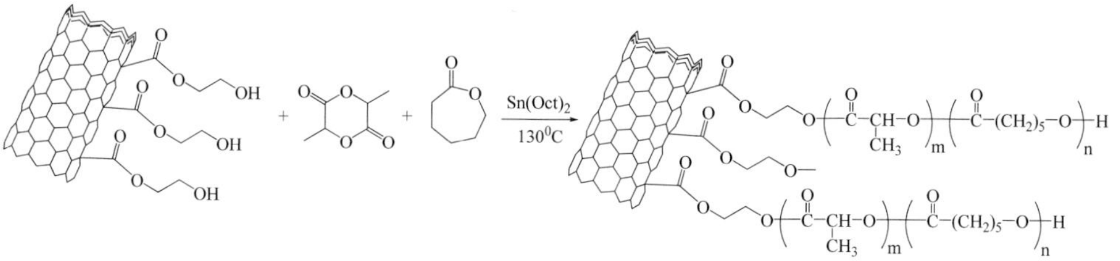

Figure 8.16 Copolymerization of L-lactide and ε-caprolactone onto MWNT-OH [31].

energy compared to PLLA/MWNT composite, indicating that the former was thermally more stable.

Feng et al. [30] covalently grafted PLLA via in-situ ROP of lactide. The grafted polymer was controlled by adjusting the feed ratio of monomer to MWNTs. The nanocomposite materials exhibited superparamagnetic performance and were aligned under low magnetic field.

Biodegradable copolymer of L-lactide and ε-caprolactone was grafted on MWNT-OH surface by ROP using stannous octanoate as the catalyst in order to reinforce the mechanical properties of the copolymer (Figure 8.16) [31]. The MWNT-g-poly(L-lactide-co-ε-caprolactone)/poly(L-lactide-co-ε-caprolactone) nanocomposites showed an increase in tensile strength and a decrease in elastic modulus compared to the neat copolymer.

8.2.2
Surface-Initiated Anionic Polymerization

The living nature of anionic polymerization [32] and its ability to afford polymers with low polydispersity and high molecular weights in a controlled way, even at low initiator concentrations, have made it an attractive choice in grafting polymers from CNTs. This chapter will deal with the anionic polymerization of linear monomers such as styrene and (meth)acrylates.

Viswanathan et al. [33] were first to grow polymer chains from nanotubes via anionic polymerization. They introduced carbanions on the surface of SWNTs by adding sec-butyllithium (sec-BuLi) to the sp^2 carbons and used them as initiating sites for the polymerization of styrene. The mutual electrostatic repulsion between individual nanotubes was able to exfoliate the SWNT bundles. Styrene was also polymerized in the solution with the free sec-BuLi initiator to linear chains, resulting in nanocomposite materials with low grafting efficiency (~10 wt.%).

Using the same approach, Chen et al. [34] showed that the carbanions on the surface of SWNTs can also initiate the polymerization of tert-butyl acrylate (tBA) but not that of methyl methacrylate (MMA). The authors suggested that a possible reason for this behavior was the different sensitivity of the two monomers toward the carbanions. The living nature of the polymerization was confirmed by initia-

Figure 8.17 Anionic polymerizations of methyl acrylate and methyl methacrylate from SWNTs and SWNT-g-(PtBA), respectively [34].

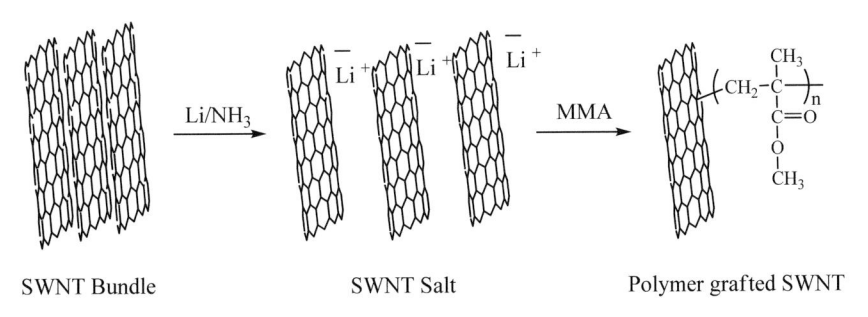

Figure 8.18 Anionic polymerization of methyl methacrylate from SWNT salts [35].

tion of the polymerization of MMA, from PtBA chains and the synthesis of a diblock copolymer (PtBA-b-PMMA) from SWNTs (Figure 8.17).

In a similar work, Liang et al. [35] used debundled nanotube salts as anionic initiators for the polymerization of MMA from SWNTs. It was shown that addition of Li/NH$_3$ to SWNTs (Figure 8.18) lead to SWNTs salts that have the ability to exfoliate SWNTs bundles and grow polymer chains from their surface.

In order to achieve higher grafting efficiency and higher molecular weights, our group used a different approach [36] to grow polymers from MWNTs via anionic polymerization in a custom-made glass apparatus (Figure 8.11a). MWNTs were covalently functionalized with 4-hydroxyethyl benzocyclobutene (BCB-EO) and 1-benzocylcobutene-1′-phenylethylene (BCB-PE) through [4 + 2] cycloaddition. The presence of these precursor initiator moieties was confirmed by FT-IR, Raman spectroscopy, and TGA. Alkoxy- and alkyl-anions were generated from MWNTs-g-(BCB-EO)$_n$ and MWNTs-g-(BCB-PE)$_n$, by addition of triphenylmethane and sec-BuLi, for the polymerization of ethylene oxide (THF) and styrene (benzene), respectively. The initiation of ethylene oxide and styrene from the surface alkoxy and alkyl anions, respectively, was found to be slow due to heterogeneous nature of the reaction. The surface-grown polymers were obtained in high conversion, forming MWNTs-g-(BCB-PEO)$_n$ and MWNTs-g-(BCB-PS)$_n$ containing only a small fraction of MWNTs < 1 wt.%. The produced nanocomposites were characterized by FT-IR, 1H NMR, Raman spectroscopy, DSC, TGA, and TEM. The TEM images showed the presence of thick layers of polymer around the surface. The formation of diblock copolymer, MWNTs-g-(BCB-PS-b-PI)$_n$ confirmed the living nature of these polymerizations.

8.3
Conclusions

The modification of CNTs with polymers has greatly benefitted from the application of ring-opening and anionic polymerizations via the surface-initiating technique. Both approaches have enabled the grafting on the CNT surfaces of a variety of well-defined polymers, with precisely controlled structure, molecular weight, and low polydispersity.

References

1 Iijima, S. (1991) *Nature*, **354**, 56.
2 Cooper, C.A., Young, R.J., and Halsall, M. (2001) *Compos. Part A.*, **32A**, 401.
3 Gao, G., Cagin, T., and Goddard, W.A. (1998) *Nanotechnology*, **9**, 184.
4 Ajayan, P.M., Stephan, O., Colliex, C., and Trauth, D. (1994) *Science*, **265**, 1212.
5 Tsubokawa, N. (2005) *Polym. J.*, **37**, 637.
6 Homenick, C.M., Lawson, G., and Adronov, A. (2007) *Polym. Rev.*, **47**, 265.
7 Spitalsky, Z., Tasis, D., Papagelis, K., and Galiotis, C. (2010) *Progr. Polym. Sci.*, **35**, 357.
8 Kamber, N.E., Jeong, W., Waymouth, R.M., Pratt, R.C., Lohmeijer, B.G.G., and Hedrick, J.L. (2007) *Chem. Rev.*, **107**, 5813.
9 Yoon, K.R., Kim, W.J., and Choi, I.S. (2004) *Macromol. Chem. Phys.*, **205**, 1218.
10 Xu, Y., Gao, C., Kong, H., Yan, D., Jin, Z.Y., and Watts, P.C.P. (2004) *Macromolecules*, **37**, 8846.
11 Zhou, L., Gao, C., and Xu, W. (2009) *Macromol. Chem. Phys.*, **210**, 1011.
12 Qu, L., Veca, M.L., Lin, Y., Kitaygorodisky, A., Chen, B., McCall, M.A., Connell, J.W., and Sun, Y.P. (2005) *Macromolecules*, **38**, 10328.
13 Yang, M., Gao, Y., Li, H., and Adronov, A. (2007) *Carbon*, **45**, 2327.
14 Buffa, F., Hu, H., and Resasco, D. (2005) *Macromolecules*, **38**, 8258.
15 Zeng, H., Gao, C., and Yan, D. (2006) *Adv. Funct. Mater.*, **16**, 812.
16 Castro, M., Lu, J., Bruzaud, S., Kumar, B., and Feller, J.-F. (2009) *Carbon*, **47**, 1930.
17 Ruelle, B., Peeterbroeck, S., Gouttebaron, R., Godfroid, T., Monteverde, F., Dauchot, J.-P., Alexandre, M., Hecq, M., and Dubois, P. (2007) *J. Mater. Chem.*, **17**, 157.
18 Yang, Y., Tsui, C.P., Tang, C.Y., Qiu, S., Zhao, Q., Cheng, X., Sun, Z., Li, R.K.Y., and Xie, X. (2010) *Eur. Polym. J.*, **46**, 145.
19 Adeli, M., Mirab, N., Alavidjeh, M.S., Sobhani, Z., and Atyabi, F. (2009) *Polymer*, **50**, 3528.
20 Xiang, L., Zhang, Z., Yu, P., Zhang, J., Su, L., Ohsaka, T., and Mao, L. (2008) *Anal. Chem.*, **80**, 6587.

21 Priftis, D., Sakellariou, G., Hadjichristidis, N., Penott, E.K., Lorenzo, A.T., and Muller, A.J. (2009) *J. Polym. Sci. Part. A: Polym. Chem.*, **47**, 4379.

22 Priftis, D., Sakellariou, G., Baskaran, D., Mays, J.W., and Hadjichristidis, N. (2009) *Soft Matter*, **5**, 4272.

23 Priftis, D., Sakellariou, G., Mays, J.W., and Hadjichristidis, N. (2010) *J. Polym. Sci. Part. A: Polym. Chem.*, **48**, 1104.

24 Yao, Y., Li, W., Wang, S., Yan, D., and Chen, X. (2006) *Macromol. Rapid Commun.*, **27**, 2019.

25 Li, J., He, W.-D., Yang, L.-P., Sun, X.-L., and Hua, Q. (2007) *Polymer*, **48**, 4352.

26 Tang, H., and Zhang, D. (2010) *J. Polym. Sci. Part. A: Polym. Chem.*, **48**, 2340.

27 Chen, G.-X., Kim, H.S., Park, B.H., and Yoon, J.-S. (2007) *Macromol. Chem. Phys.*, **208**, 389.

28 Kim, H.-S., Park, B.H., Yoon, J.-S., and Jin, H.-J. (2007) *Eur. Polym. J.*, **43**, 1729.

29 Kim, H.-S., Chae, Y.S., Park, B.H., Yoon, J.-S., Kang, M., and Jin, H.-J. (2008) *Curr. Appl. Phys.*, **8**, 803.

30 Feng, J., Cai, W., Sui, J., Li, Z., Wan, J., and Chakoli, A.N. (2008) *Polymer*, **49**, 4989.

31 Chakoli, A.N., Wan, J., Feng, J.T., Amirian, M., Sui, J.H., and Cai, W. (2009) *Appl. Surf. Sci.*, **256**, 170.

32 Hadjichristidis, N., Pitsikalis, M., Pispas, S., and Iatrou, H. (2001) *Chem. Rev.*, **101**, 3747.

33 Viswanathan, G., Chakrapani, N., Yang, H., Wei, B., Chung, H., Cho, K., Ryu, C.Y., and Ajayan, P.M. (2003) *J. Am. Chem. Soc.*, **125**, 9258.

34 Chen, S., Chen, D., and Wu, G. (2006) *Macromol. Rapid Commun.*, **27**, 882.

35 Liang, F., Beach, K., Sadana, A.K., Cantu, Y.I., Tour, J.M., and Billups, W.E. (2006) *Chem. Mater.*, **18**, 4764.

36 Sakellariou, G., Ji, H., Mays, J.W., and Baskaran, D. (2008) *Chem. Mater.*, **20**, 6217.

9
Grafting of Polymers on Nanotubes by Atom Transfer Radical Polymerization

Chao Gao

9.1
Introduction

Functionalization of nanotubes, especially carbon nanotubes (CNTs) [1, 2], with polymers has attracted increasing attention over the past years for the purposes of (1) improvement of solubility/dispersibility/wettability of nanotubes, (2) integration of both properties of polymers and nanotubes, (3) fabrication of novel hybrid nanomaterials and nanodevices, and (4) preparation of high-performance composites [3–6]. Both covalent and noncovalent methodologies have been employed to functionalize nanotubes. As to covalent modification, three grafting approaches have been developed, including "grafting to," "grafting from," and "random grafting" (Figure 9.1).

The "grafting to" approach means attaching as-prepared or commercially available polymers to nanotubes, followed by removal of unreacted polymers by filtering or centrifuging. The "random-grafting" approach denotes *in situ* polymerization of monomers through the conventional polycondensation or free radical polymerization techniques and the *in situ* formed macromolecules or macroradicals could couple onto nanotube surfaces randomly. The "grafting from" approach involves surface-initiated *in situ* polymerization in which initiating sites should be immobilized onto nanotube surfaces in advance. Hence, the "grafting from" approach normally needs controlled/living polymerizations such as controlled radical polymerizations (CRPs) [7], anionic polymerization, cationic polymerization, and ring-opening polymerization, etc.. This chapter will focus on polymer-functionalized nanotubes by CRPs. It is known that CRPs for surface-initiated polymerization mainly include atom transfer radical polymerization (ATRP) [8–11], reversible-addition fragmentation chain transfer (RAFT) polymerization [12–17], and nitroxide-mediated radical polymerization (NMRP) [18–20]. Since ATRP has been the most intensively addressed technique to functionalize nanotubes, it will be highlighted herein, whereas the results associated with other CRPs will be only mentioned briefly.

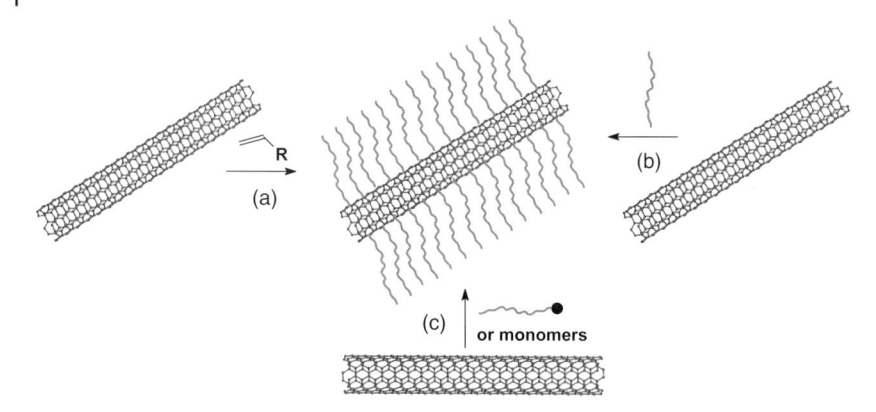

Figure 9.1 Synthesis strategies for polymer-grafted CNTs: (a) grafting from, (b) grafting to, (c) random grafting.

9.2
Grafting of Polymers on CNTs by ATRP

ATRP, a CRP based on a copper halide/nitrogen-based ligand catalyst, was first defined by Matyjaszewski and coworkers in 1995 [11]. Because of the merits such as good controllability upon molecular structure and topology, wide applicability for vinyl monomers including styrenes, acrylates and methacrylates, easily commercial availability of initiators and catalysts, excellent versatility in the synthesis of well-defined functional polymers and nanostructured materials, and high tolerance to water and polar organic solvents, ATRP has received much and sustained attention during the last 15 years, and has been developed into one of the most powerful and successful CRPs [8]. In this regard, numerous research papers and review articles have been published [8–11, 21–24].

Due to the commercial availability of functional building block of ATRP initiators such as 2-bromopropionyl bromide and 2-bromoisobutyryl bromide, surface-based macroinitiators can be readily accessed by simple esterification reaction. Therefore, ATRP has been widely employed to modify solid surfaces such as silica substrates, gold particles, polymer films, and CNTs, affording various polymer brushes with desired structures, properties, morphologies, functions, and applicable values [7]. This section only concerns the surface functionalizations of CNTs.

9.2.1
CNT-Based Macroinitiators

The most important step for ATRP approach to functionalization of CNTs is to introduce ATRP-initiating sites to CNT surfaces. The initiating sites-linked CNTs can serve as macroinitiators to initiate ATRP in the presence of monomer and catalyst/ligand, grafting polymer chains on CNTs, as depicted in Figure 9.2. Generally, the used catalyst is CuBr or CuCl, and the corresponding ligand is N,N,N',N'',N''-pentamethyldiethylenetriamine (PMDETA) or 2,2′-bipyridyl (bpy).

9.2 Grafting of Polymers on CNTs by ATRP

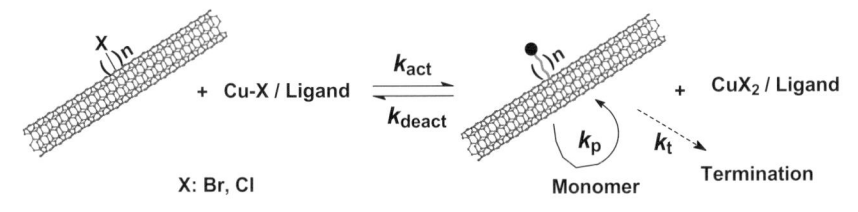

Figure 9.2 CNT-based ATRP mechanism with CuX (X: Br, Cl) as catalyst.

Several groups have reported different routes to prepare CNT-based macroinitiators via either oxidation of CNTs or addition to carbon–carbon double bonds of CNTs, as illustrated in Figure 9.3 [5]. Strong acid-oxidation followed by acylation is a highly efficient route to access the CNT-based macroinitiator. Gao and coworkers [25–27] prepared 2-bromo-2-methylpropionyl-functionalized multiwalled carbon nanotubes (MWCNTs) (CNT-Br-1) through a four-step process: (1) oxidization of pristine MWCNTs with concentrated HNO_3 for ~24 h [25, 26] or the mixture of HNO_3 and H_2SO_4 (1 : 3 by volume) for a short time (~100 min) [27] to introduce carboxylic groups onto MWCNT surfaces, (2) convertion of carboxyl groups into acyl chloride with excess of thionyl chloride, (3) preparation of hydroxyl groups-functionalized CNTs (CNT-OH) by reaction of acyl cholide with excess of glycol, (4) reaction of CNT-OH and 2-bromo-2-methylpropionyl bromide to make CNT-Br-1 in the presence of triethylamine. The bromine initiator density calculated from corresponding thermal gravimetric analysis (TGA) data achieved around 0.55 mmol of bromine groups per gram of CNT-Br-1, 0.72 mmol per gram of neat MWCNTs, or 8.6 initiator groups per 1000 carbons [27]. The main advantages of this route lie in the fact that (1) the chemicals are inexpensive and commercially avalabile, facilitating large-scale production, (2) the obtained CNT-OH could be used as macroinitiator to initiate ring-opening polymerization, (3) the process is highly reproducible, and (4) the initiating group density is sufficiently high, favoring for the construction of high density brushes. Later, Narain et al. obtained 2-bromo-2-methylpropionyl-functionalized single-walled carbon nanotubes (SWCNTs) (CNT-Br-2) through a similar protocol to the case of CNT-Br-1 except that glycol was replaced with 2-aminoethanol [28].

Alternatively, 2-bromopropionyl-functionalized SWCNTs (CNT-Br-3) [29] or 2-bromo-2-methylpropionyl-functionalized MWCNTs (CNT-Br-4) [30] was prepared through a three-step protocol via reaction of acyl chloride-functionalized SWCNTs (SWCNT-COCl) or MWCNT-COCl with 2-hydroxyethyl 2′-bromopropionate or 2-hydroxyethyl-2-bromoisobutyrate, respectively.

Another strategy to introduce functional groups onto CNTs is the addition reaction based on the C=C bonds of CNTs. For instance, Adronov et al. [31] prepared 2-bromo-2-methylpropionyl-functionalized SWCNTs (CNT-Br-5) by 1,3-dipolar cycloaddition reaction [32, 33] of 4-hydroxyphenyl glycine and octyl aldehyde with shortened SWCNTs in DMF at 130 °C for 5 days, followed by an esterification with 2-bromoisobutyryl bromide in DMF at 70 °C. By electrochemical reduction of diazonium salt, BF_4^- $^+N_2$–C_6H_4–CH_2CH_2–Br, Chehimi et al. introduced brominated aryl-initiating groups on aligned MWCNTs (CNT-Br-6) [34]. Paik et al.

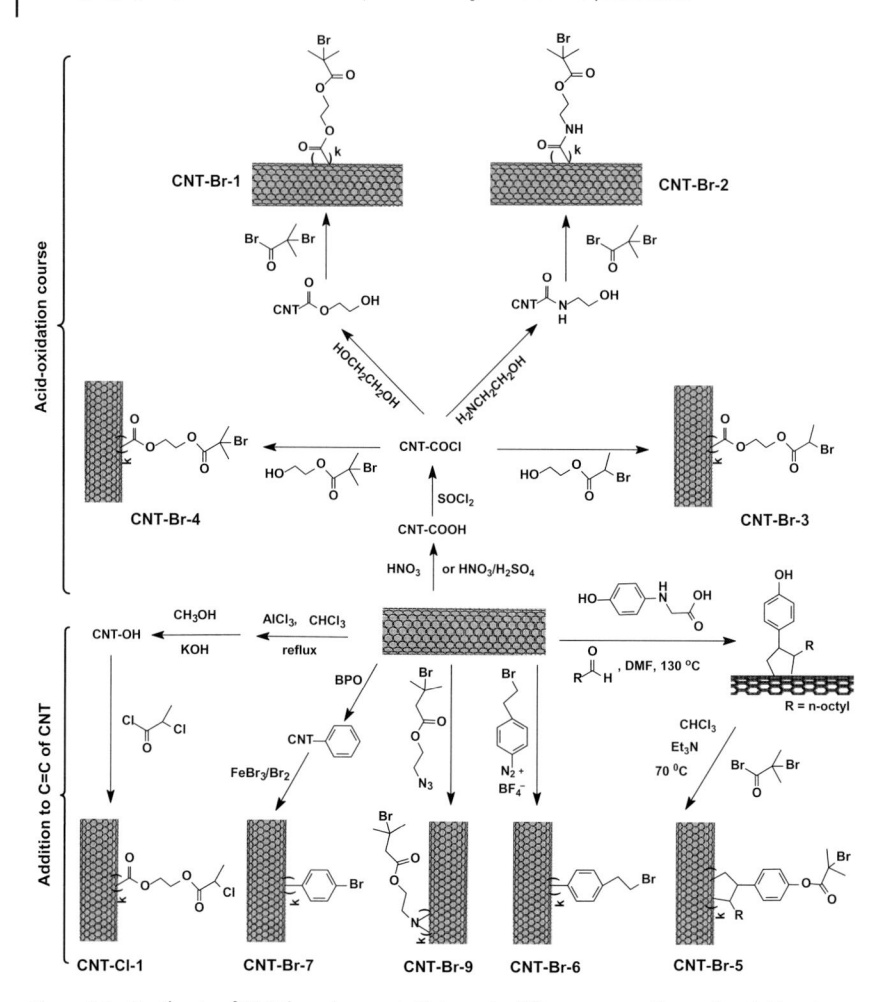

Figure 9.3 Synthesis of CNT-based macroinitiators via different routes for surface-initiated ATRP [5].

synthesized 2-chrolopropyl-linked SWCNTs (CNT-Cl-1) by a three-step process [35]: (1) direct electrophilic addition of chloroform to SWCNTs using AlCl$_3$ as the catalyst, (2) hydrolysis in the presence of KOH to form hydroxyl groups, (3) esterification of 2-chrolopropyl chloride with the hydroxyl groups linked on the SWCNTs. N-doped MWCNTs-based macroinitiator (CNT-Br-7) was prepared by reaction with benzoil peroxide (BPO) in toluene at 105 °C for 5 h, followed by bromination at 55 °C with Br$_2$ in CCl$_4$ using FeBr$_3$ as the catalyst [36]. Similarly, CNT-Cl-2 and CNT-Br-8 could be prepared by treatment of purified SWCNTs with peroxy organic acids containing Cl or Br substituents, m-chloroperbenzoic acid (MCPBA), and 2-bromo-2-methylperpropionic acid (BMPPA), respectively [37].

Most recently, Gao et al. developed a facile approach to functionalization of MWCNTs and SWCNTs via nitrene chemistry, and functional groups such as hydroxyl, carboxyl, and amino could be immobilized on CNTs in one-step electrophilic [2 + 1] cycloaddition reaction [38]. So macroinitiator CNT-Br-9 could be synthesized by one-step reaction between pristine CNTs and 2-azidoethyl-2-bromo-2-methylpropanoate. The Br density achieved 0.6 mmol g^{-1} of functionalized CNTs or 8.2 groups per 1000 carbons, which is close to the value of CNT-Br-1. Subsequent ATRP of styrene, methyl methyacrylate (MMA), and 3-azido-2-hydroxypropyl methacrylate (GMA-N$_3$), affording various polymer-grafted CNTs [38].

9.2.2
Linear Homopolymer-Grafted CNTs

9.2.2.1 CNT-Initiated ATRP of (Meth)acrylates

Polymer arms can be grafted from CNT surfaces via the ATRP "grafting from" approach using CNT-based macroinitiators. Most of the ATRP-active monomers including (meth)acrylates, styrenes, and acrylamides have been tried to perform *in situ* polymerization, growing various arms on CNTs covering from hydrophobic to hydrophilic, from oil-soluble to water-soluble, from acidic to basic, and from common to functional polymers [5]. The monomer structures are shown in Scheme 9.1, and the polymerization conditions and selected results are summarized in Table 9.1.

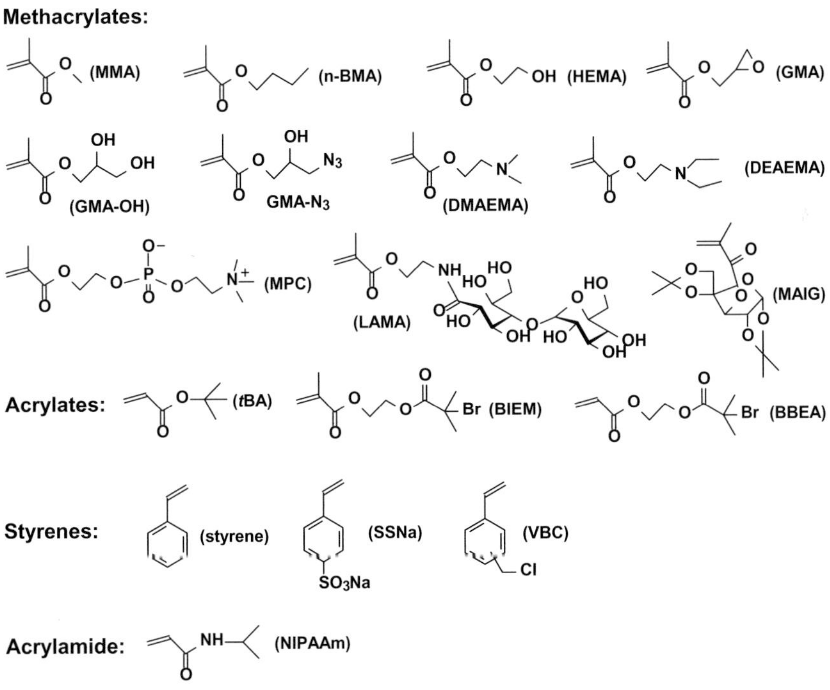

Methacrylates:

(MMA), (n-BMA), (HEMA), (GMA)

(GMA-OH), GMA-N$_3$, (DMAEMA), (DEAEMA)

(MPC), (LAMA), (MAIG)

Acrylates: (tBA), (BIEM), (BBEA)

Styrenes: (styrene), (SSNa), (VBC)

Acrylamide: (NIPAAm)

Scheme 9.1 The monomers used in the *in situ* polymerization for grafting polymer from CNTs by ATRP technique [5].

Table 9.1 Selected conditions and results of ATRP in the presence of CNT-based macroinitiator [5].

Macroinitiator	Monomer	Catalyst/ligand	Solvent/temperature/time	Highest $f_{wt.\%}$	Reference
CNT-Br-1	MMA	CuBr/PMDETA	DMF/60°C/30h	82	[25]
	MMA+HEMA	CuBr/PMDETA	DMF/60°C/20h	54.5	[25]
	DMAEMA	CuBr/PMDETA	THF/60°C/48h	80	[39]
	DEAEMA	CuBr/PMDETA	MeOH/60°C/48h	67	[40, 41]
	GMA-OH	CuBr/PMDETA	MeOH/40°C/48h	90	[27]
	GMA	CuBr/PMDETA	diphenyl ether/50°C/48h	82	[42]
	MAIG	CuBr/HMTETA	ethyl acetate/60°C/29h	71	[43]
	BIEM	CuBr/(PPh$_3$)$_2$NiBr$_2$	ethyl acetate/100°C/4.5h	38	[43]
	MAIG+BIEM	CuBr/(PPh$_3$)$_2$NiBr$_2$	ethyl acetate/100°C/4.5–29.5h	40–53	[43]
	tBA	CuBr/PMDETA	DMF/60°C/48h	75	[44]
	Styrene	CuBr/PMDETA	diphenyl ether/100°C/50h	77.9	[26]
	Styrene+tBA	CuBr/PMDETA		70	[45]
	SSNa	CuBr/PMDETA	DMF/130°C/30h	68	[44]
	NIPAAm	CuBr/PMDETA	Water/RT/48h	84	[46]
CNT-Br-2	MPC	CuBr/bpy	MeOH/25°C/24h	54	[28]
	MPC	CuBr/bpy	H$_2$O/25°C/12h	59	[28]
	LAMA	CuBr/bpy	H$_2$O/25°C/12h	67	[28]
	LAMA	CuBr/bpy	NMP/25°C/12h	70	[28]
CNT-Br-3	n-BMA	CuCl/bpy	DCB/60°C/19h	69	[29]
CNT-Br-4	Styrene	CuBr/bpy	DCB/110°C/14h	75	[47]
	MMA	CuBr/PMDETA	Toluene/90°C/24h	70.9	[30]
	Styrene	CuBr/PMDETA	Toluene/100°C/24h	33.0	[30]
	MMA+styrene	CuBr/PMDETA	Toluene/100°C/24h	24.6	[30]
	Styrene	CuBr/PMDETA	Bulk/80°C/24h	60	[48]
	CH$_2$CHCN	CuBr/PMDETA	Bulk/70°C/24h	40	[48]
	Styrene+CH$_2$CHCN	CuBr/PMDETA	Bulk/80°C/24h	70	[48]
CNT-Br-5	BBEA	CuBr/PMDETA	Toluene/100°C/24h	80	[49]
	MMA	CuBr/bpy	DMF+H$_2$O/RT/24h	–	[31]
	tBA	CuBr/bpy	DMF+H$_2$O/RT/24h	–	[31]

Table 9.1 Continued.

Macroinitiator	Monomer	Catalyst/ligand	Solvent/temperature/time	Highest $f_{wt.\%}$	Reference
CNT-Br-6	MMA	CuCl+CuCl$_2$/PMDETA	Bulk/90 °C/6 h	–	[34]
	Styrene	CuBr/PMDETA	Bulk/110 °C/16 h	–	[34]
CNT-Br-7	Styrene	CuBr/dNbpy	toluene/110 °C/18 h	40	[36]
CNT-Br-8	MMA	CuBr/PMDETA	DCB/60 °C/24 h	30	[37]
CNT-Br-9	GMA-N$_3$	CuBr/PMDETA	THF/25 °C/24 h	58.6	[38]
	Styrene	CuBr/PMDETA	Bulk/80 °C/24 h	33.3	[38]
	MMA	CuBr/PMDETA	THF/40 °C/24 h	50.5	[38]
CNT-Cl-1	Styrene	CuCl/PMDETA	Acetone/60 °C/96 h	6 nm	[35]

Gao and coworkers [25] presented controlled functionalization by ATRP for the first time in 2004. Using CNT-Br-1 as the initiator and CuBr/PMDETA as the catalyst system, poly(methyl methacrylate) (PMMA) was successfully grafted from MWCNTs [25]. The resulting polymer-functionalized MWCNTs were characterized by transmission electron microscopy (TEM), scanning electron microscopy (SEM), hydrogen-nuclear magnetic resonance (^{1}H NMR), Fourier transform infrared spectroscopy (FTIR), TGA, and Raman spectroscopy. Remarkably, core–shell nanostructures were clearly observed by TEM with CNT as the core and the grafted polymer layer as the shell, declaring that (1) the high dense and even grafting efficiency was achieved, (2) the whole nanotube outer surfaces including tube ends and body were uniformly grafted with polymers, rather than only or mainly distributed on the ends that were previously thought to be the most highly reactive locations, (3) the direct evidence of visualization was available for the polymer grafting, laying the foundation for the characterization of polymer-covered nanotubes. Furthermore, the polymer layer thickness can be readily adjusted from ~3.8 to ~14 nm (see Figure 9.4), and the corresponding polymer content was changed from 31.9 to 82.0 wt.% when the mass feed ratio of MMA to CNT-Br-1 was varied from 1/1 to 10/1. The PMMA-grafted MWNTs were well dispersible in weak polar solvents such as chloroform and THF [25]. At the same time, Ford [29] and Adronov et al. [31] independently developed the in situ ATRP strategy to functionalize SWCNTs using CNT-Br-4 and CNT-Br-5 as macroinitiators, respectively. The successful exploration of ATRP approach to controlled functionalization of CNTs triggered the fast advancement of this field, and various polymers were grafted on CNTs by ATRP and other CRP techniques subsequently.

Interestingly, PMMA-grafted MWCNTs brushes (MWCNT-g-PMMA, 85 wt.% polymer) could be soluble in THF, and self-assemble into suprastructures on solid

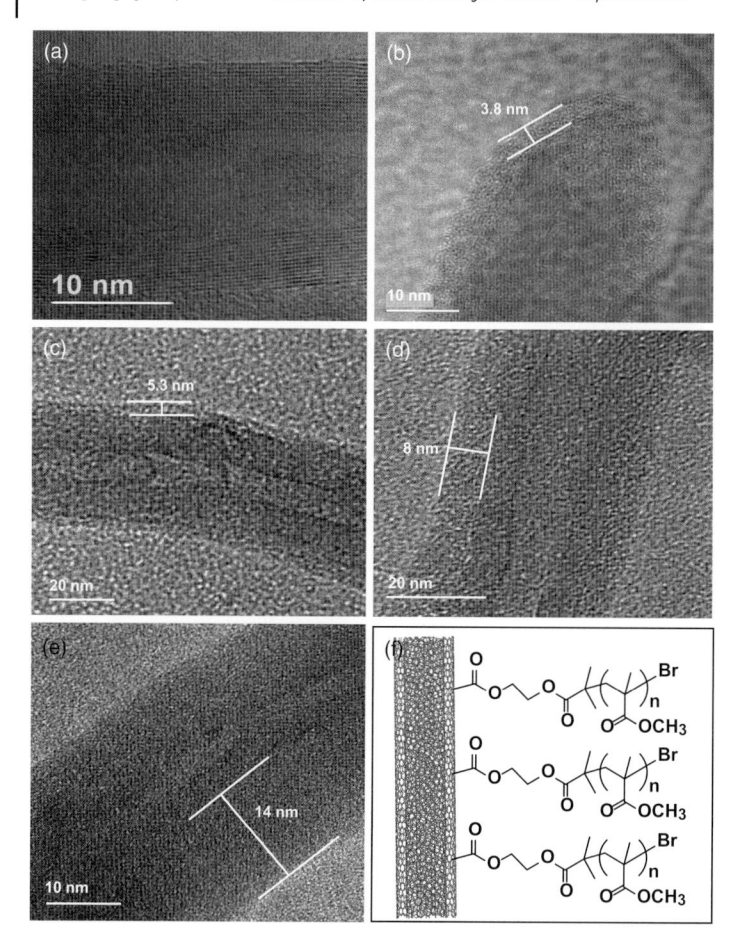

Figure 9.4 TEM images of pristine MWCNT (a) and PMMA-grafted MWCNTs (b–e) prepared by *in situ* ATRP technique, and schematic illustration of PMMA-grafted MWCNTs (f) [25].

surfaces such as gold, mica, silicon, quartz, or carbon films [50]. The combination of SEM, AFM and TEM measurements confirmed the morphology of the assembled structures, and revealed the assembly mechanism. With decreasing concentration of the MWCNT-g-PMMA from 3 to 0.1 mg ml^{-1}, the assembled structures changed from cellular and basketwork-like forms to multilayered cellular networks and individual needles (Figure 9.5). It is speculated that phase separation during evaporation of the solvent drives the MWCNT-g-PMMA nanohybrids to assemble and form the suprastructured objects, and the rigid MWCNTs stabilize the formed structures, as demonstrated by TEM and SEM measurements (see Figure 9.6). The self-assembly behavior of polymer-grafted CNTs is possibly a general phenomenon for highly soluble core–shell nanostructures.

Using CNT-Br-1 as macroinitiator, Gao and coworkers extended the PMMA grafting protocol to other monomer systems, including 2-(dimethylamino)ethyl

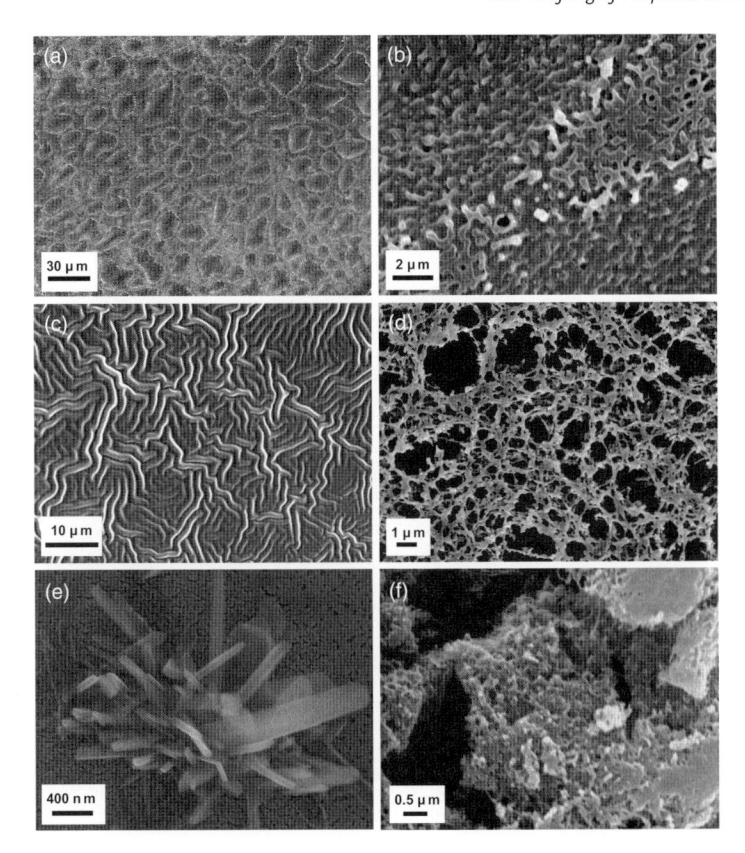

Figure 9.5 Representative SEM images of the self-assembled patterns or structures of the MWCNT-g-PMMA on a gold surface with concentrations of 3 (a, b), 2 (c), 1 (d), and 0.1 (e) mg ml^{-1}, and SEM image of the as-prepared MWCNT-g-PMMA bulk material (f). Reprinted with permission from [50].

methacrylate (DMAEMA) [39], 2-(diethylamino)ethyl methacrylate (DEAEMA) [40, 41], glycerol monomethacrylate (GMA-OH) [27], glycidyl methacrylate (GMA) [42], 3-O-methacryloyl-1,2:5,6-di-O-isopropylidene-D-glucofuranose (MAIG) [43], and tert-butyl acrylate (tBA) [44]. The structure and morphology of polymer-functionalized MWCNTs were also analyzed by FTIR, NMR, TGA, Raman spectroscopy, AFM, SEM, and TEM. The same conclusions can be drawn as in the case of PMMA-grafted MWCNTs: (1) core–shell nanostructures can be clearly observed by high-resolution TEM with the polymer layer as the shell and CNT as the core for the polymer-grafted CNTs, especially when the grafted polymer amount is higher than 50 wt.%, (2) the grafted polymer content can be readily adjusted by the feed ratio of monomer to CNT-Br-1 and the reaction time, (3) the solubility and dispersibility of CNTs are highly improved due to the polymer grafting. These results showed the good reproducibility and high generality of the ATRP methodology.

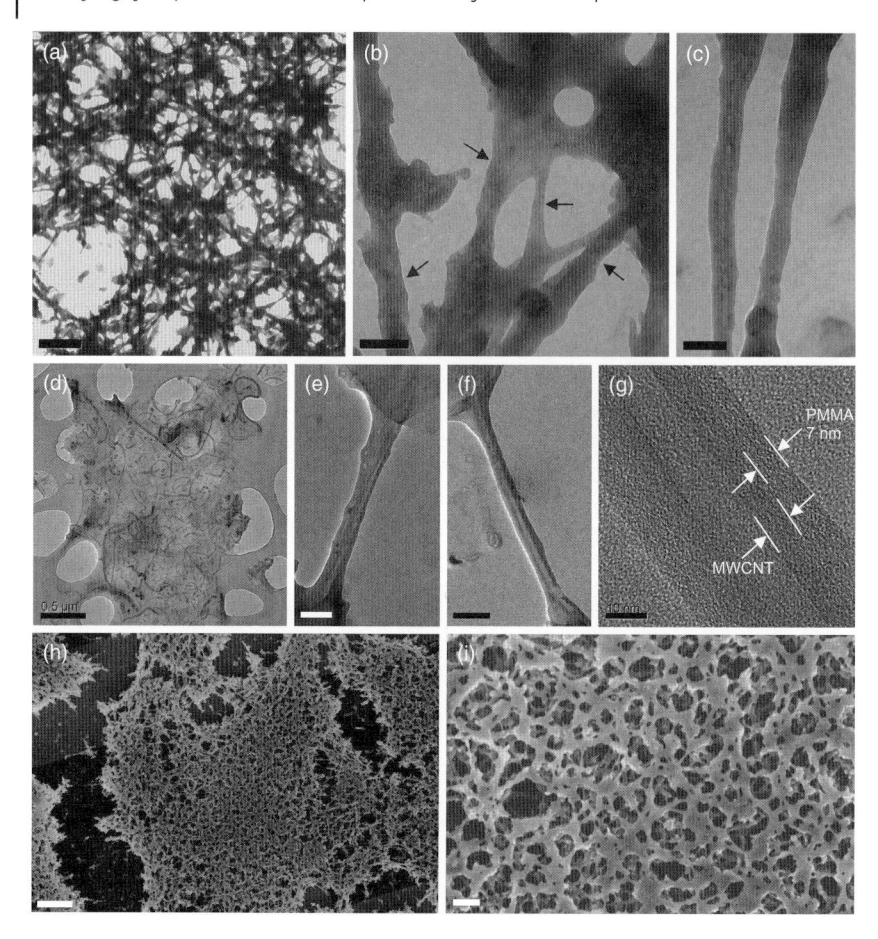

Figure 9.6 Low-voltage (100 kV) TEM images of MWCNT-*g*-PMMA on a carbon-coated TEM sample grid (a–c), high-voltage (200 kV) TEM image of MWCNT-*g*-PMMA on a multipore lacey carbon TEM sample grid (d), and high-resolution high-voltage TEM images of MWCNT-*g*-PMMA on the carbon-coated TEM sample grid (e–g), and SEM images of the same sample measured by TEM (shown in image a) of assembling on the TEM grid (h, i). The concentration of solution is 1 mg MWCNT-*g*-PMMA per 1 ml THF. The scale bars are 3 μm (a), 200 nm (b), 50 nm (c), 0.5 μm (d), 50 nm (e), 50 nm (f), 10 nm (g), 5 μm (h), and 1 μm (i). Reprinted with permission from [50].

PDEAEMA-grafted MWCNTs showed an interesting pH-responsive solubility/dispersibility because of the pH-responsive PDEAEMA shell [40]. Polyelectrolyte-grafted CNTs can also be used as nanoplatforms to fabricate hybrid nanostructures and nanomaterials by electrostatic attraction (Figure 9.7). Gao *et al.* [39, 41] grafted PDEAEMA or PDMAEMA on MWCNTs by ATRP and used as templates to load negative iron oxide (Fe_3O_4) nanoparticles or CdTe quantum dots (QDs), obtaining supraparamagnetic or fluorescent CNTs (Figure 9.8). The magnetic CNTs were adsorbed onto sheep red blood cells so that individual cells could be manipulated

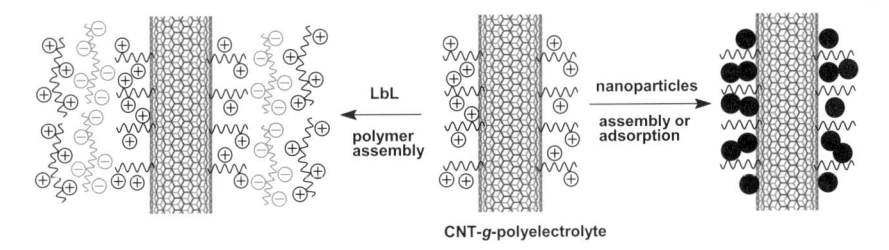

Figure 9.7 Polyelectrolyte-grafted CNTs used as nanosubstrate for multilayered polymer assembly or nanoparticle loading [5].

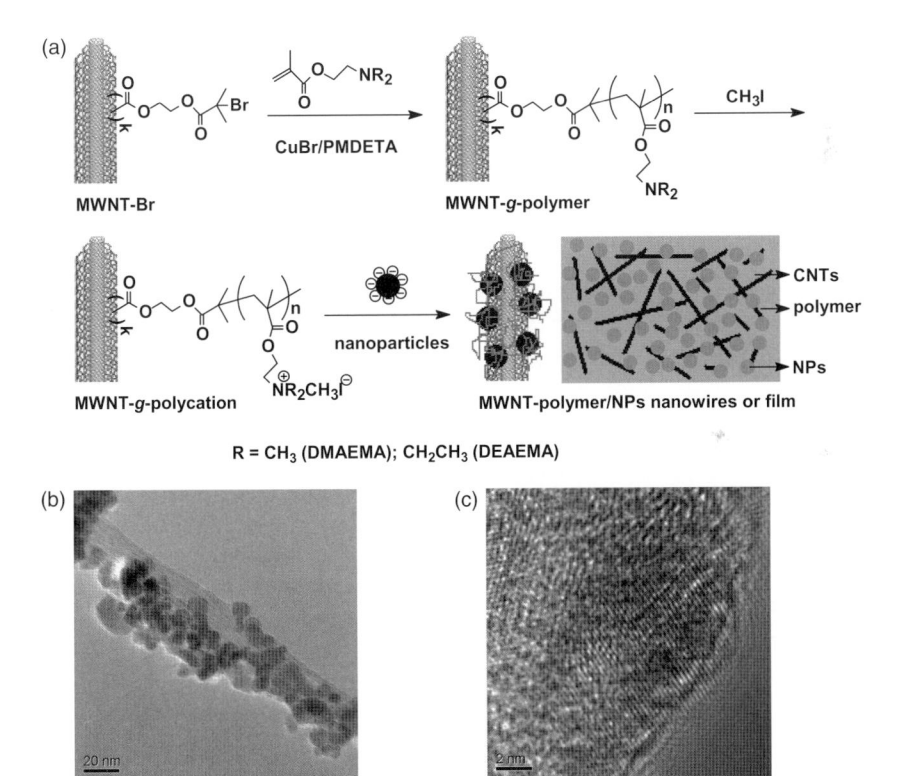

Figure 9.8 (a) Synthesis of cationic polymer-grafted MWCNTs, and assembly of nanoparticles on the polycation-functionalized MWCNTs. (b, c) TEM images of PDEAEMA-grafted MWCNT/Fe_3O_4 hybrid at different magnifications [5, 39–41].

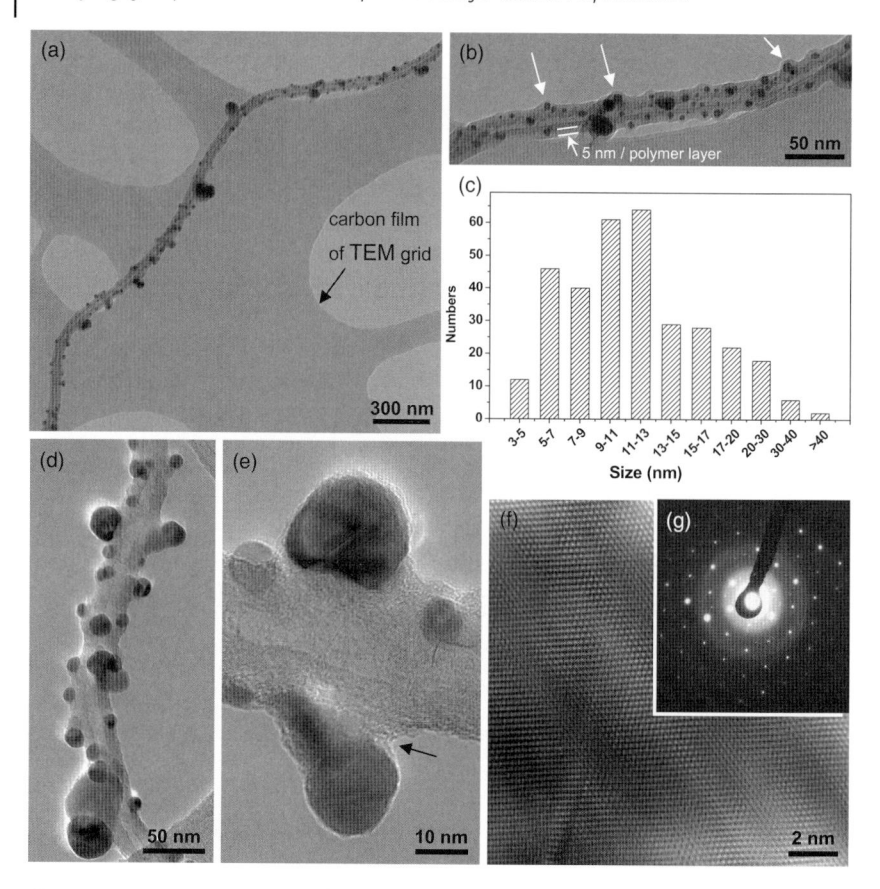

Figure 9.9 TEM measurements of nanohybrids of silver nanoparticles/MWCNT-g-PAA (60 wt.% of PAA, 2 h reaction). (a) TEM image at 100 kV. (b) Higher magnification of (a) displaying the respective phases of polymer layer, silver nanoparticles and CNTs. (c) Size distribution of the produced silver nanoparticles. (d) TEM image at 200 kV. (e) Higher magnification of (d) showing the polymer interlayer between silver nanoparticle and nanotube wall. (f) High resolution and high magnification TEM image of a silver nanoparticle, presenting the silver crystal lattice. (g) Selected area electron diffraction (SAED) patterns showing the face-centered cubic (fcc) crystal structure of the silver single nanocrystals. Reprinted with permission from [51].

in a magnetic field, opening the door of biomanipulation with magnetic nanotubes [41].

Poly(acrylic acid) (PAA)-grafted CNTs, derived from the hydrolysis of tBA-grafted CNTs, could be applied as a charged nanosubstrate to coat polymer with opposite charges via layer-by-layer (LbL) assembly approach (Figure 9.7) [44]. Moreover, PAA-grafted CNTs are good templates for depositing silver and other metal nanocrystals. As shown in Figure 9.9, Ag nanoparticles with mean diameter of ~10 nm could be evenly grown on CNT surfaces in water without any additional

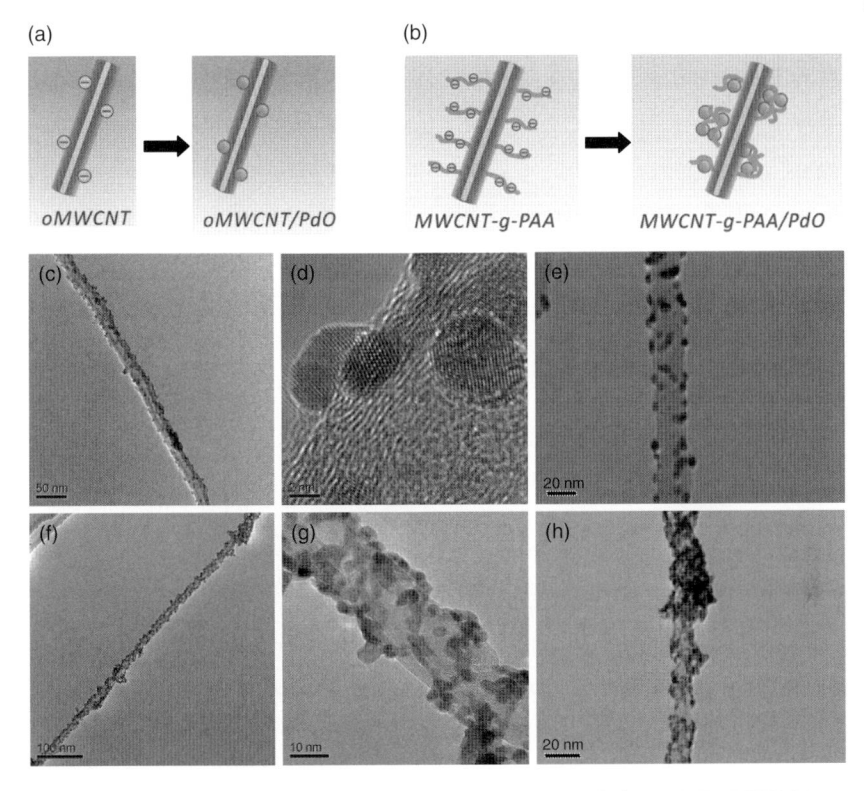

Figure 9.10 (a) Direct decoration of PdO nanoparticles on oxidized MWCNTs (oMWCNTs) from aqueous solution of Pd(NO₃)₂. (b) Decoration of PdO nanoparticles on PAA-grafted MWCNTs (MWCNT-g-PAA) from Pd(NO₃)₂ aqueous solution. (c–e) TEM images of oMWCNT/PdO. (f–h) TEM images of MWCNT-g-PAA/PdO [52].

chemcials and physical treatments [51]. Similarly, uniform PdO nanoparticles with diameter 2.3–2.6 nm can be loaded on CNT surfaces by simple decomposition of Pd(NO₃)₂ in water at room temperature, and the PAA grafting made the deposited PdO nanoparticles highly dense (Figure 9.10) [52].

Alternatively, hydrophilic multihydroxyl poly(GMA-OH)-grafted MWCNTs could be converted into multicarboxyl polymer-functionalized CNTs by reaction with succinic anhydride and then used as templates to efficiently sequestrate metal ions such as Ag^+, Co^{2+}, Ni^{2+}, Au^{3+}, La^{3+}, and Y^{3+}, giving rise to MWCNT-polymer/metal hybrid nanocomposites, nanowires, or necklace-like nanostructures, depending on the grafted polymer content and the nature of the captured metal [27]. Scheme 9.2 shows the synthesis protocol. SEM and TEM studies combined with energy dispersive spectroscopy (EDS) analyses confirmed the structure and elements of the novel hybrid nanoobjects.

Using CNT-Br-4 as the macroinitiator, Baskaran successfully grafted PMMA from the MWCNTs by ATRP with CuBr/PMDETA as the catalyst/ligand in toluene

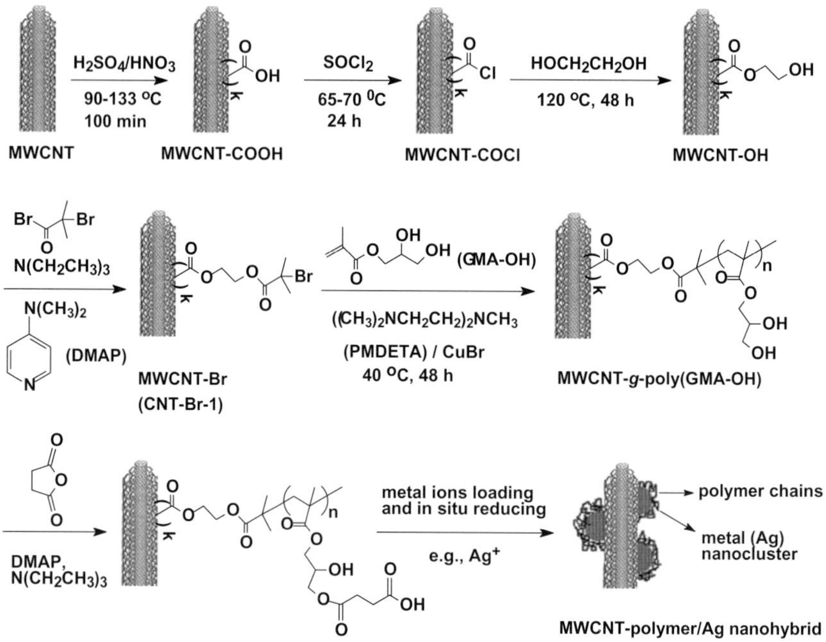

Scheme 9.2 Functionalization of MWCNTs with poly(GMA-OH) by ATRP, convertion of the hydroxyl groups poly(GMA-OH) into carboxylic groups, and metal sequestration/reduction by the grafted polyacid chains [27].

at 90 °C [30]. The amount of grafted PMMA approached 70.9 wt.%. The covalent linkage between PMMA and MWCNTs were demonstrated by FTIR, NMR, Raman spectroscopy, TGA, SEM, and TEM measurements. The glass transition temperature (T_g) of PMMA grafted on CNTs is higher than that of free PMMA by about 15–30 °C due to the covalent anchoring [30].

Chehimi et al. functionalized aligned MWCNTs with PMMA at 90 °C for 6 h using CNT-Br-6 as the initiator and CuCl+CuCl$_2$/PMDETA as the catalyst system [34]. The grafted polymer layer was observed with HRTEM. X-ray photoelectron spectroscopy (XPS) analysis confirmed the presence of PMMA by its characteristic C1s and valence band features. Liu et al. prepared PMMA-grafted SWCNTs at 60 °C for 24 h using CNT-Br-8 as the initiator and CuBr/PMDETA as the catalyst system, and they found that the grafted polymer content increased from 17 to 30 wt.% by the addition of ethyl 2-bromoisobutyrate as a coinitiator [37].

Functionalization of CNTs with biocompatible and biodegradable polymers is one of the most interesting topics in this field, because the resulting nanocomposites promise great potential in the applications of bionanotechnology. Narain et al. grafted poly(2-methacryloyloxyethyl phosphorylcholine) (polyMPC) and poly(lactobionamidoethyl methacrylate) (polyLAMA) from the SWCNT surfaces by ATRP of MPC and LAMA at 25 °C for 12–24 h with CNT-Br-2 as the initiator and CuBr/bpy as the catalyst system, and the grafted polyMPC and polyLAMA contents

Scheme 9.3 Grafting linear glycopolymer from surfaces of MWCNTs by ATRP. Reprinted with permission from [43]. © American Chemical Society.

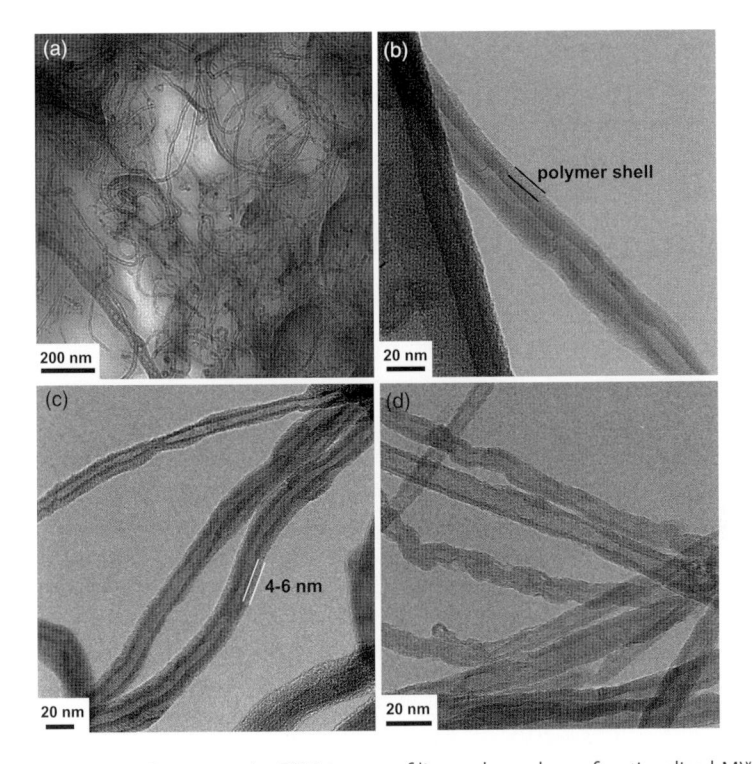

Figure 9.11 Representative TEM images of linear glycopolymer-functionalized MWCNTs at 29 (a) and 10.5 h (b), deprotected glycopolymer-functionalized MWCNTs at 29 h (c), and MWCNT-Br (d). Reprinted with permission from [43]. © American Chemical Society.

reached 59 wt.% and 70 wt.%, respectively [28]. Gao and Müller et al. functionalized MWCNTs with glycopolymer by ATRP of MAIG using CNT-Br-1 as the initiator, followed by hydrolysis of the PMAIG chains in 80% formic acid for 48 h (see Scheme 9.3), and the grafted polymer content achieved 71 wt.% [43]. Figure 9.11 shows the representative TEM images of glycopolymer-functionalized MWCNTs, revealing the even grafting effect and core–shell structure for the individual functionalized nanotubes.

9.2.2.2 CNT-Initiated ATRP of Styrenes

Styrene and its derivates is another class of ATRP-active monomers, and also widely used in the surface-initiated polymerizations. Using CNT-Br-1 as the macroinitiator, Gao and coworkers prepared polystyrene-grafted MWCNTs (MWCNT-g-PS) in diphenyl ether at 100 °C for 50 h with a catalyst system of CuBr/PMDETA [26]. TGA measurements showed that the grafted PS content increased from 28.6 wt.% to 77.9 wt.% when the feed ratio of monomer to CNT-Br-1 was increased from 1/1 to 10/1 by weight. To probe the information of grafted PS, the chemcial hydrolysis was conducted for MWCNT-g-PS samples by KOH/ethanol in THF to collect detached PS. The numer-average molecular weight (M_n) of the detached PS from the nanotube surfaces increased from 5000 to 11 000, indicating that the increased polymer amount was resulted from the increase of molecular weight. It is noteworthy that the polydispersity index (PDI) of the detached polymer was rather broad (1.77–3.57) and became broader with the grafted polymer content, which suggested that part of chains were terminated during polymerization. This is likely due to the high density of grafted chains that may cause coupling-termination between closed chains. The thermal decomposition temperature (T_d) of the PS chains can be dramatically improved by about 50–80 °C due to the covalent linkage with CNTs, and the T_g of grafted PS is also higher than that of detached PS by about 10–20 °C. The different thermal behaviors between grafted and free polymers provide further evidence for the covalent grafting. As shown in Figure 9.12a and b, pure PS spheres (indicated by arrows) could be found in the hydro-

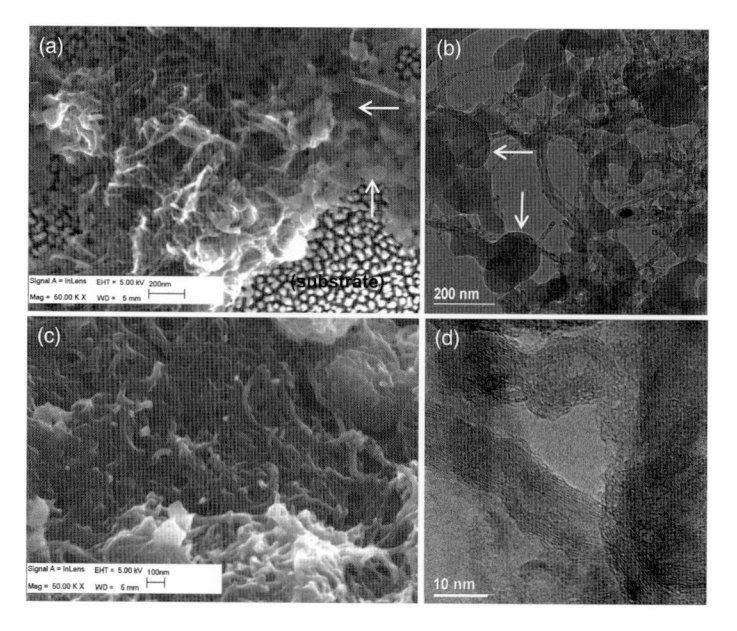

Figure 9.12 SEM (a) and TEM (b) images of detached sample of MWCNT-g-PS mixed with PS (the arrows denote the mixed PS spheres), SEM image of MWCNT-g-PS in which free PS was removed completely (c), and TEM image of thermally decomposed MWCNT-g-PS at 500 °C, indicating the destruction of tubular structures (d) [26].

lyzed MWCNTs if the detached polymer was not removed completely, whereas MWCNT-g-PS exhibited even morphology of polymer-coated nanotubes (Figure 9.12c). This comparison demonstrated that the mixed polymer can be removed from the product by sufficient washing and centrifuging. In addition, the thermally decomposed MWCNT-g-PS was also investigated by TEM and SEM observations, revealing that the covalently functionalized moieties on the MWCNTs have strong negative influence on the thermal degradation and morphology of CNTs (Figure 9.12d), and the more the functionalized moieties, the stronger the influence.

The macroinitiator of CNT-Br-1 can also initiate ATRP of sodium 4-styrenesulfonate (SSNa) in DMF at 130 °C [44]. The content of grafted polySSNa varied from 25 to 68 wt.% when the molar feed ratio of SSNa to CNT-Br-1 increased from 20/1 to 100/1. TEM observations also showed the featured core–shell structure for polySSNa-grafted MWCNTs. UV/Vis spectra demonstrated that the polySSNa-grafted MWCNTs are highly soluble in water.

Ford and coworkers prepared SWCNT-g-PS in DCB at 110 °C for 14 h with CuBr/bpy as the catalyst [47]. The grafted PS content increased with increasing the reaction time, and approached 75 wt.% at 14 h with a monomer conversion of 76%. According to the Raman and near-IR spectra, the polymer grafting did not change the sidewall structures of SWCNTs considerably.

With CNT-Br-4 as the macroinitiator, Baskaran [30] and Ryu et al. [48] independently prepared MWCNT-g-PS by in situ ATRP using the same catalyst of CuBr/PMDETA under different reaction conditions (see Table 9.1), and they also extended the surface-initiated polymerization to copolymerization of MMA/styrene [30] and styrene/acrylonitrile [48], respectively. In the presence of CNT-Cl-1, Paik et al. obtained SWCNT-g-PS with 6 nm thickness of PS shell by using CuCl/PMDETA catalyst in acetone at 60 °C for 96 h [35]. With CNT-Br-7 as macroinitiator and CuBr/4,4′-dinonyl-2,2′-bipyridine (dNbpy) as catalyst system, N-doped CNTs could also be grafted with PS (3–5 nm thickness of polymer layer with fraction of ~40 wt.%) [36].

9.2.2.3 CNT-Initiated ATRP of Acrylamides

Generally, the controllability of acrylamides by ATRP is not so effective as that of methacrylates and styrenes in homogeneous solution. Nevertheless, CNT-initiated ATRP of N-isopropylacrylamide (NIPAAm) was reported with good control over the grafted polymer content [46]. The representative HRTEM images of MWCNT-g-PNIPAAm are given in Figure 9.13, showing the high and even grafting efficiency with uniform core–shell structures. It is well known that PNIPAAm is a temperature-sensitive material that has a lower critical solution temperature (LCST) of 32–34 °C due to its hydrophilic/hydrophobic phase transition in water. The combination of NMR, UV/vis, and DSC analyses confirmed that the MWCNT-g-PNIPAAm nanohybrids are still thermosensitive, with an LCST of 34 °C. Interestingly, the thermoresponsive behavior of hydrophilic/hydrophobic phase transition for MWCNT-g-PNIPAAm is highly reversible and very fast, as probed by multicycle scanning on DSC (Figure 9.14). This promises the potential application of this kind of smart nanowires in the micro/nano switching and other relevant fields.

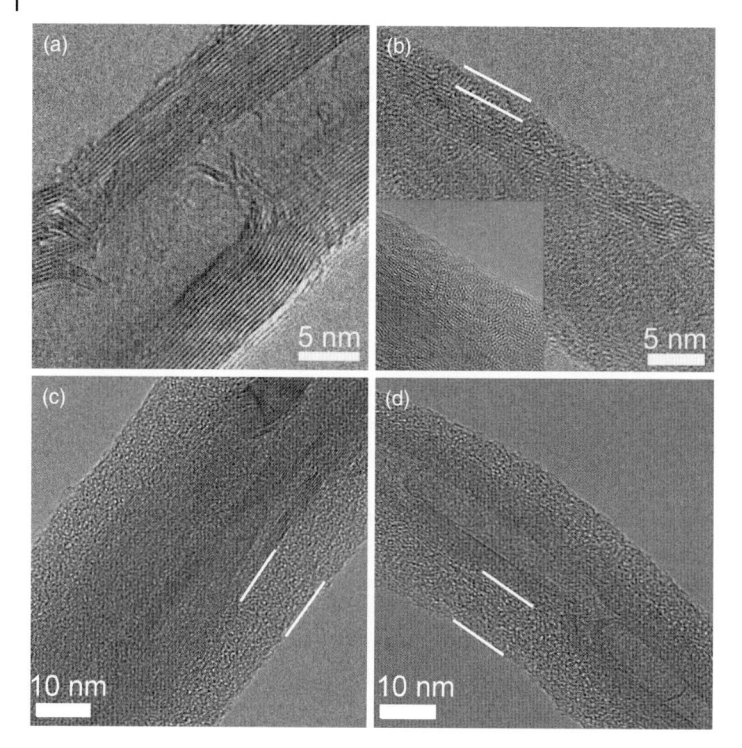

Figure 9.13 HRTEM images of pristine MWCNT (a), PNIPAAm-grafted MWCNT with polymer contents of 51 wt.% (b), 68 wt.% (c), and 84 wt.% (d). Inset of image (b): local amplification of image b. The marked polymer shells in images (b–d) have thicknesses of 3, 11, and 14 nm, respectively. Reprinted with permission from [46]. © American Chemical Society.

Figure 9.15a shows the AFM images of MWCNT-g-PNIPAAm at 20 °C and 50 °C, revealing that individually dispersed nanorods were evenly separated from one another at 20 °C, whereas polygonal aggregates with a magnitude of one micrometer were observed at 50 °C. The thermosensitive mechanism is schematically illustrated in Figure 9.15b. At temperatures below the LCST, the predominantly intermolecular hydrogen bonding between the grafted PNIPAAm chains and water molecules causes a loosely coiled conformation of PNIPAAm chains and a high surface free energy. This makes the whole nonocylinder hydrophilic and highly soluble or dispersible in water. As the temperature is increased to the LCST and above, the hydrogen bonds between the grafted PNIPAAm chains and water molecules is mostly broken and replaced by that between C=O and N–H groups of the PNIPAAm chains, leading to a compact, collapsed chain conformation with a low surface free energy, which makes the PNIPAAm chains hydrophobic. Then the hydrophobic association among inter- and intrananocylinders drives the individual cylinders to assemble and become separated from water.

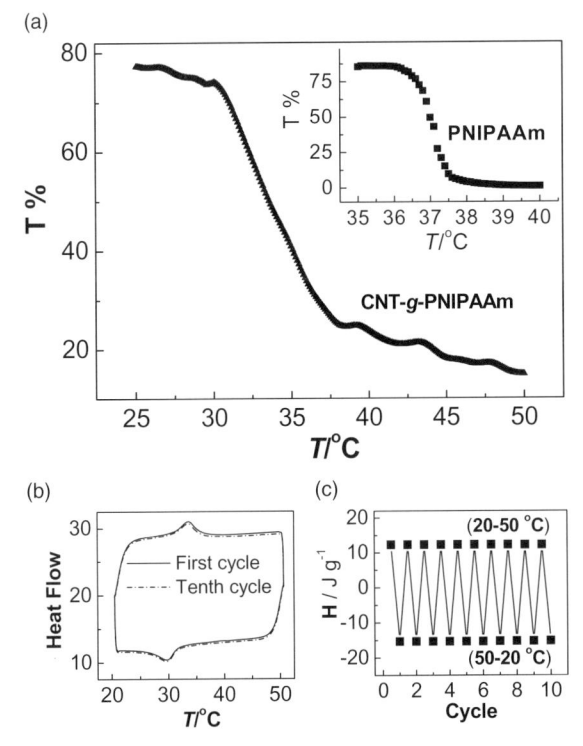

Figure 9.14 (a) Transmittance of PNIPAAm-grafted MWCNTs (84 wt.% of polymer) or neat PNIPAAm ($M_n = 15\,200$, PDI = 1.45) (inset) in water as a function of temperature with a concentration of 1 mg ml^{-1}. (b) DSC curves of the first and the tenth cycles for the PNIPAAm-grafted MWCNTs with a sample concentration of ca. 1 mg ml^{-1}. (c) Enthalpy of transition for the PNIPAAm-grafted MWCNTs as a function of cycle in DSC measurements. Reprinted with permission from [46]. © American Chemical Society.

Alternatively, PNIPAAm can be grafted on aligned MWCNTs, offering a smart surface that displayed temperature-sensitive suprahydrophilic and suprahydrophobic transition behavior [53].

Likewise, other carbon surfaces such as carbon fibres [54], carbon spheres [55], carbon black [56–61], multilayered carbon nanoonions (CNOs) [62], and graphene nanosheets [63–65] could be grafted with polymers via the *in situ* ATRP technique.

9.2.2.4 Controllability of CNT-Initiated ATRP

To achieve a controlled functionalization is the basic purpose for the surface-initiated controlled/living polymerization. For CNT cases, two techniques have been employed to control the grafted polymer amount through the manners of feed ratio and reaction kinetics (time).

Through the feed ratio of monomer to CNT-based macroinitiator (R_{feed}), the grafted polymer fraction by weight (f_{wt}%) could be well controlled, as demonstrated

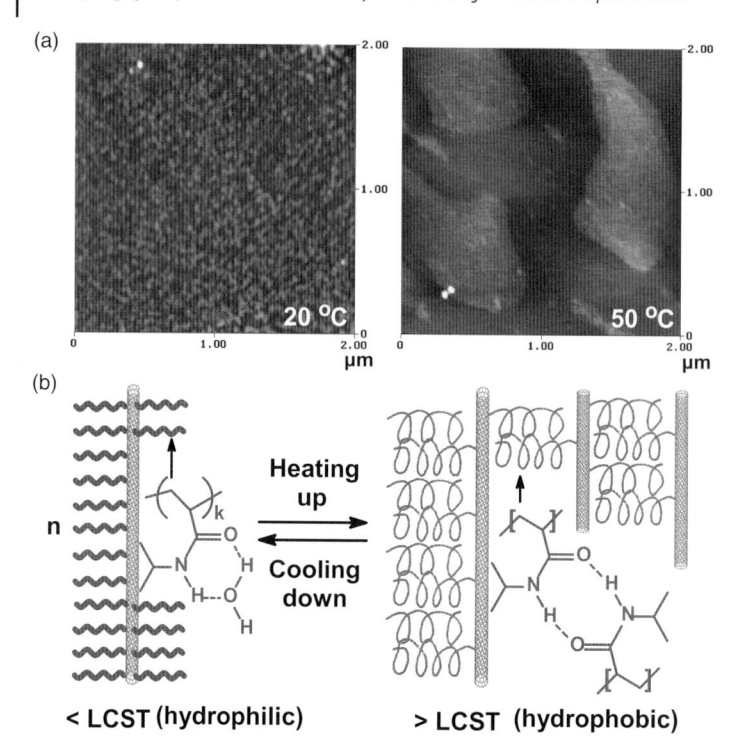

Figure 9.15 (a) AFM images of MWCNT-g-PNIPAAm at 20 °C (left) and 50 °C (right). (b) Schematic mechanism for the reversible self-assembly of the PNIPAAm-grafted MWCNT nanowires. The AFM-tested samples were prepared as follows: The sample of MWCNT-g-PNIPAAm dissolving in water (ca. 0.5 mg ml^{-1}) at 20 °C was dropped onto a mica plate, getting one sample (20 °C); the solution was then heated to 50 °C and dropped onto a mica plate, obtaining another sample (50 °C). Reprinted with permission from [46]. © American Chemical Society.

by Gao and other researchers [25–27]. In the first example of polymer-grafting on MWCNTs by ATRP in 2004, Gao and coworkers had demonstrated that the grafted polymer amount increased with increasing R_{feed} [25]. This conclusion had been corroborated with other examples mentioned above. Later, Gao et al. investigated the controllability in details in the case of poly(GMA-OH)-grafted MWCNTs with or without free initiator (or sacrificial initiator) [27]. In the absence of free initiator, poly(GMA-OH) with f_{wt}% 50–90% was grown on MWCNTs when R_{feed} was varied from 1.0/1 to 8.8/1; in the prescence of free initiator of ethyl 2-bromoisobutyrate, the f_{wt}% was increased from 46% to 88% when R_{feed} of monomer/CNT-Br-1/ethyl 2-bromoisobutyrate was adjusted from 15.9/1/3.8 to 51.5/1/0.86. The free polymer initiated by the free initiator was also collected to detect the molecular parameters by GPC measurements. The comparative experiments showed that (1) the presence of free initiator has no significant effect on the nanotube-surface initiating polymerizations, (2) both the polymer content grafted on the nanotubes and M_n of free polymer (3000–17 000) can be efficiently controlled by adjusting the feed ratio of monomer to co-initiators, and (3) the PDI of the free polymer increased when either the feed ratio or the molecular weight was increased. Figure 9.16 shows the TGA weight loss curves of poly(GMA-OH)-grafted MWCNTs without (Figure 9.16a) and with free initiator (Figure 9.16b), and the relationship between $f_{wt}\%$ and the molar feed ratio (R_{mole}) (Figure 9.16c).

Figure 9.17 shows the typical SEM images of poly(GMA-OH)-functionalized MWCNTs with different polymer contents. For the samples with relatively low polymer fraction (<50 wt.%), rodlike structures can be clearly observed. With increasing the polymer content, the nanotube structure gets more and more blurred, and only the continuous bulk polymer phase can be observed for the samples of solid powder when the polymer fraction is very high (>70 wt.%) [27]. Around 12–14 nm thickness of polymer layer coated on the tube convex surface can be observed under TEM for the sample with 88 wt.% of polymer (Figure 9.17h).

With the case of MAIG, Gao et al. showed that the grafted polymer amount can also be controlled through the polymerization kinetics with or without sacrificial initiator [43]. In both manners a linear dependence of molecular weight on conversion was obtained, and the polymer amounts grafted on MWCNTs could be well controlled in a wide range by the reaction time and monomer conversion, as shown in Figure 9.18. It is noteworthy that coupling was found in the GPC curves of free polymer when the conversion of monomer reached ca. 45–50%, whereas no coupling occurred despite of very high conversion of this monomer (>80%) for its ATRP in the absence of CNTs.

In the cases of SWCNTs, Ford et al. first investigated the kinetics for ATRP of n-BMA initiated by the SWCNT-Br (CNT-Br-3) by the addition of MBP as a coinitiator [29]. Linear kinetic plots of $\ln[M_0/M]$ (herein M_0 and M represent initial and measured concentrations of monomer, respectively) versus reaction time and M_n versus monomer conversion were obtained, and the PDI of the free polymer is low ($M_w/M_n < 1.6$). Moreover, the grafted polymer amount also increased linearly with M_n of the free polymer, indicating the growth of polymer from SWCNTs is also well controlled.

The controlled functionalization of CNTs with polymer paves the way for the fabrication of CNT-based devices and high-performance composites with tailor-made structures and properties.

9.2.3
Linear Block-Copolymer-Grafted CNTs

Due to the "living" character of ATRP, the terminal groups of polymer-grafted CNTs could initiate ATRP of another monomer, grafting block copolymer on CNTs. Gao and coworkers first prepared PHEMA-block-PMMA-grafted MWCNTs by sequential ATRP of MMA and HEMA in DMF at 60 °C with catalyst system of CuBr/PMDETA [25]. Figure 9.19 shows a typical TEM image of the product, indicating the core–shell structure of polymer-grafted CNTs as well.

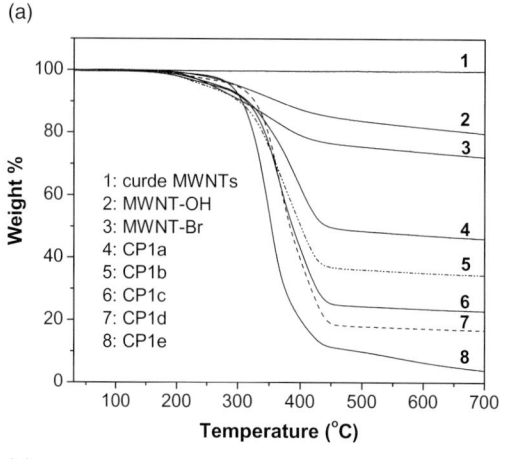

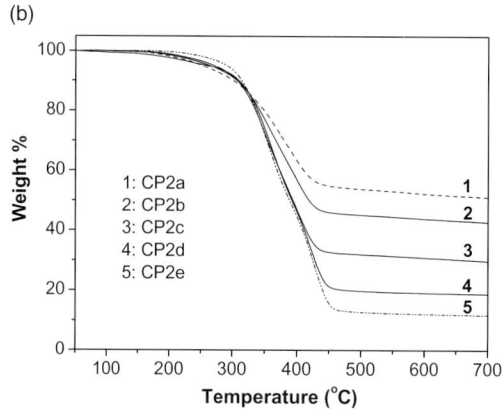

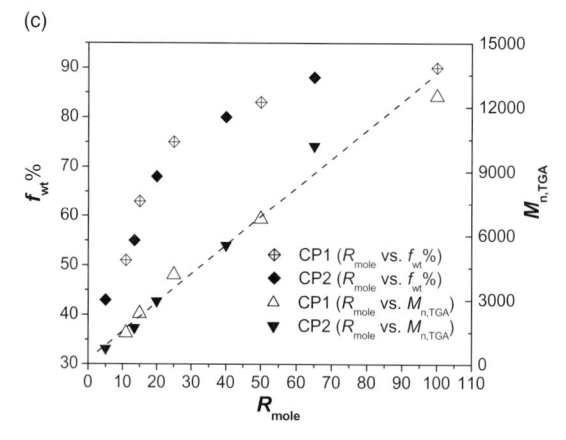

Figure 9.16 Thermogravimetric analysis (TGA) of weight loss curves under nitrogen of pristine MWCNTs, hydroxyl-functionalized MWCNTs (MWNT-OH), MWCNT-based macroinitiator (MWNT-Br), and MWCNT-g-poly(GMA-OH) without free initiator (a, CP1 series) and with free initiator (b, CP2 series). (c) The weight fraction of grafted poly(GMA-OH) calculated from the TGA data (f_{wt}%) and the average molecular weight of the poly(GMA-OH) calculated from TGA ($M_{n,TGA}$) as a function of molar feed ratio (R_{mole}) for the CNT-initiated polymerizations in the presence and absence of free initiator [27]. © American Chemical Society.

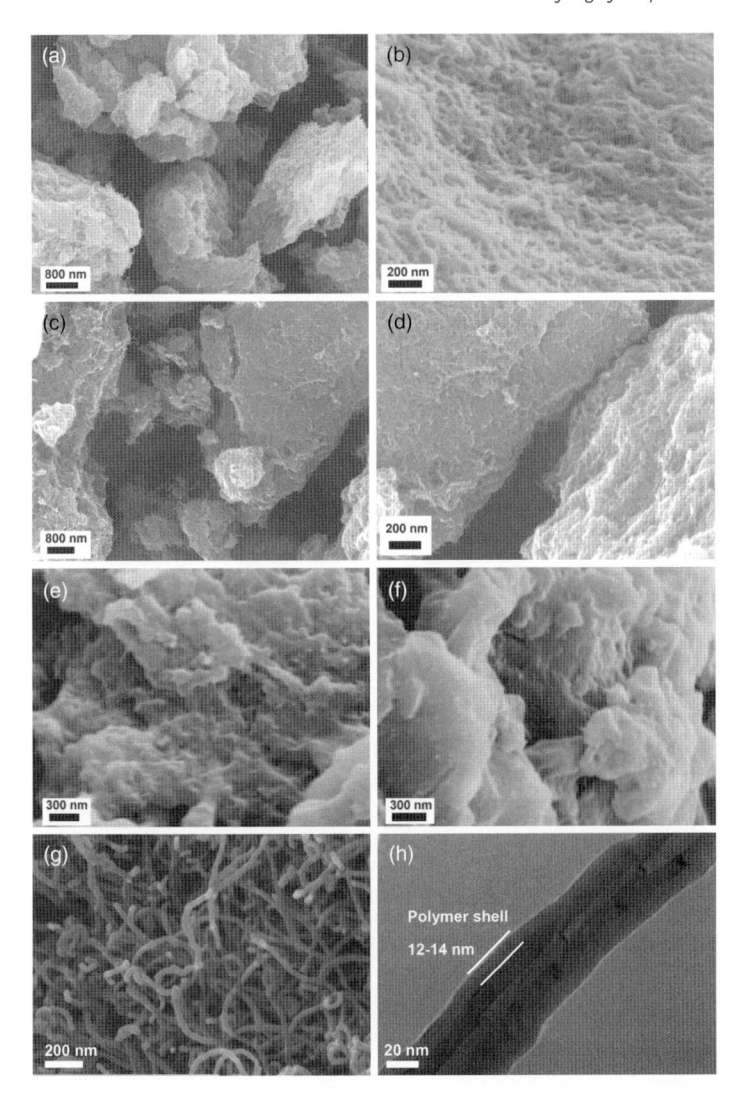

Figure 9.17 Representative SEM images for samples of MWCNT-g-poly(GMA-OH) with 46 wt.% (a, b), 68 wt.% (c, d), 80 wt.% (e), and 88 wt.% (f) of polymer, using CNT-Br-1 and EBiB as coinitiators. As a comparison, the SEM images of pristine MWCNTs are also shown (g). The samples for the SEM testing were prepared by directly loading the solid powder of samples on carbon film substrates. Thus, the tubelike structure cannot be observed clearly in the images of e and f because of the covering of the high content of polymer. (h) TEM image of the MWCNT-g-poly(GMA-OH) with 88 wt.% of polymer, showing the core–shell structure. Reprinted with permission from [27]. © American Chemical Society.

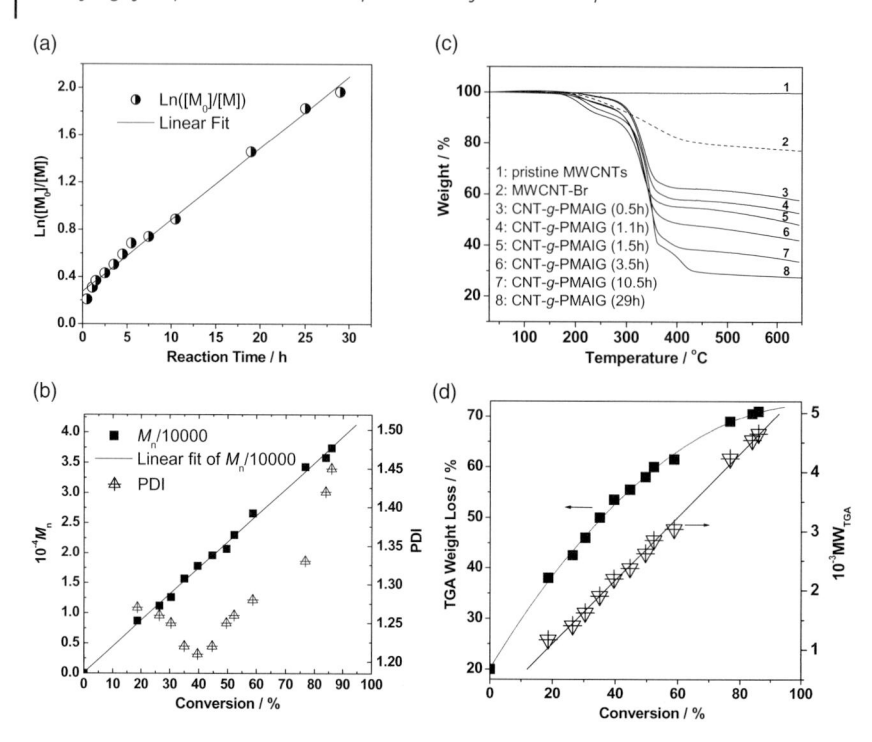

Figure 9.18 The first-order time-conversion plot (a), and apparent number-average molecular weight (M_n) and PDI of the free polymer as a function of monomer conversion (b). TGA weight loss curves for the pristine MWCNTs, MWCNT-Br, and linear glycopolymer (PMAIG)-grafted MWCNTs obtained at different reaction time (c), and the content and average molecular weight of the polymer grafted on MWCNTs, calculated from corresponding TGA data, as a function of monomer conversion (d). Reprinted with permission from [43]. © American Chemical Society.

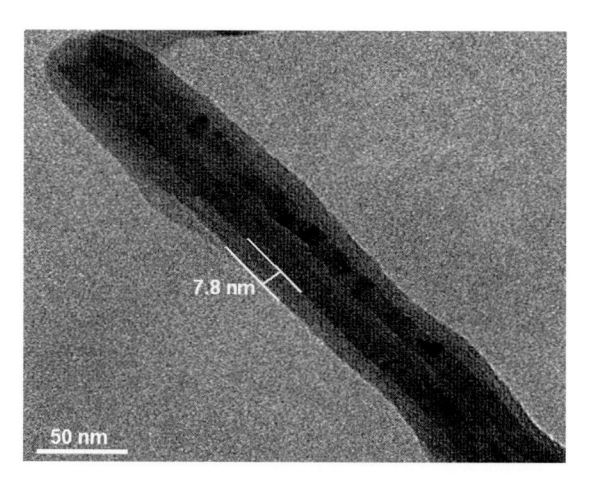

Figure 9.19 A representative TEM image of PHEMA-*b*-PMMA-grafted MWCNT. Reprinted with permission from [25]. © 2004, American Chemical Society. © American Chemical Society.

Scheme 9.4 Synthesis of PS-block-PAA-grafted MWCNTs by sequential ATRP of styrene and tBA [45].

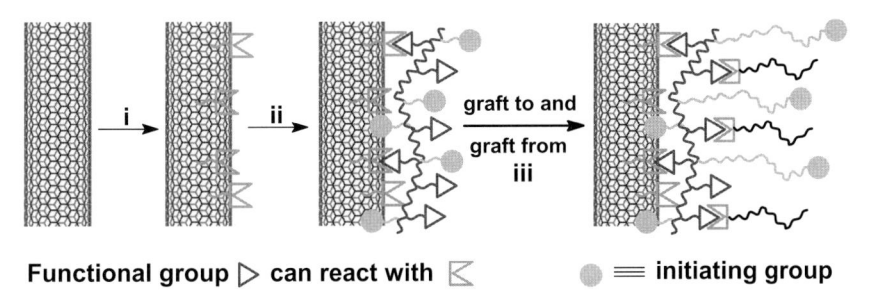

Figure 9.20 Gemini-grafting strategy to functionalize CNTs by a combination of "grafting to" and "grafting from" approaches. (i) premodification of pristine CNTs to introduce functional groups on CNTs, (ii) linkage of polymer with two types of functional groups on CNTs, (iii) grafting polymer on CNTs via either one-pot orthogonal multigrafting or sequential multigrafting [66].

Similarly, amphiphilic block copolymer, PS-block-PAA, was successfully grafted from MWCNTs by sequential ATRP of styrene and tBA followed by hydrolysis of PtBA block, as shown in Scheme 9.4 [45]. The resulting product could self-assemble into a Janus film at the interface of chloroform and water.

Recently, Gao and coworkers presented a novel "Gemini grafting" strategy to graft amphiphilic polymer brushes on CNTs by a combination of ATRP grafting-from and click-grafting to approaches [66]. As shown in Figure 9.20, a multifunctional polymer chain with both bromine and azido groups was firstly attached to CNT surfaces by the azide-alkyne click chemistry, giving rise to multifunctional CNTs. Subseqnent ATRP and click chemistry reaction allowed the hydrophobic poly(n-butyl methacrylate) (PnBMA) or PS and hydrophilic poly(ethylene oxide) (PEG) grafted on CNTs, respectively. Both MWCNTs and SWCNTs were employed in the experiments. The resulting amphiphilic brushes-grafted CNTs were

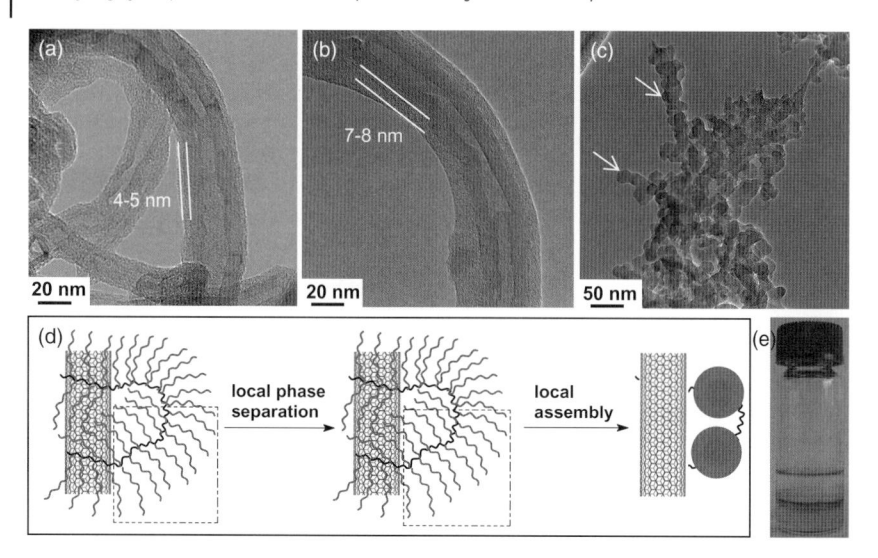

Figure 9.21 TEM images of PnBMA-grafted MWCNTs with azido groups (MWCNT-Az-PnBMA) (a), MWCNTs grafted with both PnBMA and PEG brushes (MWNT-PnBMA-PEG) (b, c). (d) Cartoon for the local phase separation and assembly of amphiphilic polymer brushes into Janus polymer structures on CNTs as shown in image c (marked by arrows). (e) Photograph of MWCNT-PnBMA-PEG dispersed in a mixed solvent of water (upper layer) and chloroform (bottom layer). Reprinted with permission from [66]. © American Chemical Society.

characterized by SEM, TEM (Figure 9.21a–c), TGA, Raman, and FTIR spectrometers. In addition, such compounds could self-assemble into micelles-like particles attached on CNTs (Figure 9.21c and d) and Janus film at the oil/water interface as shown in Figure 9.21e. This versatile Gemini-grafting strategy could be extended to other substrates to fabricate amphiphilic brushes with desired structures and properties.

Besides, Liu et al. prepared MWCNT-based macroinitiator through the addition reaction between ATRP initiator, 1-bromoethylbenzene or macroinitiator, bromine-terminated PS, and CNTs with BrCu/PMDETA as the catalyst system at 80 °C for 24 h [67]. Further ATRP of styrene or NIPAAm on the functionalized CNTs afforded PS-grafted CNTs or V-shaped amphiphilic polymer PS-PNIPAAm-grafted CNTs, which showed Janus behavior at the interface of water and chloroform at 20 °C and moved into chloroform phase at 50 °C.

9.2.4
Hyperbranched Polymer-Grafted CNTs by a Combination of ATRP and SCVP

Besides linear polymers, hyperbranched polymers (HPs) can also be grafted on CNTs by either "grafting to" or "grafting from" approach. HPs are highly branched macromolecules with three-dimensional dendritic architecture [68–71]. Because of their unique characters such as high solubility, low viscosity, and abundant functional groups as well as their cost-effective availability over dendrimers, HPs have received increasing attention during the past two decades [72–80]. Based on the multifunctional groups, HP-grafted nanomaterials could act as highly reactive platforms and highly robust templates for the fabrication of complex functional nanodevices, magnetic catalysts, dye-adsorbents, and drug carriers [81–83]. HPs have been covalently grafted on MWCNTs by ring-opening polymerization [84, 85], polycondensation of AB_2 monomers or monomer pairs [86–91], or self-condensing vinyl polymerization (SCVP) [43, 49].

The SCVP approach to HPs was first reported by Fréchet and coworkers through polymerization of AB*-type monomer [92]. The AB* monomer generally contains a vinyl group and a functional group that can initiate the polymerization of the vinyl moiety. Thus, AB* is also coined as inimer (inimer = *ini*tiator + mono*mer*) [93, 94]. If the AB* inimer possesses a halogen atom that can initiate ATRP of AB* itself, HPs can be prepared by a combination of SCVP and ATRP. Likewise, HP-grafted CNTs could be prepared in the prescence of CNT-based macroinitiator.

Gao *et al.* obtained HP-grafted MWCNTs with inimer of 2-(2-bromoisobutyryloxy) ethyl methacrylate (BIEM) via SCVP-ATRP process using CuBr/bis (triphenylphosphine)nickel(II) bromide as catalyst system and CNT-Br-1 as macroinitiator in ethyl acetate at 100 °C for 4.5 h [43]. The f_{wt}% of HPs approached 38%. This process was then extended to self-condensing vinyl copolymerization (SCVCP) of BIEM and MAIG with different monomer ratios, affording polymer-grafted CNTs with degree of branching (DB) from 0.49 to 0.21. After deprotection of MAIG, hyperbranched glycopolymer-functionalized CNTs was obtained. Scheme 9.5 shows the reaction process. Kinetic studies were carried out for SCVCP with different ratios of MAIG to BIEM (γ). In the cases of lower γ (0.5 and 1), the apparent molecular weight and PDI increased with conversion; in the cases of higher γ (2.5 and 5), the molecular weight also increased with conversion at the beginning, and leveled off after certain conversion (ca. 90%). Figure 9.22 shows the results with γ 0.5 for the free polymer separated from the reaction system, indicating the typical SCVP GPC elution curves with broad peaks accompanying with oligomers.

Alternatively, another AB* inimer, 2-((bromobutyryl)oxy)ethyl acrylate (BBEA), could be polymerized via SCVP in the presence of CNT-Br-4 in toluene at 100 °C with catalysts system of CuBr/PMDETA [49]. It was found that the f_{wt}% increased with the reaction time, and achieved up to 80% at 24 h. The SCVCP of BBEA and *t*BA was successfully conducted. The resulting HP-coated MWCNTs showed good dispersibility in organic solvents such as THF and chloroform.

9.3
Functionalization of CNTs by Other CRPs

Similar to the ATRP methodology, polymers could be grafted on CNTs by other CRPs such as NMRP and RAFT polymerization if the corresponding initiating

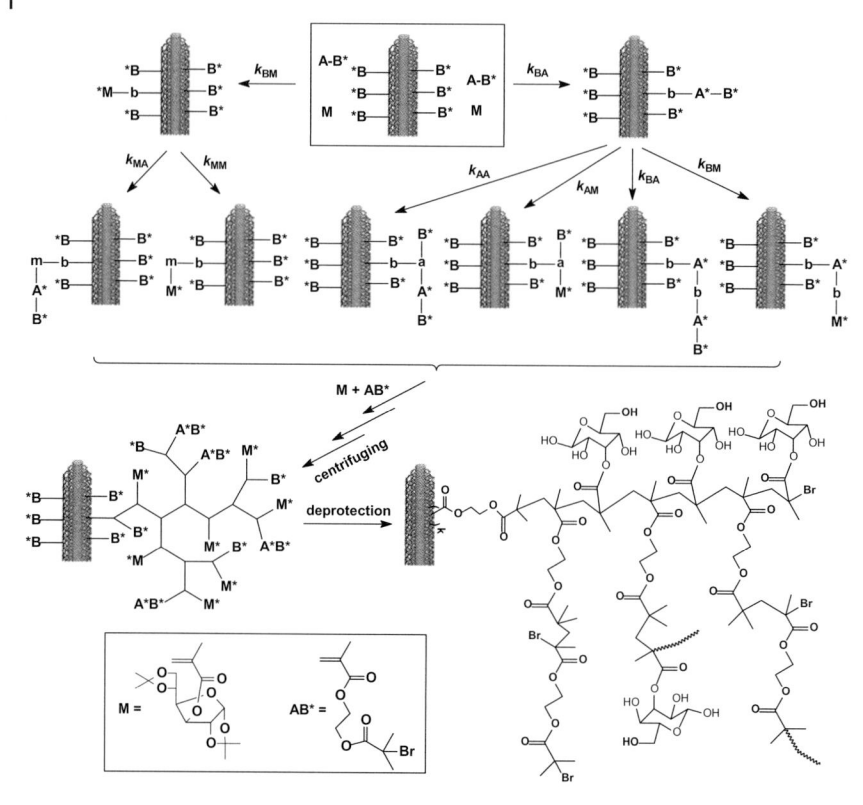

Scheme 9.5 Synthetic strategy for grafting hyperbranched glycopolymer from surfaces of MWCNTs by SCVCP of inimer (AB*) and monomer (M) via ATRP. Reprinted with permission from [43]. © American Chemical Society.

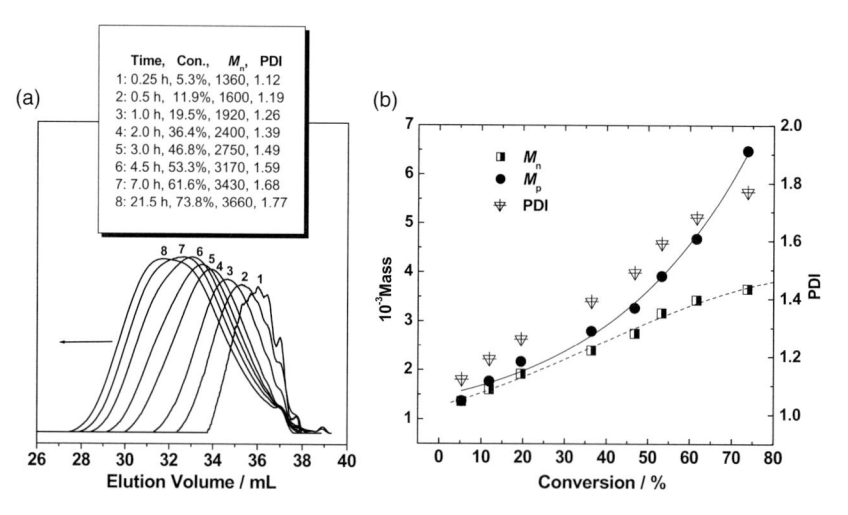

Figure 9.22 GPC curves of the free HP collected from the SCVCP system of MAIG and BIEM with $\gamma = 0.5$ at different conversion or reaction time (a), and corresponding molecular weight (M_n = numer-average molecular weight, M_p = peak molecular weight) and PDI of the polymer as a function of conversion (b) [43].

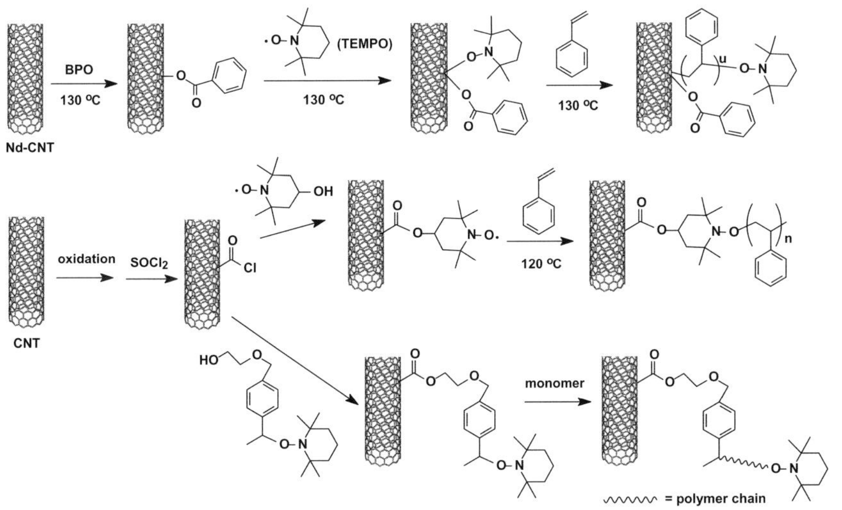

Scheme 9.6 Preparation of 2,2,6,6-tetramethylpiperidinyl-1-oxyl (TEMPO)-functionalized CNTs (macroinitiator of NMRP) and polymer-grafted CNTs by NMRP [5].

group of 2,2,6,6-tetramethylpiperidinyl-1-oxyl (TEMPO) or chain-transfer agent (CTA) was introduced on CNTs priorly. Schemes 9.6 and 9.7 show the preparation of CNT-based macroinitiator for NMRP and macroCTA for RAFT polymerization respectively, as well as the polymer grafting process [5]. The most reported cases of NMRP on CNTs are aromatic monomers such as styrene, vinyl pyridine, and SSNa, and the grafted polymer amount is relatively low (~30–60 wt.%) [95–98]. As a comparison, the main three types of monomers including (meth)acrylates, styrenes, and acrylamides have been successfully polymerized via the RAFT process in the presence of CNT-based macroCTA, and the grafted polymer amount could be very high (up to 87 wt.%) and well controlled by polymerization time [99–108]. Moreover, CNT-initiated block copolymerizations were also conducted by subsequent addition of different monomers through RAFT polymerization, indicating the versatility, generality, and flexibility of RAFT polymerization.

9.4
Grafting of Polymers on Other Nanotubes by ATRP

Besides CNTs, boron nitride nanotubes (BNNTs) are another kind of nanotubes with unique attributes. BNNTs exhibited very similar mechanical properties and thermal conductivity to CNTs, and they were demonstrated with notably higher chemical stability and resistance to oxidation [109]. Significantly, BNNTs are transparent to visible light due to their constant wide band gap (around 5.2–5.8 eV). According to the theoretical predictions, BNNTs may show very high thermal conductivities, while being electrically insulating, promising BNNTs could be used

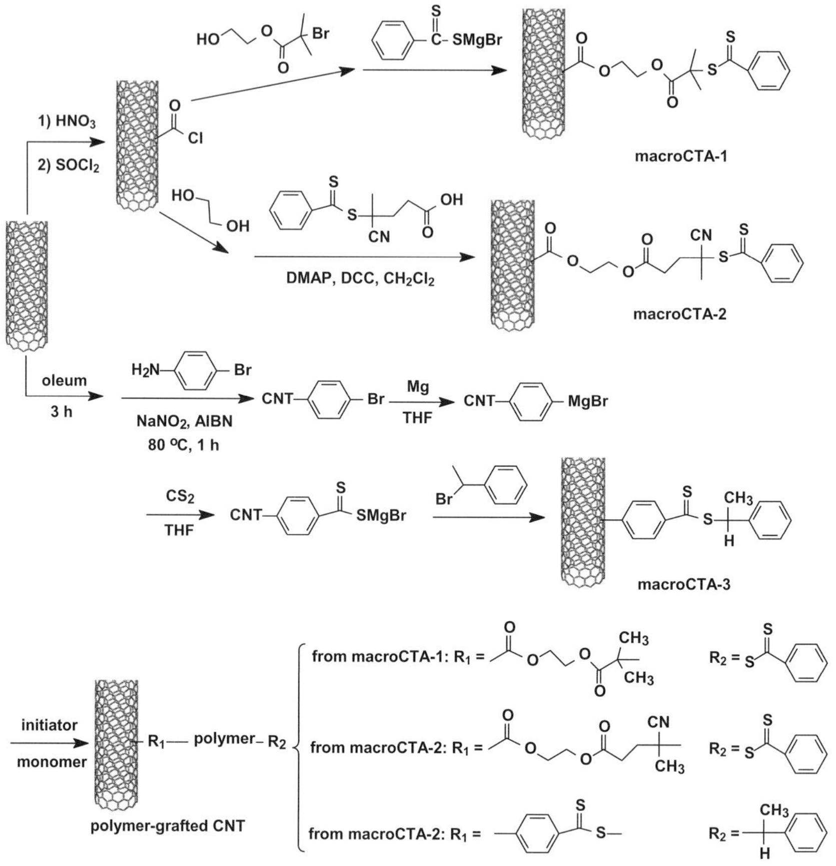

Scheme 9.7 Preparation of thiocarbonylthio group-functionalized CNTs (macroCTA) and polymer-grafted CNTs by RAFT polymerization [5].

as nanofillers for insulating polymeric composites with good thermal conductivity and transparence [109].

Based on the amino groups of BNNTs, Zhi et al. introduced ATRP-initiating groups on BNNTs by reaction with chloroacetyl chloride at 150 °C for 120 h, and then grafted PS and PMMA by ATRP technique with catalyst system of CuCl/4,4′-dinonyl-2,2′-dipyridyl (DIDIPY) in xylene at 130 °C for 48 h [110], as shown in Scheme 9.8. It was found that the grafted polymer thickness on BNNTs was not uniform but ranged from 0.8 to 5 nm, which was likely caused by the defect-based functionalization process. Such polymer-coated BNNTs could be used to prepare carbon-covered BNNTs by high-temperature calcining (960 °C) in Ar gas.

Halloysite (formula: $Al_2Si_2O_5(OH)_4 \cdot 2H_2O$, 1:1 layer aluminosilicate), a kind of natural clay material with hollow tubule structures, had also been modified with polymers by ATRP technique. Yang and coworkers grafted cross-linked PS on halloysite surfaces through *in situ* ATRP of styrene and divinylbenzene with the

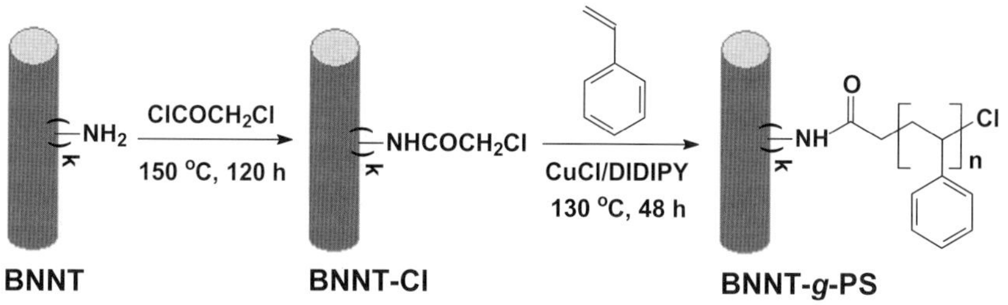

Scheme 9.8 Grafting polystyrene on BNNTs by ATRP technique [110].

catalyst system of CuBr/PMDETA in toluene at 110 °C, and then obtained polymeric nanotubes and nanowires after removal of halloysite template by HF/HCl treatment [111]. The halloysite-based macroinitiator was prepared by modification of H_2O_2-treated halloysite with 3-aminopropyltrimethoxysilane (APS) at 120 °C in toluene for 12 h, followed by reaction with 2-bromoisobutyryl bromide in dry dichloromethane containing dry triethylamine at 25 °C for 10 h. TGA measurements showed that the grafted polymer content increased with the reaction time, and GPC measurements for the PS cleaved from the halloysite also indicated the increasing tendency of molecular weights. Liu and coworkers grafted HPs on 2-bromoisobutyric acid-modified halloysite nanotubes by SCVP-ATRP of inimer, 2-((bromoacetyl)oxy)ethyl acrylate, with the catalysts system of CuBr/bpy at 90 °C for 12 h [112]. The grafted HP amount was around 30 wt.%.

In addition, one-dimensional attapulgite [113], zinc oxide nanowires [114], and silicon nanowires [115] had been used as substrates for polymer grafting by ATRP. After dissolving the substrates, polymer nanotubes could be obtained [115].

9.5
Conclusions

ATRP had been demonstrated as a powerful tool for the controlled functionalization of CNTs and other nanotubes. Through ATRP, various kinds of polymers have been successfully grafted on CNTs, covering from hydrophilic to hydrophobic, from water-soluble to oil-soluble, from acidic to basic, from temperature-responsive to pH-sensitive, and from polyelectrolytes to sugar-contained biocompatible polymers. After even grafting polymer on nanotubes, core–shell nanostructures could be observed by TEM as a characteristic of high-efficiency grafting with polymer layer as shell and CNT as cylindrical core. Notably, all of the CNT surfaces including tube ends and main body could be coated with polymer chains for CNTs with defects, especially for those made by chemical-vapor deposition method. Up to date, the grafted polymer amount could be controlled by the means of feed ratio and reaction time (kinetics). However, to achieve high-performance composites with desired properties, the control over polymeric arm density is also crucial. Regrettably, such reports were rarely published. The polymer-grafted CNTs could be used as templates and building blocks to load other compounds such as nanocrystals and semiconductors (QDs) for the purposes of catalysts, magnets, photoluminescence, sensors, supramolecular self-assembly, and so forth, promising the applications of such CNTs.

Acknowledgments

Financial supports from the National Natural Science Foundation of China (No. 50473010, No. 50773038, and No. 20974093), National Basic Research Program of China (973 Program) (No. 2007CB936000), Qianjiang Talent Foundation of Zhejiang Province (No. 2010R10021), the Fundamental Research Funds for the Central Universities (No. 2009QNA4040), and the Foundation for the Author of National Excellent Doctoral Dissertation of China (No. 200527) are kindly acknowledged. I thank the coauthors very much for their contributions to our publications related to this topic.

References

1 Iijima, S. (1991) *Nature*, **354**, 56.
2 Iijima, S. and Ichihashi, T. (1993) *Nature*, **363**, 603.
3 Sahoo, N.G., Rana, S., Cho, J.W., Li, L., and Chan, S.H. (2010) *Prog. Polym. Sci.*, **35**, 837.
4 Spitalsky, Z., Tasis, D., Papagelis, K., and Galiotis, C. (2010) *Prog. Polym. Sci.*, **35**, 357.
5 Gao, C. (2010) *Encyclopedia of Nanoscience and Nanotechnology*, vol. X (ed. H.S. Nalwa) American Scientific Publishers, Stevenson Ranch, CA, pp. 1–51.
6 Tasis, D., Tagmatarchis, N., Bianco, A., and Prato, M. (2006) *Chem. Rev.*, **106**, 1105.
7 Barbey, R., Lavanant, L., Paripovic, D., Schuewer, N., Sugnaux, C., Tugulu, S., and Klok, H.A. (2009) *Chem. Rev.*, **109**, 5437.
8 Matyjaszewski, K. and Tsarevsky, N.V. (2009) *Nat. Chem.*, **1**, 276.
9 Pintauer, T. and Matyjaszewski, K. (2008) *Chem. Soc. Rev.*, **37**, 1087.
10 Braunecker, W.A. and Matyjaszewski, K. (2007) *Prog. Polym. Sci.*, **32**, 93.
11 Wang, J.S. and Matyjaszewski, K. (1995) *Macromolecules*, **28**, 7901.
12 Moad, G., Rizzardo, E., and Thang, S.H. (2009) *Aust. J. Chem.*, **62**, 1402.
13 Stenzel, M.H. (2009) *Macromol. Rapid Commun.*, **30**, 1603.
14 Moad, G., Rizzardo, E., and Thang, S.H. (2006) *Aust. J. Chem.*, **59**, 669.
15 Moad, G., Rizzardo, E., and Thang, S.H. (2005) *Aust. J. Chem.*, **58**, 379.
16 McCormack, C.L. and Lowe, A.B. (2004) *Acc. Chem. Res.*, **37**, 312.
17 Chiefari, J., Chong, Y.K., Ercole, F., Krstina, J., Jeffery, J., Le, T.P.T., Mayadunne, R.T.A., Meijs, G.F., Moad, C.L., Moad, G., Rizzardo, E., and Thang, S.H. (1998) *Macromolecules*, **31**, 5559.
18 Brinks, M.K. and Studer, A. (2009) *Macromol. Rapid Commun.*, **30**, 1043.
19 Ghannam, L., Parvole, J., Laruelle, G., Francois, J., and Billon, L. (2006) *Polym. Int.*, **25**, 1199.
20 Hawker, C.J., Bosman, A.W., and Harth, E. (2001) *Chem. Rev.*, **101**, 3661.
21 Tsarevsky, N.V. and Matyjaszewski, K. (2007) *Chem. Rev.*, **107**, 2270.
22 Matyjaszewski, K. and Xia, J.H. (2001) *Chem. Rev.*, **101**, 2921.
23 Ouchi, M., Terashima, T., and Sawamoto, M. (2009) *Chem. Rev.*, **109**, 4963.

24 Kamigaito, M., Ando, T., and Sawamoto, M. (2001) *Chem. Rev.*, **101**, 3689.
25 Kong, H., Gao, C., and Yan, D. (2004) *J. Am. Chem. Soc.*, **126**, 412.
26 Kong, H., Gao, C., and Yan, D. (2004) *Macromolecules*, **37**, 4022.
27 Gao, C., Vo, C.D., Jin, Y.Z., Li, W.W., and Armes, S.P. (2005) *Macromolecules*, **38**, 8634.
28 Narain, R., Housni, A., and Lane, L. (2006) *J. Polym. Sci., Part A: Polym. Chem.*, **44**, 6558.
29 Qin, S., Qin, D., Ford, W.T., Resasco, D.E., and Herrera, J.E. (2004) *J. Am. Chem. Soc.*, **126**, 170.
30 Baskaran, D., Mays, J.W., and Bratcher, M.S. (2004) *Angew. Chem. Int. Ed.*, **43**, 2138.
31 Yao, Z., Braidy, N., Botton, G.A., and Adronov, A. (2003) *J. Am. Chem. Soc.*, **125**, 16015.
32 Georgakilas, V., Kordatos, K., Prato, M., Guldi, D.M., Holzinger, M., and Hirsch, A. (2002) *J. Am. Chem. Soc.*, **124**, 760.
33 Georgakilas, V., Tagmatarchis, N., Pantarotto, D., Bianco, A., Briand, J.P., and Prato, M. (2002) *Chem. Commun.*, 3050.
34 Matrab, T., Chancolon, J., L'hermite, M.M., Rouzaud, J.-N., Deniau, G., Boudou, J.-P., Chehimi, M.M., and Delamar, M. (2006) *Colloids Surf. A*, **287**, 217.
35 Choi, J.H., Oh, S.B., Chang, J., Kim, I., Ha, C.-S., Kim, B.G., Han, J.H., Joo, S.-W., Kim, G.-H., and Paik, H.-J. (2005) *Polym. Bull.*, **55**, 173.
36 Fragneaud, B., Masenelli-Varlot, K., Gonzalez-Montiel, A., Terrones, M., and Cavaillé, J.-Y. (2006) *Chem. Phys. Lett.*, **419**, 567.
37 Liu, M., Yang, Y., Zhu, T., and Liu, Z. (2007) *J. Phys. Chem. C*, **111**, 2379.
38 Gao, C., He, H., Zhou, L., Zheng, X., and Zhang, Y. (2009) *Chem. Mater.*, **21**, 360.
39 Li, W., Gao, C., Qian, H., Ren, J., and Yan, D. (2006) *J. Mater. Chem.*, **16**, 1852.
40 Li, W., Kong, H., Gao, C., and Yan, D. (2005) *Chin. Sci. Bull.*, **50**, 2276.
41 Gao, C., Li, W., Morimoto, H., Nagaoka, Y., and Maekawa, T. (2006) *J. Phys. Chem. B*, **110**, 7213.
42 Li, W.W. (2007) Master Thesis (supervisor: Chao Gao), Shanghai Jiao Tong University, China.
43 Gao, C., Muthukrishnan, S., Li, W., Yuan, J., Xu, Y., and Müller, A.H.E. (2007) *Macromolecules*, **40**, 1803.
44 Kong, H., Luo, P., Gao, C., and Yan, D. (2005) *Polymer*, **46**, 2472.
45 Kong, H., Gao, C., and Yan, D. (2004) *J. Mater. Chem.*, **14**, 1401.
46 Kong, H., Li, W., Gao, C., Yan, D., Jin, Y., Walton, D.R.M., and Kroto, H.W. (2004) *Macromolecules*, **37**, 6683.
47 Qin, S., Qin, D., Ford, W.T., Resasco, D.E., and Herrera, J.E. (2004) *Macromolecules*, **37**, 752.
48 Shanmugharaj, A.M., Bae, J.H., Nayak, R.R., and Ryu, S.H. (2007) *J. Polym. Sci. Part A: Polym. Chem.*, **45**, 460.
49 Hong, C.-Y., You, Y.-Z., Wu, D., Liu, Y., and Pan, C.-Y. (2005) *Macromolecules*, **38**, 2606.
50 Gao, C. (2006) *Macromol. Rapid Commun.*, **27**, 841.
51 Gao, C., Li, W., Jin, Y.Z., and Kong, H. (2006) *Nanotechnology*, **17**, 2882.
52 He, H.K. and Gao, C. (2010) *Molecules*, **15**, 4679.
53 Sun, T., Liu, H., Song, W., Wang, X., Jiang, L., Li, L., and Zhu, D. (2004) *Angew. Chem. Int. Ed.*, **43**, 4663.
54 Liu, P. and Su, Z.X. (2005) *Polym. Int.*, **54**, 1508.
55 Jin, Y.Z., Gao, C., Kroto, H.W., and Maekawa, T. (2005) *Macromol. Rapid Commun.*, **25**, 1133.
56 Liu, T.Q., Jia, S., Kowalewski, T., Matyjaszewski, K., Casado-Portilla, R., and Belmont, J. (2003) *Langmuir*, **19**, 6342.
57 Pyun, J., Kowalewski, T., and Matyjaszewski, K. (2003) *Macromol. Rapid Commun.*, **24**, 1043.
58 Liu, T.Q., Casado-Portilla, R., Belmont, J., and Matyjaszewski, K. (2005) *J. Polym. Sci. Part A: Polym. Chem.*, **43**, 4695.
59 Liu, T.Q., Jia, S.J., Kowalewski, T., Matyjaszewski, K., Casado-Portilla, R., and Belmont, J. (2006) *Macromolecules*, **39**, 548.

60 Yang, Q., Wang, L., Xiang, W.D., Zhou, J.F., and Tan, Q.H. (2007) *J. Polym. Sci. Part A: Polym. Chem.*, **45**, 3451.

61 Yang, Q., Wang, L., Huo, J., Ding, J.H., and Xiang, W.D. (2010) *J. Appl. Polym. Chem.*, **117**, 824.

62 Zhou, L., Gao, C., Zhu, D., Xu, W., Chen, F.F., Palkar, A., Echegoyen, L., and Kong, E.S.-W. (2009) *Chem. Eur. J.*, **15**, 1389.

63 Yang, Y.F., Wang, J., Zhang, J., Liu, J.C., Yang, X.L., and Zhao, H.Y. (2009) *Langmuir*, **25**, 11808.

64 Fang, M., Wang, K.G., Lu, H.B., Yang, Y.L., and Nutt, S. (2010) *J. Mater. Chem.*, **20**, 1982.

65 Lee, S.H., Dreyer, D.R., An, J.H., Velamakanni, A., Piner, R.D., Park, S., Zhu, Y.W., Kim, S.O., Bielawski, C.W., and Ruoff, R.S. (2010) *Macromol. Rapid Commun.*, **31**, 281.

66 Zhang, Y., He, H., and Gao, C. (2008) *Macromolecules*, **41**, 9581.

67. Liu, Y.-L. and Chen, W.-H. (2007) *Macromolecules*, **40**, 8881.

68 Flory, P.J. (1952) *J. Am. Chem. Soc.*, **74**, 2718.

69. Kim, Y.H. and Webster, O.W. (1990) *J. Am. Chem. Soc.*, **112**, 4592.

70 Kim, Y.H. and Webster, O.W. (1992) *Macromolecules*, **25**, 5561.

71 Kim, Y.H. (1998) *J. Polym. Sci. Part A: Polym. Chem.*, **36**, 1685.

72 Gao, C. and Yan, D. (2004) *Prog. Polym. Sci.*, **29**, 183.

73 Gao, C. (2009) Chapter 2: hyperbranched polymers and functional nanoscience, in *Novel Polymers and Nanoscience* (ed. M. Adeli), Transworld Research Network, India.

74 Wilms, D., Stiriba, S.-E., and Frey, H. (2010) *Acc. Chem. Res.*, **43**, 129.

75 Calderon, M., Quadir, M.A., Sharma, S.K., and Haag, R. (2010) *Adv. Mater.*, **22**, 190.

76 Voit, B.I. and Lederer, A. (2009) *Chem. Rev.*, **109**, 5924.

77 Voit, B. (2005) *J. Polym. Sci. Part A: Polym. Chem.*, **43**, 2679.

78 Mori, H. and Müller, A.H.E. (2003) *Top. Curr. Chem.*, **228**, 1.

79 Jikei, M. and Kakimoto, M. (2001) *Prog. Polym. Sci.*, **26**, 1233.

80 Inoue, K. (2000) *Prog. Polym. Sci.*, **25**, 453.

81 Zhou, L., Gao, C., and Xu, W. (2010) *Langmuir*, **26**, 11217.

82 Zhou, L., Gao, C., Hu, X., and Xu, W. (2010) *ACS Appl. Mater. Interfaces*, **2**, 1211.

83 Zhou, L., Gao, C., and Xu, W. (2010) *ACS Appl. Mater. Interfaces*, **2**, 1483.

84 Zhou, L., Gao, C., and Xu, W. (2009) *Macromol. Chem. Phys.*, **210**, 1011.

85 Xu, Y., Gao, C., Kong, H., Yan, D., Jin, Y., and Watts, P.C.P. (2004) *Macromolecules*, **37**, 8846.

86 Gao, C., Jin, Y.Z., Kong, H., Whitby, R.L.D., Acquah, S.F.A., Chen, G.Y., Qian, H., Hartschuh, A., Silva, S.R.P., Henley, S., Fearon, P., Kroto, H.W., and Walton, D.R.M. (2005) *J. Phys. Chem. B*, **109**, 11925.

87 Yuan, W., Jiang, G.H., Che, J.F., Qi, X.B., Xu, R., Chang, M.W., Chen, Y., Lim, S.Y., Dai, J., and Chan-Park, M.B. (2008) *J. Phys. Chem. C*, **112**, 18754.

88 Yang, Y.K., Xie, X.L., Yang, Z.F., Wang, X.T., Cui, W., Yang, J.Y., and Mai, Y.W. (2007) *Macromolecules*, **40**, 5858.

89 Choi, J.Y., Han, S.W., Huh, W.S., Tan, L.S., and Baek, J.B. (2007) *Polymer*, **48**, 4034.

90 Choi, J.Y., Oh, S.J., Lee, H.J., Wang, D.H., Tan, L.S., and Baek, J.B. (2007) *Macromolecules*, **40**, 4474.

91 Zheng, Y.P., Zhang, J.X., Yu, P.Y., Liu, L.L., and Gao, Y. (2009) *J. Compos. Mater.*, **43**, 2771.

92 Fréchet, J.M.J., Henmi, M., Gitsov, I., Aoshima, S., Leduc, M.R., and Grubbs, R.B. (1995) *Science*, **269**, 1080.

93 Müller, A.H.E., Yan, D.Y., and Wulkow, M. (1997) *Macromolecules*, **30**, 7015.

94 Yan, D.Y., Müller, A.H.E., and Matyjaszewski, K. (1997) *Macromolecules*, **30**, 7024.

95 Dehonor, M., Masenelli-Varlot, K., González-Montiel, A., Gauthier, C., Cavaillé, J.Y., Terrones, H., and Terrones, M. (2005) *Chem. Commun.*, 5349.

96 Zhao, X.D., Fan, X.H., Chen, X.F., Chai, C.P., and Zhou, Q.F. (2006) *J. Polym. Sci. Part A: Polym. Chem.*, **44**, 4656.

97. Zhao, X.D., Lin, W.R., Song, N.H., Chen, X.F., Fan, X.H., and Zhou, Q.F. (2006) *J. Mater. Chem.*, **16**, 4619.
98. Fan, D.Q., He, J.P., Tang, W., Xu, J.T., and Yang, Y.L. (2007) *Eur. Polym. J.*, **43**, 26.
99. Cui, J., Wang, W.P., You, Y.Z., Liu, C.H., and Wang, P.H. (2004) *Polymer*, **45**, 8717.
100. Hong, C.Y., You, Y.Z., and Pan, C.Y. (2005) *Chem. Mater.*, **17**, 2247.
101. Xu, G.Y., Wu, W.T., Wang, Y.S., Pang, W.M., Wang, P.H., Zhu, G.R., and Lu, F. (2006) *Nanotechnology*, **17**, 2458.
102. Hong, C.Y., You, Y.Z., and Pan, C.Y. (2006) *J. Polym. Sci., Part A: Polym. Chem.*, **44**, 2419.
103. You, Y.Z., Hong, C.Y., and Pan, C.Y. (2006) *Nanotechnology*, **17**, 2350.
104. Hong, C.Y., You, Y.Z., and Pan, C.Y. (2006) *Polymer*, **47**, 4300.
105. Xu, G.Y., Wu, W.T., Wang, Y.S., Pang, W.M., Zhu, Q.R., Wang, P.H., and You, Y.Z. (2006) *Polymer*, **47**, 5909.
106. Xu, G.Y., Wu, W.T., Wang, Y.S., Pang, W.M., Zhu, Q.R., and Wang, P.H. (2007) *Nanotechnology*, **18**, Art. No. 145606.
107. Pei, X.W., Hao, J.C., and Liu, W.M. (2007) *J. Phys. Chem. C*, **111**, 2947.
108. Wang, G.J., Huang, S.Z., Wang, Y., Liu, L., Qiu, J., and Li, Y. (2007) *Polymer*, **48**, 728.
109. Zhi, C.Y., Bando, Y., Tang, C.C., Huang, Q., and Golberg, D. (2008) *J. Mater. Chem.*, **18**, 3900.
110. Zhi, C., Bando, Y., Tang, C., Kuwahara, H., and Golberg, D. (2007) *J. Phys. Chem. C*, **111**, 1230.
111. Li, C., Liu, J., Qu, X., Guo, B., and Yang, Z. (2008) *J. Appl. Polym. Sci.*, **110**, 3638.
112. Mu, B., Zhao, M., and Liu, P. (2008) *J. Nanopart. Res.*, **10**, 831.
113. Liu, P. and Wang, T. (2007) *Ind. Eng. Chem. Res.*, **46**, 97.
114. Rupert, B.L., Mulvihill, M.J., and Arnold, J. (2006) *Chem. Mater.*, **18**, 5045.
115. Mulvihill, M.J., Rupert, B.L., He, R., Hochbaum, A., Arnold, J., and Yang, P.D. (2005) *J. Am. Chem. Soc.*, **127**, 16040.

10
Polymer Grafting onto Carbon Nanotubes via Cationic Ring-Opening Polymerization
Ye Liu and Decheng Wu

10.1
Introduction

Carbon nanotubes (CNTs) have unique structures including high aspect ratios, high surface areas, and stable nanosized tubes. These features render CNTs promising for various applications [1–6]. However, suitable functionalization of CNTs is a prerequisite for most of applications, especially for biotechnology such as preparation of good-delivery systems for drugs, genes, and proteins, and the functionalization of CNTs play a vital role in improving solution dispersity and biocompatibility of CNTs, facilitating drugs loading, and enhancing targeting capability [1, 3, 5, 7–10].

One important way to functionalize CNTs is grafting polymers onto the sidewalls of CNTs [1, 3, 10–15]. The polymers grafting can be realized via grafting-from or grafting-to approach. In the grafting-to approach, polymers are linked to the sidewalls of CNTs, but it is difficult to graft dense polymer chains onto the sidewalls of CNTs due to steric hindrance among high-molecular-weight polymer chains and the sidewalls of CNTs. However, it is feasible to obtain dense polymer grafting via the grafting-from approach, in which polymers are grown from the sidewalls of CNTs. In the grafting-from approach, polymerization of monomers is performed in the presence of CNTs attached with certain groups, which can be initiator or chain transfer agent of the polymerization. In principle, all types of polymerization, that is, free radical polymerization, ionic chain polymerization, ring-opening polymerization, and condensation polymerization, can be applied for grafting polymers onto the sidewalls of CNTs via the grafting-from approach.

In comparison to other types of polymerization, ring-opening polymerization of cyclic monomers, which can be divided into cationic, anionic, and radical ring-opening polymerization, could produce polymers with heteroelements in the main chains [16]. Some of the polymers obtained via ring-opening polymerization are well-known materials for various biorelated applications. Therefore, it is meaningful to explore ring-opening polymerization to graft polymers onto sidewalls of

---

Surface Modification of Nanotube Fillers, First Edition. Edited by Vikas Mittal.
© 2011 Wiley-VCH Verlag GmbH & Co. KGaA. Published 2011 by Wiley-VCH Verlag GmbH & Co. KGaA.

CNTs via the grafting-from approach. In this chapter, polymer-grafting CNTs are obtained by performing cationic ring-opening polymerization of cyclic monomers in the presence of amino groups or hydroxyl groups functionalized CNTs in one pot. First cationic ring-opening polymerization of cyclic monomers in the presence of hydroxyl compounds or amino compounds in one pot is discussed in terms of the mechanisms and the applications for preparation of new materials, which is fundamental to understanding and preparation of polymer-grafted CNTs via cationic ring-opening polymerization; then the preparation of two types of polymer-grafted CNTs, that is, polyethylenimine (PEI)-grafted multiwalled CNTs (PEI-g-MWNTs) and hyperbranched polyether-grafted multiwalled CNTs (MWNT-HP), is described; and finally the application of PEI-g-MWNTs for gene delivery is discussed.

10.2
Cationic Ring-Opening Polymerization of Cyclic Monomers in the Presence of Chain Transfer Reagents in One Pot

Cationic ring-opening polymerization of different types of cyclic monomers containing heteroatoms including cyclic ether and cyclic amines can be realized [16]. Transfer reaction occurs when hydroxyl compounds and amino compounds are presented in cationic ring-opening polymerization of cyclic ethers and cyclic amines, respectively. The presence of chain transfer agents can reduce the amounts of cyclic products formed by the chain transfer reaction to polymer chains. In addition, telechelic polymers and new copolymers can be obtained by performing cationic ring-opening polymerization of cyclic monomers with certain chain transfer reagents. Recently functional polyamines are obtained by cationic ring-opening polymerization of aziridine in the presence of amino compounds, which will be described below [17, 18].

Cationic ring-opening polymerization of cyclic monomers in the presence of chain transfer reagents might be performed in two ways [19, 20]. One way is to add cyclic monomers to the solution containing initiator and chain transfer reagents slowly to keep the monomer concentration in the solution close to zero. In this way, the polymerization is via active monomer (AM) mechanism as described in Scheme 10.1, and the polymerization can be taken as living polymerization with the molecular weight of polymers controlled by the molar ratio of monomer to the chain transfer reagent, and no cyclic oligomer is formed. But the polymerization process is tedious and is also difficult to realize. In contrast, it is feasible to perform cationic ring-opening polymerization with cyclic monomers, initiator, and chain transfer agents in one pot. However, the polymerization mechanisms of the one-pot polymerization become complicated. The mechanisms of two typical types of cationic ring-opening polymerizations in the presence of chain transfer agents in one pot are described below, and typical examples of applying this type of polymerization for the preparation of new materials are illustrated.

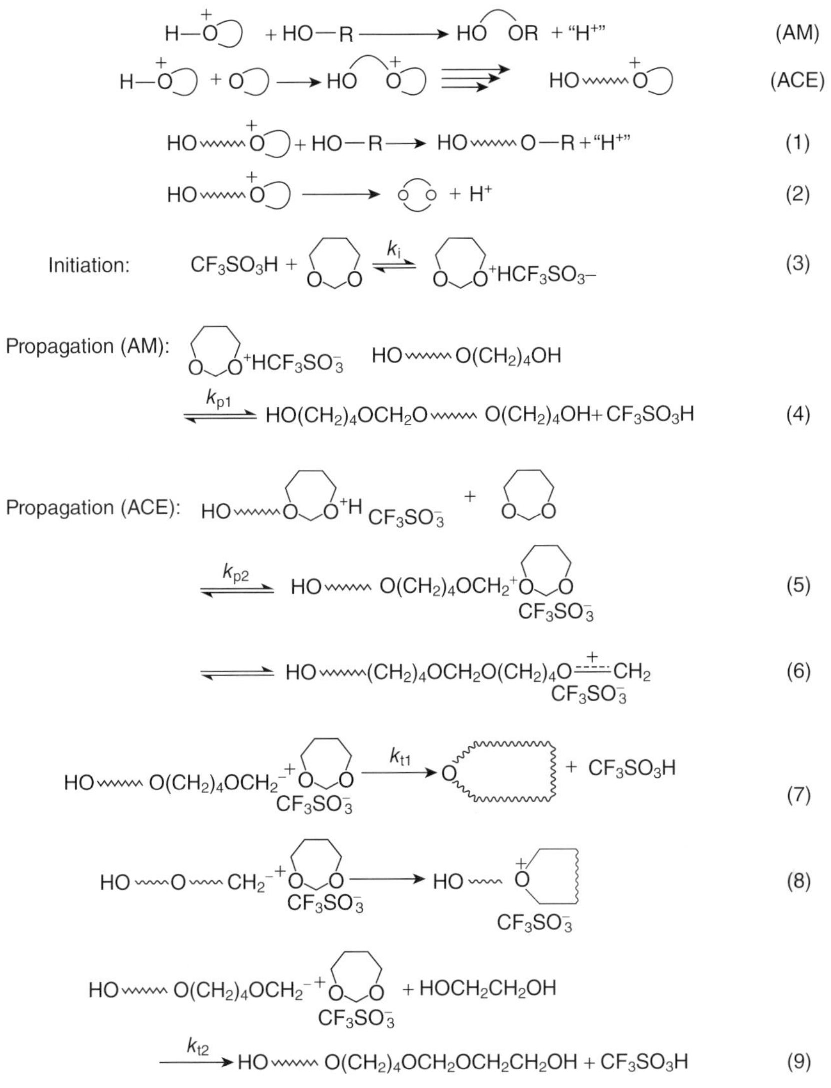

10.2 Cationic Ring-Opening Polymerization of Cyclic Monomers

Scheme 10.1 Possible mechanisms of propagation.

$$H{-}\overset{+}{O}{\frown} + HO{-}R \longrightarrow HO\frown OR + \text{``}H^{+}\text{''} \quad \text{(AM)}$$

$$H{-}\overset{+}{O}{\frown} + O{\frown} \longrightarrow HO\frown \overset{+}{O}{\frown} \rightleftharpoons HO\text{www}\overset{+}{O}{\frown} \quad \text{(ACE)}$$

$$HO\text{www}\overset{+}{O}{\frown} + HO{-}R \longrightarrow HO\text{www}O{-}R + \text{``}H^{+}\text{''} \quad (1)$$

$$HO\text{www}\overset{+}{O}{\frown} \longrightarrow \overset{}{O}{\frown} + H^{+} \quad (2)$$

Initiation: $\quad CF_3SO_3H + \overset{}{O}{\frown} \underset{}{\overset{k_i}{\rightleftharpoons}} \overset{}{O}{\overset{+}{O}}H\,CF_3SO_3^{-} \quad (3)$

Propagation (AM): $\overset{}{O}{\overset{+}{O}}H\,CF_3SO_3^{-} \quad HO\text{www}O(CH_2)_4OH$

$$\underset{}{\overset{k_{p1}}{\rightleftharpoons}} HO(CH_2)_4OCH_2O\text{www}O(CH_2)_4OH + CF_3SO_3H \quad (4)$$

Propagation (ACE): $HO\text{www}O\overset{+}{\underset{}{O}}H\,CF_3SO_3^{-} + \overset{}{O}{\frown}$

$$\underset{}{\overset{k_{p2}}{\rightleftharpoons}} HO\text{www}O(CH_2)_4OCH_2\overset{+}{O}{\frown} \quad (5)$$
$$CF_3SO_3^{-}$$

$$\rightleftharpoons HO\text{www}(CH_2)_4OCH_2O(CH_2)_4O\overset{+}{\cdots}CH_2 \quad (6)$$
$$CF_3SO_3^{-}$$

$$HO\text{www}O(CH_2)_4OCH_2\overset{+}{\underset{CF_3SO_3^{-}}{O}}{\frown} \xrightarrow{k_{t1}} \overset{}{O}{\frown}\text{rings} + CF_3SO_3H \quad (7)$$

$$HO\text{www}O\text{www}CH_2\overset{+}{\underset{CF_3SO_3^{-}}{O}}{\frown} \longrightarrow HO\text{www}\overset{+}{O}{\frown}\text{rings} \quad (8)$$
$$CF_3SO_3^{-}$$

$$HO\text{www}O(CH_2)_4OCH_2\overset{+}{\underset{CF_3SO_3^{-}}{O}}{\frown} + HOCH_2CH_2OH$$

$$\xrightarrow{k_{t2}} HO\text{www}O(CH_2)_4OCH_2OCH_2CH_2OH + CF_3SO_3H \quad (9)$$

AM = activated monomer mechanism
ACE = active chain end mechanism

10.2.1
Mechanism of Cationic Ring-Opening Polymerization of Cyclic Monomers in the Presence of Chain Transfer Agents in One Pot

Cationic ring polymerization of cyclic monomers in the presence of chain transfer agents in one pot can be divided into two categories: one is that additional chain transfer agents are added such as hydroxyl compounds or amino compounds; and the other type is that the hydroxyl groups or amino groups are contained in the monomers such as hydroxyl groups in glycidol and amines in aziridine. The mechanisms of two typical polymerizations in the respective categories are discussed below.

10.2.1.1 Cationic Ring-Opening Polymerization of 1,3-Dioxepane (DOP) in the Presence of Ethylene Glycol (EG) in One Pot [19]

Cationic polymerization of DOP initiated with triflic acid in the presence of EG in one pot is adapted as a model of cationic polymerization of cyclic monomers in the presence of external chain transfer reagent. For these systems, there are two possible mechanisms of propagation: (i) an AM mechanism and (ii) an active chain-end (ACE) mechanism as described in Scheme 10.1. In the ACE, the chain transfer reaction could occur with hydroxyl groups of the external transfer agents as indicated in Eq. (10.1),[1] and with the internal end-hydroxyl group as shown in Eq. (10.2) forming cyclic products. The results from the kinetic study of cationic ring-opening polymerization of DOP in the presence of EG in one pot provide understanding of the mechanism.

The initiation of the cationic polymerization of DOP using triflic acid could be described by Eq. (10.3). When the polymerization proceeds according to AM, the propagation reaction could be described by Eq. (10.4). When the polymerization proceeds according to ACE, the propagation steps could be described by Eqs. (10.5) and (10.6). For DOP, cyclic oxonium ion is the major active species, which is used to represent the chain propagation center. During the propagation reaction, two types of chain transfer reactions should be considered: (i) intramolecular chain transfer reactions as described by Eqs. (10.7) and (10.8) (reaction as shown by Eq. (10.8) is preferred at the initial stage of the polymerization); (ii) intermolecular chain transfer reaction as described by Eq. (10.9) occurs when EG is presented. Triflic acid reproduced from the reactions shown in Eqs. (10.4), (10.7), and (10.9) reinitiates polymerization.

When the reaction reaches steady state without polymerization termination, Eq. (10.10) should be valid if the polymerization proceeds just following AM:

$$k_i[\text{DOP}][\text{I}] = k_{p1}[\text{M*}][\text{EG}] \tag{10.10}$$

where k_i is the rate constant of the initiation reaction, k_{p1} is the rate constant of the propagation reaction, [DOP] is the amount concentration of monomer, [I] is the amount concentration of initiator, [M*] is the amount concentration of propagation of active species, and [EG] represents the amount concentration of polydiol and glycol added. [EG] equals the initial concentration of EG because only one polydiol molecule of higher molecular weight is produced when each glycol molecule reacts with one oxonium ion. Thus, the polymerization rate R_p is

$$R_p = k_{p1}[\text{M*}][\text{EG}] = k_i[\text{DOP}][\text{I}] \tag{10.11}$$

---

1) For Eqs. (10.1)–(10.9), see Scheme 10.1.

On the other hand, the polymerization rate R_p is determined by determining reaction as described by Eq. (10.5) when the polymerization proceeds just following ACE:
Therefore,

$$R_p = k_{p2}[M^*][DOP] \tag{10.12}$$

where k_{p2} is the rate constant of propagation. At steady-state conditions, Eq. (10.13) should be valid:

$$k_i[DOP][I] = k_{t1}[M^*] + k_{t2}[M^*][EG] \tag{10.13}$$

where k_{t1} and k_{t2} are the rate constant of reactions described by Eqs. (10.7) and (10.8), respectively.
Thus,

$$[M^*] = k_i[DOP][I]/(k_{t1} + k_{t2}[EG]) \tag{10.14}$$

Combination of Eqs. (10.12) and (10.14) yields

$$R_p = k_{p2} \times k_i[DOP]^2[I]/(k_{t1} + k_{t2}[EG]) \tag{10.15}$$

When AM and ACE mechanisms coexist in the polymerization, the rate of polymerization is

$$R_p = (k_{p1}[EG] + k_{p2}[DOP])[M^*] \tag{10.16}$$

At steady state, Eq. (10.17) should describe the process

$$k_i[DOP][I] = k_{t1}[M^*] + k_{t2}[M^*][EG] + k_{p1}[M^*][EG] \tag{10.17}$$

Therefore,

$$[M^*] = k_i[DOP][I]/(k_{t1} + k_{t2}[EG] + k_{p1}[EG]) \tag{10.18}$$

Combination of Eqs. (10.16) and (10.18) yields

$$R_p = k_{p1}k_i[DOP][I][EG]/(k_{t1} + k_{t2}[EG] + k_{p1}[EG]) + \\ k_{p2}k_i[DOP]^2[I]/(k_{t1} + k_{t2}[EG] + k_{p1}[EG]) \tag{10.19}$$

Figure 10.1 shows typical time versus conversion curves for the polymerization. It can be seen that reasonable rate and yield are obtained when the mole ratio [EG]/[I] is kept from 20 to 100. The polymerization rate, R_p, could be obtained from Figure 10.1, and Figure 10.2 shows the relationship of R_p against $[DOP]^2[I]/[EG]$. It is apparent that a good linear relationship illustrates the validity of Eq. (10.20):

$$R_p = k[DOP]^2[I]/[EG] + b \tag{10.20}$$

where k and b are constants, and [DOP] and [I] were kept to be almost constant in the kinetic measurement.

If the polymerization only proceeds according to AM, [EG] does not affect R_p as shown by Eq. (10.11). So the experimental results presented by Eq. (10.20) indicated that AM cannot be the sole mechanism of the polymerization. When only ACE is present, b in Eq. (10.20) should be removed according to Eq. (10.15).

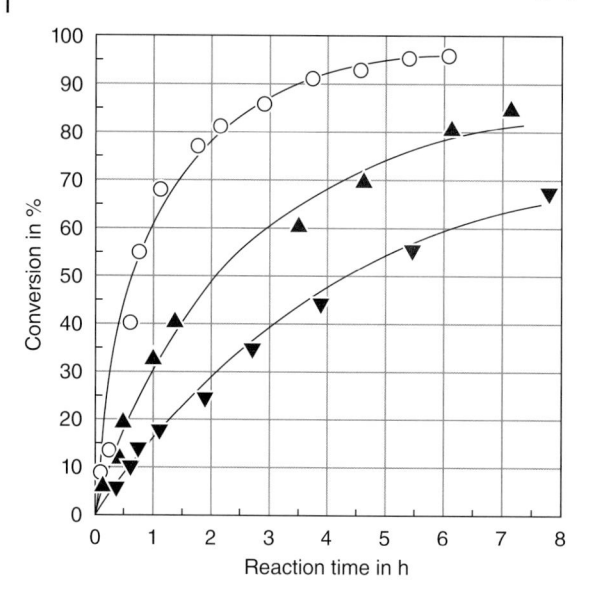

Figure 10.1 Typical time-conversion curves for the polymerization of DOP initiated by triflic acid in the presence of EG at 25 °C. (○) [DOP] = 6.21 mol/l; [EG] = 0.21 mol/l; [CF$_3$SOH] = 0.01 mol/l. (▲) [DOP] = 6.10 mol/l; [EG] = 0.61 mol/l; [CF$_3$SOH] = 0.01 mol/l. (▼) [DOP] = 5.93 mol/l; [EG] = 0.99 mol/l; [CF$_3$SOH] = 0.01 mol/l. Reproduced from Ref. [19] with permission from Wiley-VCH Verlag GmbH & Co. KGaA.

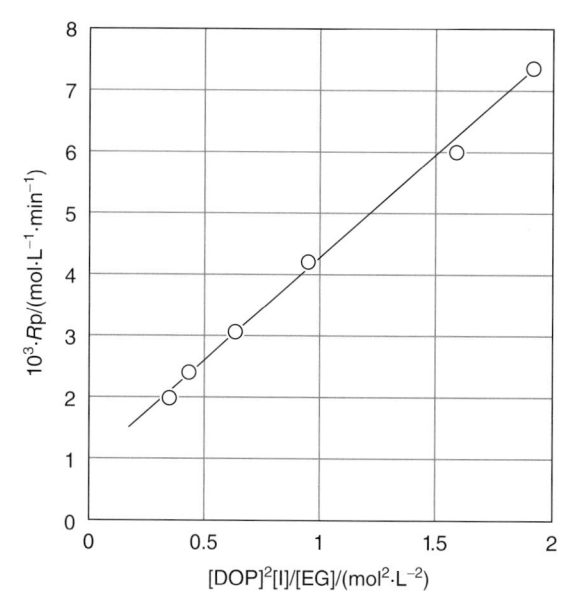

Figure 10.2 Plot of polymerization rate R_p against [DOP]^2[I]/[EG]. Polymerization conditions are the same as in Figure 10.1. Reproduced from Ref. [19] with permission from Wiley-VCH Verlag GmbH & Co. KGaA.

Therefore the polymerization proceeds according to ACE in combination with AM. Further the propagation chains formed in ACE are mainly transferred to the external chain transfer reagent, EG, due to that k_{t1} in Eqs. (10.15) and (10.19) is negligible in comparison with k_{t2}[EG] when Eq. (10.20) is compared with Eqs. (10.15) and (10.19). This is important because the dominant external chain transfer reaction could reduce the formation of cyclic oligomers, and could be explored for the preparation of new polymers when suitable chain transfer reagents are adapted. Moreover it could be stated that $k[DOP]^2[I]/[EG]$ in Eq. (10.20) represents the contribution of ACE to the polymerization, and b represents the contribution of AM. As $[DOP]^2[I]/[EG]$ decreases, the proportion of the ACE process decreases and that of the AM process increases. When $[DOP]_2[I]/[EG] = 0$, this means that the polymerization is performed in the way that monomer is added to the mixture of diol and initiator at a rate equal to the consumption of monomer, the polymerization proceeds according to AM.

10.2.1.2 Mechanism of Cationic Ring-Opening Polymerization of Glycidol [21]

Glycidol contains the oxirane ring and hydroxyl group; the hydroxyl groups in the monomers can functionalize as chain transfer reagent of cationic ring-opening polymerization of the oxirane rings. Two possible polymerization mechanisms, that is, AM and ACE, might exist. ACE always exists in the polymerization due to the condition in the one-pot polymerization. However, the presence of AM should be investigated.

As shown in Scheme 10.2, two types of repeating units in the polymer main chains could be produced in AM mechanism. One type of the unit contains secondary hydroxyl group, and the other type contains primary hydroxyl groups as substituent of the polyether main chain. When the polymerization proceeds by the ACE mechanism as shown in Scheme 10.3, the chain propagation step would form

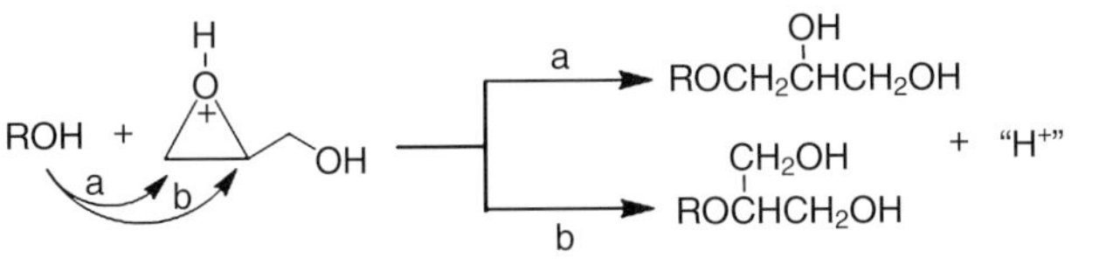

Scheme 10.2 AM mechanism in cationic ring-opening polymerization of glycidol.

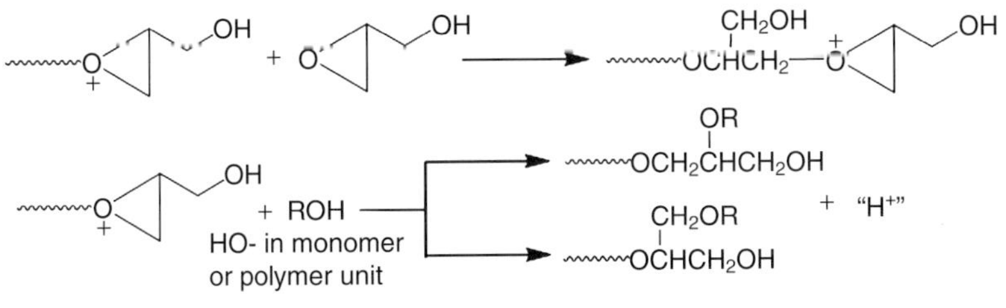

Scheme 10.3 ACE mechanism in cationic ring-opening polymerization of glycidol.

the backbone composed exclusively of –CH$_2$–CH(CH$_2$OH)–O– repeating units with primary hydroxyl as substituent of the polyether main chain although α- or β- ring-opening may lead to h–h, h–t, or t–t sequences; also only primary hydroxyl groups should be formed as substituent of the polyether chain when the chain transfer reaction involving the hydroxyl groups of polymer or monomer occurs. Therefore, analysis of the primary hydroxyl groups and the secondary hydroxyl groups in the hyperbranched polyglycerol could provide information of whether AM is present in the polymerization.

^{13}C NMR is a versatile tool to differentiate primary hydroxyl groups and secondary hydroxyl groups. Figure 10.3 is ^{13}C NMR spectrum of polyglycidol obtained from polymerization of glycidol in D$_2$O solution initiated with BF$_3$·OEt$_2$. On the basis of the assignment of signals in the ^{13}C NMR spectra of linear polyglycidol without secondary hydroxyl group, some assignments in much more complex ^{13}C NMR spectra of polyglycidol prepared by anionic polymerization of the silylated monomer followed by hydrolysis, and DEPT experiments, the peaks contributed to the structures containing primary hydroxyl group and secondary hydroxyl group, are identified as below, respectively:

a: 60.5 ppm; b: 68 ppm; c: 78 ppm   g: 72 ppm; h: 68 ppm

```
      c  b                              g   h
   —CHCH₂O—                          —CH₂CHCH₂O—
      |                                    |
     CH₂OH                                 OH
      a
```

In Figure 10.3b and c, it is obvious that units containing secondary hydroxyl group could be observed. Therefore, it is clear that AM mechanism exists.

The straightforward proof of the presence of the repeating units containing secondary hydroxyl groups comes from the studies of the ^{29}Si NMR spectra of the silylated polymers. A typical spectrum, recorded in CDCl$_3$ solution, is shown in Figure 10.4.

In addition to the signal of the unreacted excess of bis-(trimethylsilgl)acetamide (BSA) at 2.70 ppm and a small signal at 6.40 ppm attributed to (CH$_3$)$_3$Si–O–Si(CH$_3$)$_3$ (side product of the reaction between BSA and adventitious water), two signals appear at 15.70 and 16.35 ppm in the region characteristic for the ROSi(CH$_3$)$_3$. Addition of the model alcohol containing primary hydroxyl groups position, that is, CH$_3$CH$_2$OCH$_2$CH$_2$OH, leads to an increase in the intensity of the 16.35 ppm signal. This allows the following assignment:

```
     CH₂OSi(CH₃)₃                      OSi(CH₃)₃
        |                                 |
   —CH₂CHO—                        —CH₂CHCH₂O—
     16.35 ppm                         15.70 ppm
```

Integration of the signals indicates that 52% of the hydroxyl group is primary hydroxyl group and the remaining 48% is secondary hydroxyl group. The quantitative determination of the mechanism participation including the

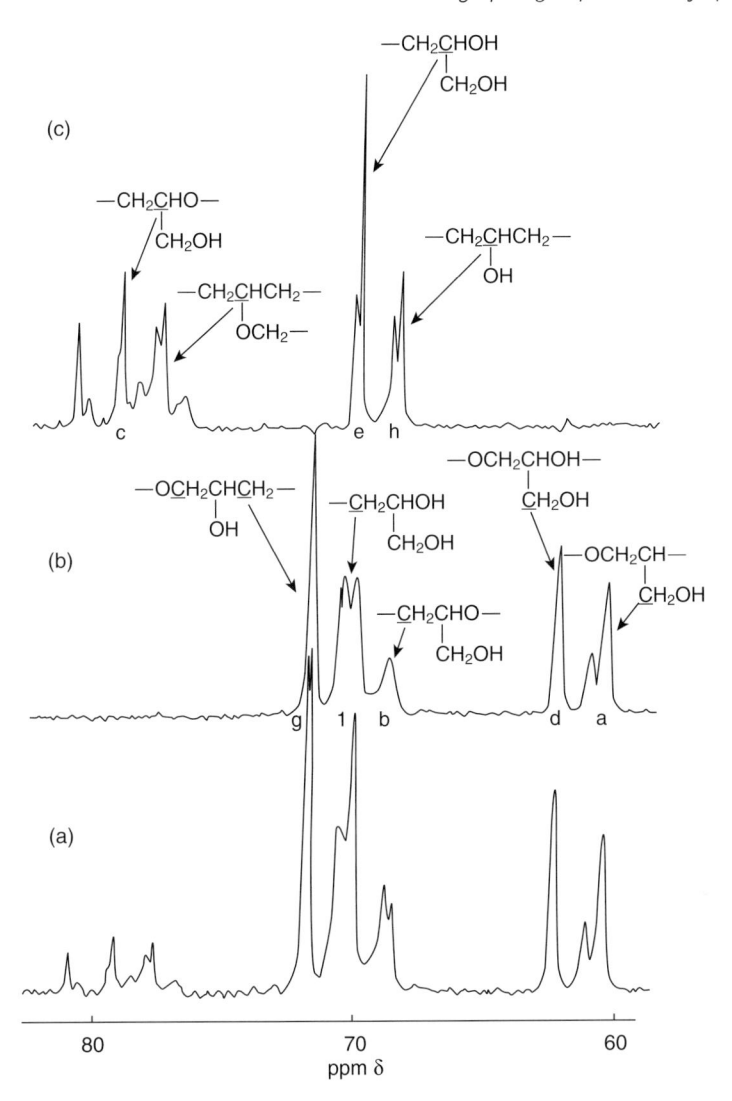

Figure 10.3 ^{13}C NMR spectra (obtained in D$_2$O, room temperature) of polyglycidol with $M_n = 1600$ obtained in the polymerization initiated with BF$_3$·OEt$_2$: (a) single pulse spectrum (2000 transients); (b) DEPT CH$_2$ subspectrum; (c) DEPT CH subspectrum. Reproduced from Ref. [21] with permission from American Chemical Society.

proportion of AM to ACE and the proportions of the reactions a and b in Eq. (10.21)[2] could not be obtained on the basis of these results; however, it is clear that AM exists in the polymerization, and the proportion is not insignificant.

2) For Eq. (10.21), see Scheme 10.2.

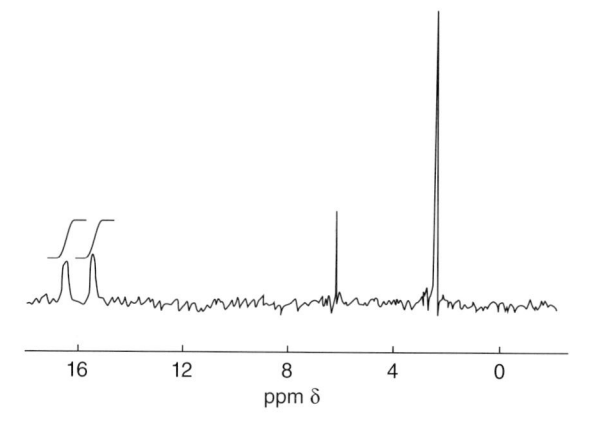

Figure 10.4 ^{29}Si NMR spectrum of polyglycidol with M_n = 7600 obtained via polymerization of glycidol in bulk initiated with HPF$_6$·OEt$_2$ following by silylation with a threefold excess of bis-(trimethylsilgl) acetamide (BSA) in dimethylacetamide (DMA) solution. The NMR spectrum was recorded in DMA at room temperature. Reproduced from Ref. [21] with permission from American Chemical Society.

10.2.2
Preparation of Functional Polyamine

Cationic ring-opening polymerization of cyclic monomers in the presence of chain transfer reagents has been explored for the preparation of new polymers. Polyamines have attracted more and more interests for safe and efficient gene delivery, which is currently the most formidable challenge in gene therapy. New types of polyamines could be prepared via performing cationic ring-opening polymerization of aziridine in the presence of amine compounds. Below are two typical examples.

Amino groups in chitosan could be explored as chain transfer reagents in cationic ring-opening polymerization of aziridine to provide hyperbranched PEI-graft-chitosan (PEI-g-chitosan) as described in Scheme 10.4 [17]. PEI-g-chitosan is obtained by performing cationic ring-opening polymerization of aziridine in the presence of water-soluble oligo-chitosan in one pot. It was demonstrated that all the amino groups of chitosan are grafted with hyperbranched PEI. Under certain condition, the molecular weight of the grafted hyperbranched PEI is ca. 206. PEI-g-chitosan has enough high molecular weight for efficient DNA delivery, but it can be expected that the single PEI units produced after degradation of the chitosan main chains will be low toxic and safe due to the low molecular weight. Therefore, PEI-g-chitosan is promising for safe and efficient gene delivery for gene therapy.

The other example is the preparation of poly(ethylene glycol)-block-PEI (PEG-b-PEI) by performing cationic ring-opening polymerization of aziridine in the presence of monoamino-terminated PEG in one pot as described in Scheme 10.5 [18]. When the molecular weight of monoamino-PEG is below 5000, this approach

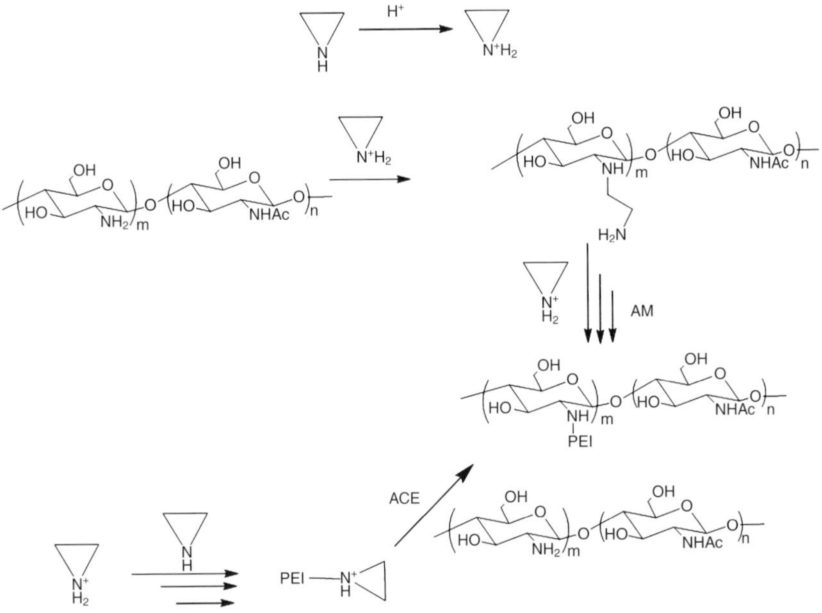

Scheme 10.4 Approach to PEI-g-chitosan.

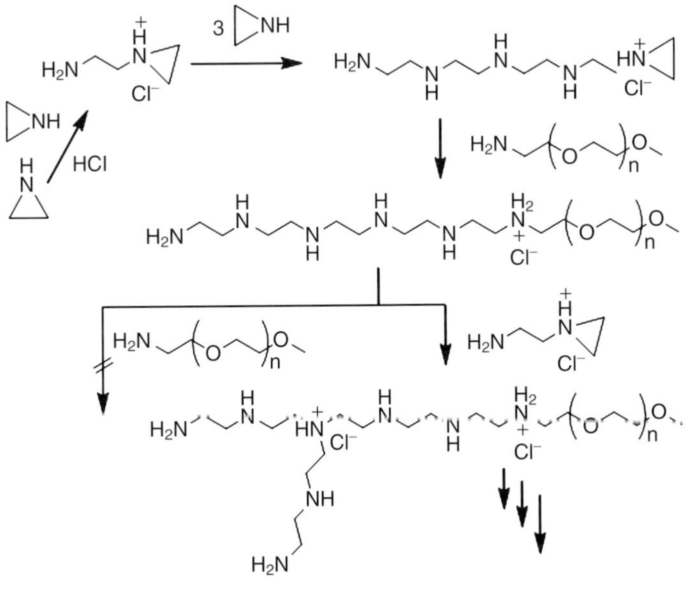

Scheme 10.5 Approach to PEG-b-PEI.

could provide block copolymer. However, monoamino-PEG of molecular weight higher than 5000 could not be blocked with PEI. The obtained PEG-*b*-PEI is demonstrated to be promising for gene delivery.

10.3
Preparation of Polymers-Grafted Carbon Nanotubes

In principle, polymers could be grafted onto the sidewalls of CNTs by performing cationic ring-opening polymerization of cyclic monomers in the presence of CNTs with hydroxyl groups or amino groups. The grafting is realized via the chain transfer of the cationic ring-opening polymerization of cyclic monomers to hydroxyl groups or the amino groups. The feasible way is performing cationic polymerization of cyclic monomers in the presence of hydroxyl group or amine group functionalized CNTs in one pot. As mentioned above, ACE and AM mechanisms exist in the polymerization; therefore, the grafting is also obtained via ACE and AM. In this approach, hyperbranched PEI and polyether are grafted to the sidewalls of CNTs.

10.3.1
Preparation of PEI-grafted CNTs [11]

As described in Scheme 10.6, PEI was grafted onto the surface of MWNTs by performing cationic polymerization of aziridine in the presence of amine-functionalized MWNTs (MWNT-NH$_2$). MWNT-NH$_2$ was obtained by introducing carboxylic acid groups onto the surface of MWNTs through refluxing MWNTs in 3.0 M nitric acid, then transforming carboxylic acid to acyl chloride by reaction with thionyl chloride, followed by reaction with ethylenediamine. The grafting of PEI should be realized via two mechanisms, that is, AM and activated chain mechanism (ACM), in which protonated aziridine monomers and the terminal iminium ion groups of propagation chains were transferred to amines on the surface of MWNTs, respectively.

The relative amount of PEI grafted onto the surface of MWNTs could be determined using TGA performed under nitrogen. As shown in Figure 10.5, MWNTs are thermally stable up to 600 °C and pure PEI degrades completely at ca. 500 °C. At 500 °C, pristine MWNTs, MWNTs-NH$_2$, and PEI-*g*-MWNTs show negligible, ca. 2.3% and 10.5% weight loss, respectively, so PEI-*g*-MWNTs contained ca. 8.2% PEI. PEI brushes render PEI-*g*-MWNTs good dispersity in aqueous solution well, and the suspension formed is stable after 6 months, but MWNTs-NH$_2$ shows poor dispersion in aqueous solution and precipitate was observed in several hours as shown in Figure 10.6.

TEM can provide direct evidence of the grafting of PEI onto the surface of MWNTs. Figure 10.7 shows TEM images of PEI-*g*-MWNTs on a holey carbon film. Individually dispersed MWNT separated from others could be observed. Further the inserted picture with a high magnification indicates that PEI is grafted onto

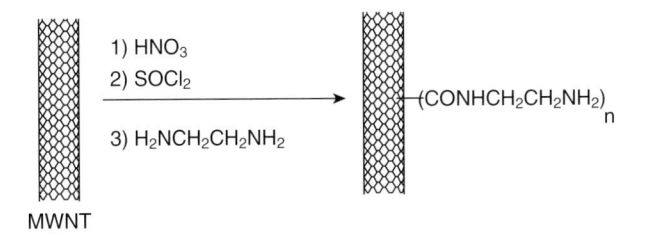

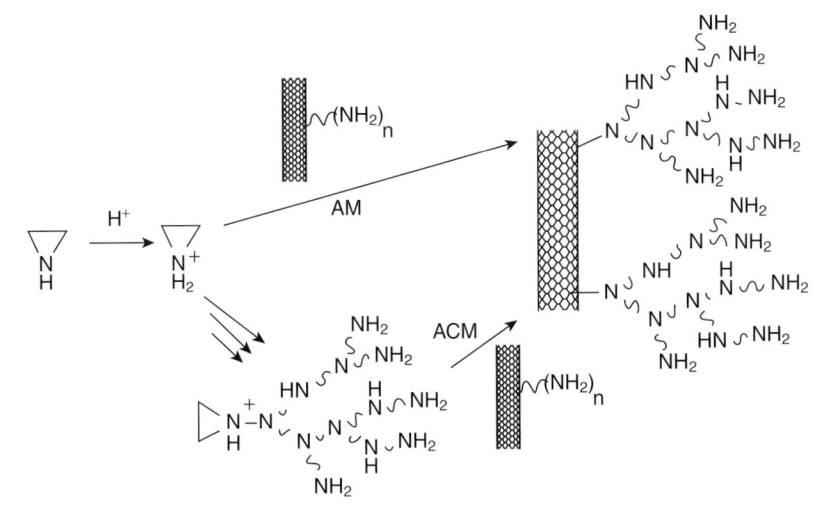

10.3 Preparation of Polymers-Grafted Carbon Nanotubes

Scheme 10.6 Approach to PEI-grafted MWNTs (PEI-g-MWNTs).

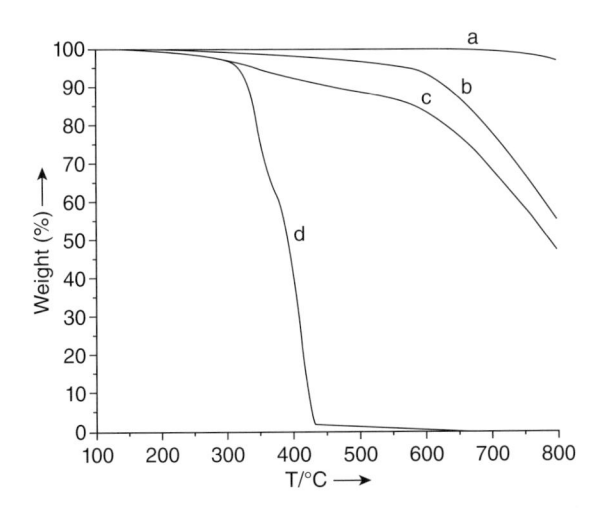

Figure 10.5 TGA curves of (a) pristine MWNTs, (b) MWNTs-NH$_2$, (c) PEI-g-MWNTs, (d) PEI under nitrogen. Reproduced from Ref. [11] with permission from Wiley-VCH Verlag GmbH & Co. KGaA.

Figure 10.6 Photographs of vials containing (a) MWNTs-NH$_2$ in aqueous solution, which precipitated in several hours, and (b) stable aqueous suspension solution of PEI-g-MWNTs, which is stable after 6 months.

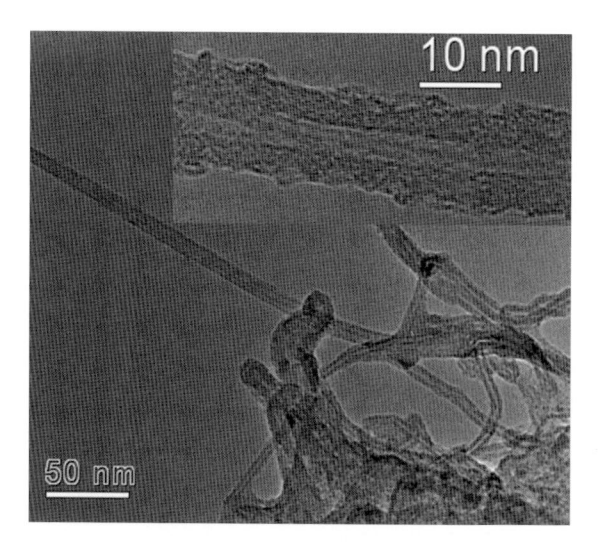

Figure 10.7 TEM images of PEI-g-MWNTs. Reproduced from Ref. [11] with permission from Wiley-VCH Verlag GmbH & Co. KGaA.

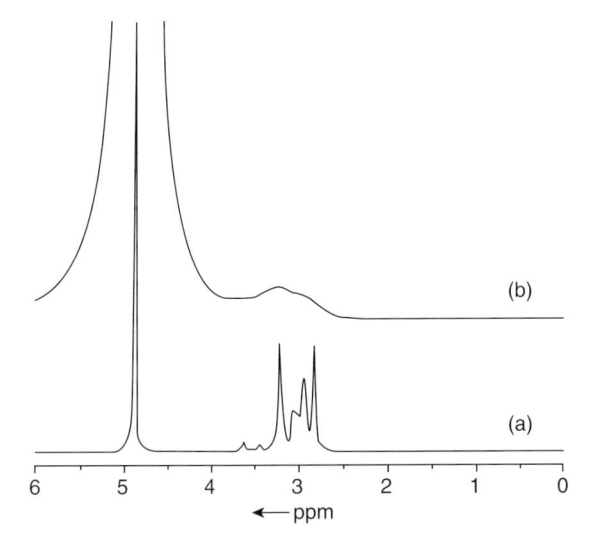

Figure 10.8 Comparison of ^{1}H NMR spectrum of (a) PEI in D$_2$O (pH 7) with (b) PEI-g-MWNTs in D$_2$O (pH 7). Reproduced from Ref. [11] with permission from Wiley-VCH Verlag GmbH & Co. KGaA.

the surface of MWNTs as lumps with different sizes instead of a uniform coating. This is due to the fact that the carboxylic acid groups, then ethylenediamine, and PEI were preferably introduced to the defects, the most active locations for chemical or physical functionalization; however, these defects tend to cluster at the bends along the surface of MWNTs grown by CVD.

The chemistry of PEI brushes grafted onto the sidewalls of MWNTs could be investigated using solution NMR. Figure 10.8 shows the comparison of ^{1}H NMR spectrum of PEI-g-MWNTs with free PEI in D$_2$O (pH 7.0). The peak at ca. 3.1 ppm for PEI-g-MWNTs is attributed to the grafted PEI, and the significantly reduced mobility of PEI chains in PEI-g-MWNTs leads to broadened peaks. Here a partial of amines of PEI are protonated (pK_a of PEI is higher than 8.0), and it is found that protonation or partial protonation of PEI is necessary for forming a stable aqueous suspension of PEI-g-MWNTs. Neutralized PEI by adjusting pH to 9 or higher leads to precipitation of dispersed PEI-g-MWNTs in several hours.

10.3.2
Preparation of Hyperbranched Polyether-Grafted CNTs [22]

As indicated in Scheme 10.7, hyperbranched polyether is obtained from cationic ring-opening polymerization of cyclic monomer 3-ethyl-3-(hydroxymethyl)oxetane (EHOX) containing hydroxyl group, which is similar to glycidol; and hyperbranched polyether is formed by chain transfer reaction to the hydroxyl group in EHOX monomer. The polymerization was performed in solution of CH$_2$Cl$_2$

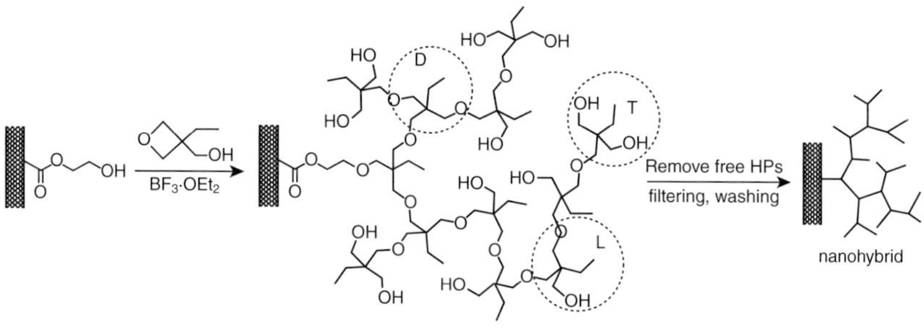

Scheme 10.7 Approach to hyperbranched polyether-grafted MWNTs MWNT-HPs.

initiated with $BF_3 \cdot OEt_2$ in the presence of hydroxyl groups attached MWNTs in one pot. The covalent attachment of hyperbranched polyethere is achieved via the chain transfer to the hydroxyl groups on the sidewall of MWNTs to produce MWNT-HP. The chain transfer reaction should be realized via either ACE or AM similar to cationic ring-opening polymerization of glycidol.

The content of hyperbranched polyEHOX in MWNT-HP could be determined using the weight loss in TGA results as for PEI-g-MWNTs. The content of hyperbranched polyEHOX could be tuned by applying different monomer/MWNT-OH feed ratios (R_{feed}), which results in hyperbranched brushes of different sizes and hence various thickness on the tubes. The quantity of polymer grown on the tubes increases slightly when the R_{feed} is lower than 20/1, increases dramatically when the R_{feed} is in a range of 20/1–30/1, and then slightly rises again when the R_{feed} is greater than 30/1. When the ratio is 5 ml of EHOX to 1 g of MWNTs-OH or 50 ml of EHOX to 1 g of MWNTs-OH, the polymer content in MWNT-HP is ca. 20% (w/w) and 87% (w/w), respectively. On the other hand, the number-average molecular weight (M_n) of the grafted HP increases with R_{feed}. So a higher feed ratio leads to a greater amount and a higher molecular weight of the grafted polymer. MWNT-HP obtained exhibits relatively good dispersibility/solubility in polar solvents such as methanol, ethanol, DMF, and DMSO. Dispersibility of the samples increases with the increasing content of the grafted polymer in MWNT-HP. The solution of MWNT-HP containing of 87% (w/w) of HP in MeOH shows 10 times higher UV-vis absorption at 600 nm than MWNT-OH.

Another important topological property of the grafted hyperbranched polymer is the degree of branching (DB). A hyperbranched polymer generally contains dendritic units (D), linear units (L), terminal units (T), and one initial unit (I), and the DB can be calculated using Fréchet's equation

$$DB = (D+T)/(D+T+L)$$

In order to get DB values, the hyperbranched polymers grafted onto the MWNT surfaces were cleaved from the sidewalls of MWNTs, and the structures were characterized using quantitative (inverse-gated) ^{13}C NMR spectra. Figure 10.9 shows a typical quantitative ^{13}C NMR spectrum of the cleaved hyperbranched

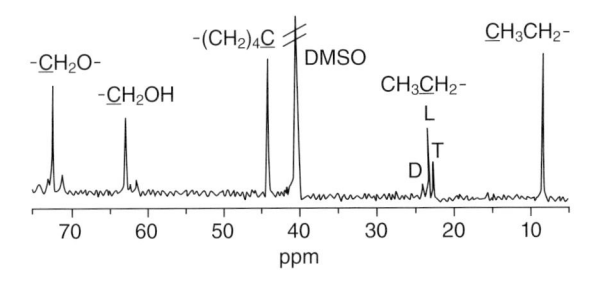

Figure 10.9 Quantitative ^{13}C NMR spectrum (DMSO-d_6) of the hyperbranched polyethers cleaved from MWNT-HP containing 87% (w/w) HP. Reproduced from Ref. [22] with permission from American Chemical Society.

polymer from MWNT sidewalls. The carbon signals of CH_3CH_2-, $C(CH_2)_4-$, $-CH_2OH$, and $-CH_2OCH_2-$ appeared at $\delta = 8.4$, 44, 62.9, and 72.3 ppm, respectively. Three adjacent peaks are found at 22.6, 23.3, and 23.9 ppm, which are assigned to the CH_3CH_2- carbon signals corresponding to terminal, linear, and dendritic units as described in Scheme 10.7, respectively. From the relative integration value of the three peaks, the DB is easily calculated using Fréchet's equation. The DB increases with the increasing molecular weight for cleaved and free HP, which accords with the theoretical prediction. In comparison, the DB of the cleaved hyperbranched polymers is lower than that of the free hyperbranched polymers although the cleaved HP has a larger M_n; and a smaller molecular weight leads to a higher disparity ($\Delta DB = DB_{free} - DB_{tree}$). This may be attributed to steric hindrance among the sidewalls of MWNTs and the grafted molecules; and a smaller molecular weight results in a stronger hindrance effect due to the limited space. However, the DB of the grafted brushes is 0.42 in MWNT-HP obtained with a feed ratio of 50 ml of EHOX to 1 g of MWNT-OH. This value is almost equal to the DB of free HPs, which approaches the maximum (<0.45, typical 0.43–0.44). Therefore, hyperbranched polymers with a high DB can be grown onto the sidewalls of MWNT.

TEM was also used to view the molecular structure of the MWNT-HPs. Figure 10.10 shows typical HRTEM images of MWNT-HP containing 87% (w/w) of HP. The core–shell structure of the 1D nanohybrids is clearly observed, that is, the hyperbranched molecular brushes are grown on the sidewalls of MWNTs with high density. It is found that a higher grafted-polymer quantity results in a thicker polymer shell. The diameter of the same functionalized MWNT is not constant but varies according to the diameter of the MWNT and the thickness of the local polymer shell instead (different diameter points are marked by the arrows in Figure 10.10a). The thickness of the HP shell shown in Figure 10.10a, c, and d is ca. 12–18, 6–8, and 7–10 nm, respectively. For the tubes grafted with a thinner polymer layer, the sheet structure of graphite is detected at a higher magnification (see Figure 10.10c and d).

For some nanotubes, more defects exist at the ends and bends mainly because of the existence of more pentagons and heptagons where then showed more

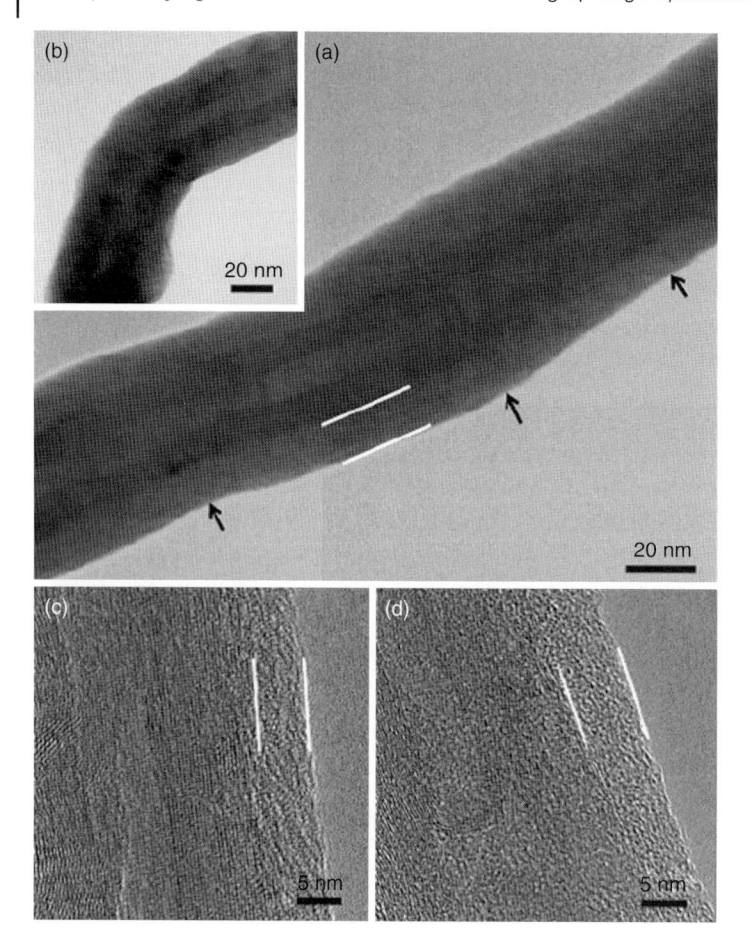

Figure 10.10 TEM images of MWNT-HP containing 87% (w/w) HP. Reproduced from Ref. [22] with permission from American Chemical Society.

reactivity. Therefore, the polymer layers are preferred to be located in these locations such as in PEI-g-MWNTs described above. However, the polymer layers at the bends or ends of MWNT-HP (one tube bend is shown in Figure 10.10b) have almost the same thickness as the rest of the tube. HP content in MWNT-HP is around 87% (w/w/) (in comparison, PEI content in PEI-g-MWNTs is less than 10% (w/w)), so the higher HP content should result in larger hyperbranched brushes, and the larger hyperbranched brushes should form more thicker and even polymer layers on the sidewalls of CNTs. In contrast, the polymer layers in PEI-g-MWNTs should be thinner and uneven, and more information about the initiation sites and probably the defect density can be reflected. In addition, CNTs from different batches or sources might have varied defect density.

The linear polymer chains generally become more rigid after linking on the sidewalls of MWNTs, as confirmed by the increase in glass transition temperature

(T_g). Similar phenomenon was observed with hyperbranched polymers. DSC measurement results reflected that T_g of HPs linked to the MWNTs is ca. 5–7 °C higher than that of cleaved HPs, which indicates that the covalent linkage of hyperbranched polymer to the surfaces of MWNTs reduces the chain flexibility.

10.4
Applications of Polymer-Grafted CNTs

Polymers-grafted CNTs obtained via cationic ring-opening polymerization is promising for many applications, and one of the most important applications is for the development of unique drug delivery systems due to their unique nanometer-sized structure and polymer chemistry. Pristine CNTs cannot be dispersed in aqueous solution, show significant toxicity, and cannot effectively immobilize active species, but grafting polymers onto sidewalls of CNTs via cationic ring-opening polymerization could improve these properties.

10.4.1
PEI-g-CNTs with Reduced Cytotoxicity and Good Capability to Enter Cells [11]

Figure 10.11 shows the cytotoxicity of PEI-g-MWNTs in 293, COS7, and HepG2 cells. Some cytotoxicity was observed when the concentration of PEI-g-MWNTs is 5 µg/ml in the three types of cells, but more than 50% of cells are still viable when the concentration of PEI-g-MWNTs reaches 100 µg/ml. In comparison with free

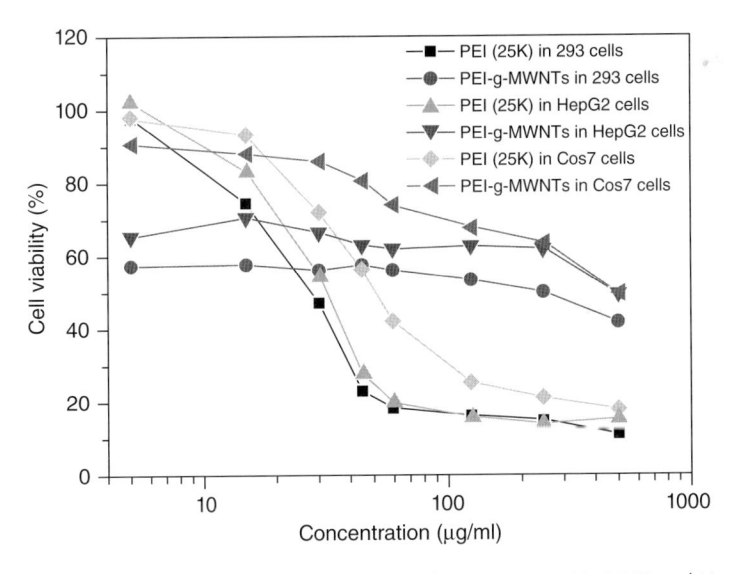

Figure 10.11 Cytotoxicity of PEI-g-MWNTs and PEI (25 K) in 293, COS7, and HepG2 cells.

PEI (M_n: 25 K), which is usually used as a benchmark of new materials for DNA delivery, PEI-g-MWNTs have much lower cytotoxicity at high concentration.

PEI-g-MWNTs have the capability to enter cells. Figure 10.12 shows confocal fluorescence micrographs of 293, COS7, and HepG2 cells after incubated with FITC-labeled PEI-g-MWNTs. When the incubation is performed at 37 °C for 1 h, Figure 10.12a, c, and e clearly indicates that PEI-g-MWNTs enter cells. There are two mechanisms suggested for the uptake mechanism of functional CNTs by cells.

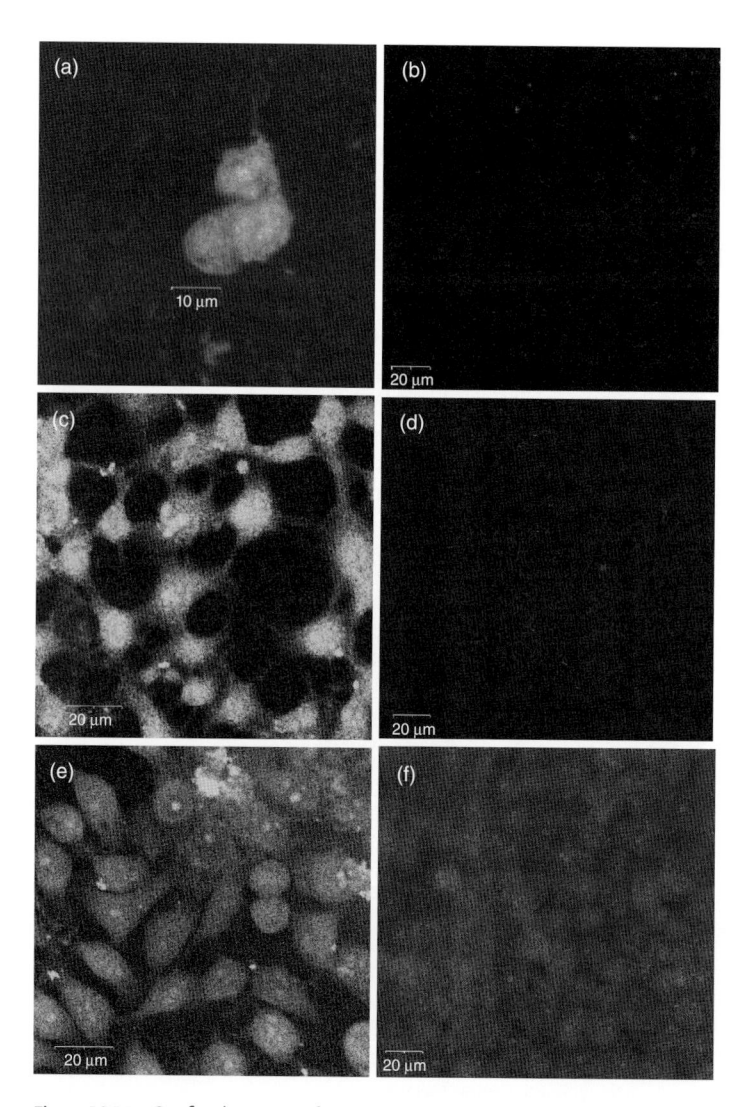

Figure 10.12 Confocal images of (a and b) 293 cells, (c and d) COS7 cells, and (e and f) HepG2 cells after incubated with the complex of cDNA (pCMV-Luc) and FITC labeled PEI-g-MWNTs for 1 h at 37 °C for (a), (c), and (e) and at 4 °C for (b), (d), and (f).

One is via an energy-independent passive diffusion process [23–25], and the other is via endocytosis [26]. Figure 10.12b, d, and f shows confocal fluorescence micrographs of 293, COS7, and HepG2 cells after incubated with FITC-labeled PEI-g-MWNTs at 4 °C for 1 h. In comparison to Figure 10.12a, c, and e, it is apparent that the fluorescent intensity is significantly lower. So the process is temperature dependent; therefore PEI-g-MWNTs enter cells via endocytosis.

10.4.2
PEI-g-MWNTs Promising for Efficient DNA Delivery [11]

Low cytotoxicity and the capability to enter cells render PEI-g-MWNTs promising for drug delivery. PEI is a well-known polymer good for gene delivery due to its high content density of amines comprised of primary amine, secondary amine, and tertiary amine, which provide so-called proton sponge effect facilitating escape of gene/PEI complexes from endosomes. Therefore good gene transfection efficiency could be obtained. It is natural to expect that PEI-g-MWNTs with PEI brushes should be good for gene delivery.

Hyperbranched PEI brushes in PEI-g-MWNTs are helpful to anchor DNA securely onto the sidewalls of MWNTs. As shown in Figure 10.13, pristine MWNTs and MWNTs attached with small molecular amines have little effect on the migration of DNA even at a high weight ratio of 100:1 (Figure 10.13b and c). Also a monolayer of PEI (25 K) adsorbed onto the surface of CNTs directly (after thorough rinsing) did not prevent the migration of DNA (Figure 10.13d). In contrast, PEI-g-MWNTs could prevent the migration of DNA totally inhibited in gel electrophoresis when the weight ratio of PEI-g-MWNTs to DNA is ca. 4:1 (Figure 10.13a). This reflects that stable complexes of PEI-g-MWNTs and DNA could be formed.

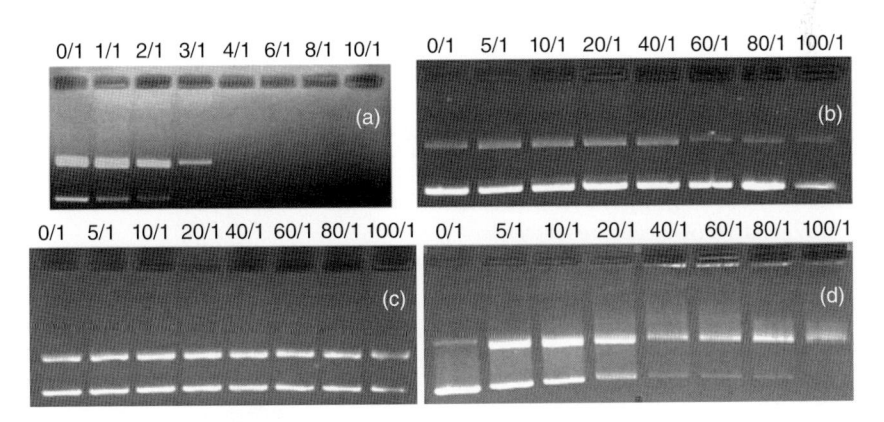

Figure 10.13 Agarose gel elctrophoresis retardation of DNA by (a) PEI-g-MWNTs, (b) pristine MWNTs, (c) MWNTs-NH$_2$, and (d) MWNTs treated with PEI (25 K) followed by thorough rinsing. Lane numbers correspond to different weight ratios of the respective materials to DNA.

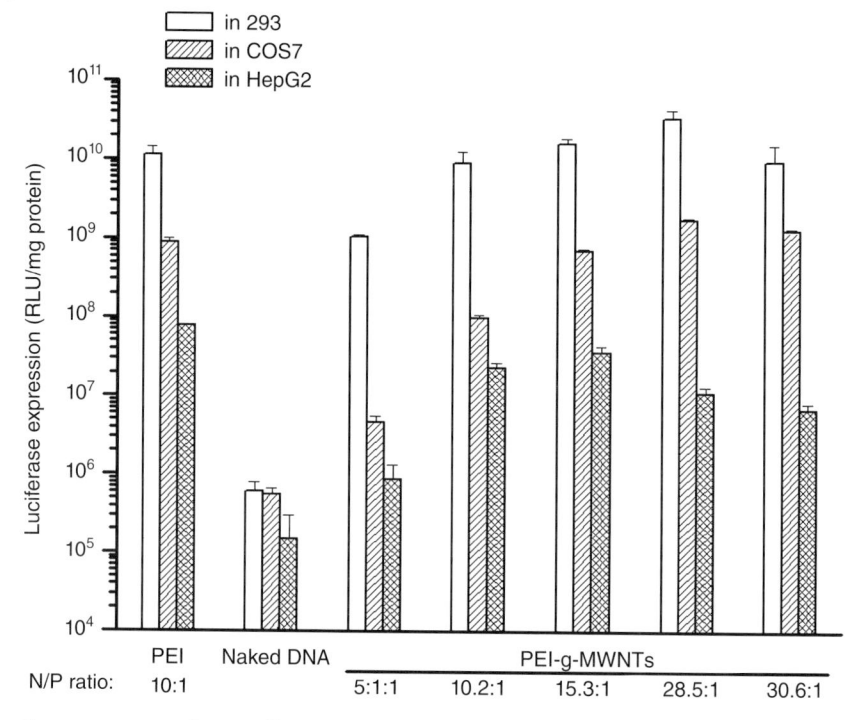

Figure 10.14 Transfection efficiency of PEI-g-MWNTs for DNA delivery relative to that of PEI (25 K) and naked DNA in 293, COS7, and HepG2 cells. The level of gene (pCMV-Luc) expression is given in relative light units (RLUs) per milligram of protein for quadruplicate runs (mean ± standard deviation (n = 4)). Reproduced from Ref. [11] with permission from Wiley-VCH Verlag GmbH & Co. KGaA.

The transfection efficiencies of PEI-g-MWNTs for DNA delivery in 293, COS7, and HepG2 cells are compared with naked DNA and PEI (25 K) in Figure 10.14. The optimal N/P ratios for PEI-g-MWNTs to DNA were about 28.5 : 1, 28.5 : 1, and 15.3 : 1 in 293, COS7, and HepG2 cells, respectively. Under these conditions, the transfection efficiencies of PEI-g-MWNTs are much higher than those naked DNA obtained under the optimal N/P ratios. It is more important that the transfection efficiency of PEI-g-MWNTs is thrice, twice, and half of those of PEI (25 K) in 293, COS7, and HepG2 cells, respectively.

The higher transfection efficiency of PEI-g-MWNTs is attributed to several factors: first is the secure immobilization of DNA onto the surface of MWNTs; therefore stable complexes of DNA and PEI-g-MWNTs could be formed and this protects DNA well from degradation; the second is the capability of PEI-g-MWNTs to enter cells as discussed above; and the third is that the proton-sponge effect of the grafted PEI brushes could facilitate the escape of the PEI-g-MWNTs/DNA complexes from endosomes. This is good for DNA to enter nucleus to realize transformation. Hence the combination of low cytotoxicity and good DNA transfection efficiency makes PEI-g-MWNTs promising for gene therapy.

References

1 Pastorin, G. (2009) *Pharm. Res.*, **26**, 746.
2 Lacerda, L., Bianco, A., Prato, M., and Kostarelos, K. (2006) *Adv. Drug Delivery Rev.*, **58**, 1460.
3 Bianco, A., Kostarelos, K., and Prato, M. (2005) *Curr. Opin. Chem. Biol.*, **9**, 674.
4 Bianco, A., Kostarelos, K., Partidos, C.D., and Prato, M. (2005) *Chem. Commun.*, 571.
5 Prato, M., Kostarelos, K., and Bianco, A. (2008) *Acc. Chem. Res.*, **41**, 60.
6 Liu, Z., Sun, X.M., Nakayama-Ratchford, N., and Dai, H.J. (2007) *ACS Nano*, **1**, 50.
7 Kostarelos, K., Bianco, A., and Prato, M. (2009) *Nature Nanotechnol.*, **4**, 627.
8 Hampel, S., Kunze, D., Haase, D., Kramer, K., Rauschenbach, M., Ritscbel, M., Leonbardt, A., Thomas, J., Oswald, S., Hoffmann, V., and Buechner, B. (2008) *Nanomedicine*, **3**, 175.
9 Liu, Z., Chen, K., Davis, C., Sherlock, S., Cao, Q.Z., Chen, X.Y., and Dai, H.J. (2008) *Cancer Res.*, **68**, 6652.
10 Tasis, D., Tagmatarchis, N., Bianco, A., and Prato, M. (2006) *Chem. Rev.*, **106**, 1105.
11 Liu, Y., Wu, D.C., Zhang, W.D., Jiang, X., He, C.B., Chung, T.S., Goh, S.H., and Leong, K.W. (2005) *Angew. Chem. Int. Ed.*, **44**, 4782.
12 Lo, M.Y., Lay, C.L., Lu, X.H., and Liu, Y. (2008) *J. Phys. Chem. B*, **112**, 13218.
13 Lu, F.S., Gu, L.R., Meziani, M.J., Wang, X., Luo, P.G., Veca, L.M., Cao, L., and Sun, Y.P. (2009) *Adv. Mater.*, **21**, 139.
14 Lay, C.L., Liu, H.Q., Tan, H.R., and Liu, Y. (2010) *Nanotechnology*, **21**, 065101.
15 Wu, W., Li, R.T., Bian, X.C., Zhu, Z.S., Ding, D., Li, X.L., Jia, Z.J., Jiang, X.Q., and Hu, Y.Q. (2009) *ACS Nano*, **3**, 2740.
16 Odian, G. (2004) *Principles of Polymerization*, John Wiley & Sons, Inc., New York.
17 Wong, K., Sun, G.B., Zhang, X.Q., Dai, H., Liu, Y., He, C.B., and Leong, K.W. (2006) *Bioconjugate Chem.*, **17**, 152.
18 Petersen, H., Martin, A.L., Stolnik, S., Roberts, C.J., Davies, M.C., and Kissel, T. (2002) *Macromolecules*, **35**, 9854.
19 Liu, Y., Wang, H.B., and Pan, C.Y. (1997) *Macromol. Chem. Phys.*, **198**, 2613.
20 Biedron, T., Kubisa, P., and Penczek, S. (1991) *J. Polym. Sci. Part A: Polym. Chem.*, **29**, 619.
21 Tokar, R., Kubisa, P., Penczek, S., and Dworak, A. (1994) *Macromolecules*, **27**, 320.
22 Xu, Y.Y., Gao, C., Kong, H., Yan, D.Y., Jin, Y.Z., and Watts, P.C.P. (2004) *Macromolecules*, **37**, 8846.
23 Chen, J.Y., Chen, S.Y., Zhao, X.R., Kuznetsova, L.V., Wong, S.S., and Ojima, I. (2008) *J. Am. Chem. Soc.*, **130**, 16778.
24 Kostarelos, K., Lacerda, L., Pastorin, G., Wu, W., Wieckowski, S., Luangsivilay, J., Godefroy, S., Pantarotto, D., Briand, J.P., Muller, S., Prato, M., and Bianco, A. (2007) *Nature Nanotechnol.*, **2**, 108.
25 Zhou, F.F., Xing, D., Wu, B.Y., Wu, S.N., Ou, Z.M., and Chen, W.R. (2010) *Nano Lett.*, **10**, 1677.
26 Kam, N.W.S., Liu, Z., and Dai, H.J. (2006) *Angew. Chem. Int. Ed.*, **45**, 577.

11
Plasma Deposition of Polymer Film on Nanotubes
Chuh-Yung Chen

11.1
Introduction

Nanocomposites are a class of composites in which the dimensions of the reinforcing phase are of the order of nanometers. Because of their nanometer size characteristics, nanocomposites possess much superior properties than conventional microcomposites [1–3]. The incorporation of single-walled (SWCNTs) and multi-walled carbon nanotubes (MWCNTs) in polymeric composites has attracted much attention recently with the main purpose to enhance the electrical, mechanical, and thermal properties for further applications [4–9]. These nanocomposite materials are explored with a view to attain improved mechanical features and tear resistance with efficient load transfer as well as to achieve certain levels of electrical conductivity through a percolation network for charge mitigation and electromagnetic shielding. Polyimides (PIs) that are widely used in microelectronics to aerospace industries have emerged as the focus of this study. Nevertheless, significant levels of electrostatic charge tend to accumulate on their insulating surfaces, thus causing local heating and premature degradation of electronic components or space structures. A surface resistivity ranged from 10^6 to 10^{10} Ω/cm can be achieved by adding SWCNT to PI while mitigating the build-up of electrostatic charge [10]. Since CNTs are generally insoluble and severely bundled, their homogeneous dispersion in the desired polymer matrices represents a significant challenge [4, 10–12].

To address this issue of the dispersion of CNTs in polymer materials, we established the "grafting modification" method, which involves functionalization of CNTs along with argon plasma treatment to increase their compatibility level with polymer matrix [13–21]. The effect of molecular cover along the CNTs needs to be considered when the property of CNTs alone was studied. However, the cover effect (covalent functionalization of CNTs that affects the electron tunneling process in the outer wall of CNTs and reduces electrical conductivity) does not apply to our case because the conductivity of CNTs alone was not studied. Starting with acid pretreatment of the CNTs under sonication, the grafting of monomer

---

Surface Modification of Nanotube Fillers, First Edition. Edited by Vikas Mittal.
© 2011 Wiley-VCH Verlag GmbH & Co. KGaA. Published 2011 by Wiley-VCH Verlag GmbH & Co. KGaA.

with polar functional groups under plasma conditions not only improves their dispersal in polymer matrices, but also enhances the electrical conductivity and the mechanical and thermal properties of the resulting nanocomposites. By grafting only a small amount of MA onto the CNTs, the modified CNTs were proven to disperse in the polymer matrix more easily, leading to superior conductivity and mechanical properties as shown by our previous results [17, 18]. This modification method is more effective in enhancing the properties of the CNTs in relation to polymers than other commonly used wet methods, such as acid-, alkyl-, and amine-based modifications.

11.2
Principle and Experiment

11.2.1
Plasma Treatment of CNT

The plasma treatment was carried out in a parallel plane electrode reactor of which the volume of the stainless steel vacuum chamber was about $6 \times 10^3 \, cm^3$. The electrodes were made of circular Cu plates (diameter = 5 cm) separated at a distance of 3 cm from each other. The chamber pressure was evacuated to 10^{-6} Torr prior to the plasma process with the purified CNTs being spread evenly inside the chamber. Argon gas at 99.99% purity was introduced into the reactor via a leak valve at a flow rate of 20 ml/min, and the vacuum pressure of the plasma reactor was maintained at 0.1 Torr by using the throttle valve. The CNTs were exposed under Ar plasma treatment for 3 min using a RF-power generator (13.56 MHz) operating at 50 W. It is known that plasma treatment can create free radicals and hence injecting any moiety immediately after the process will allow grafting of functional groups onto the CNTs surface. We have successfully grafted GMA-IDA [20], HEMA [16], MA [17, 18], and AN [19] onto the CNTs, whereby mainly the latter two will be discussed in the following sections. The MWCNTs were primarily studied for this plasma system and will be abbreviated as CNTs in the followings for simplification.

11.2.2
Maleic Anhydride System

Maleic anhydride (MA) with a concentration of 0.1 M dissolved in toluene was prepared in advance and was allowed to flow into the chamber to react with the plasma-treated CNTs at 50 °C for 3 h to complete the grafting method. After the grafting polymerization, the CNTs-grafted MA was washed repeatedly using toluene followed by separation using centrifugal method. The sediments were dried in a vacuum oven at 70 °C overnight to remove the residual solvent. The resulting products obtained were the functionalized CNTs–MA (we mention here as mCNTs).

11.2.2.1 Preparation of the mCNTs/Epoxy Nanocomposites

The pristine epoxy resin was prepared by mixing EPON 828 with a diamine curing agent at 70 °C for 30 min. The mixture was placed in a vacuum oven for 2 h for outgassing purpose, followed by a three-step thermal curing schedule (80 °C for 2 h, 120 °C for 2 h, and 140 °C for 4 h). For the preparation of the mCNTs/epoxy nanocomposites, mCNTs were first added into the diamine curing agent, and the mixture was shear mixed for 2 h and sonicated for another 15 min. EPON 828 resin was added subsequently and further shear mixed with the mCNTs/diamine mixture at 80 °C for 30 min. The resulting mixture was then outgassed in a vacuum oven for 2 h and cast into an aluminum mold. The curing cycle of mCNTs/epoxy nanocomposites is the same as that of the pristine epoxy. Figure 11.1 illustrates the procedure of functionalization of the CNTs and subsequent preparation of the mCNTs/epoxy nanocomposites.

11.2.2.2 Preparation of the mCNTs/Polyimide Nanocomposites

The nanocomposites of mCNTs and PI were prepared using a three-step method. First, mCNTs were dispersed in N,N-dimethylacetamide (DMAc) under ultrasonication for 1 h, and then 4,4′-oxydianiline (ODA) was added and stirred for 2 h to obtain a mCNTs–ODA suspension solution. Next, an equimolar amount of pyromellitic dianhydride (PMDA) was added to the mCNTs–ODA suspension solution. The mCNTs/PMDA–ODA polyamic acid (PAA) prepolymer was obtained after 6-h stirring. The entire reaction was carried out with stirring in a nitrogen-purged flask immersed in an ultrasonic bath at 25 °C. Finally, the mCNTs/PAA suspension solution was cast onto a glass plate and dried in an oven at 70 °C for 2 h, and the mCNTs/PAA film thus prepared was thermally cured in an air-circulating oven at 300 °C to obtain a solvent-free, free-standing mCNTs/PI film. A series of such mCNTs/PI nanocomposite films was prepared with mCNTs concentrations ranging from 0.0 to 3.0 wt.% (0.0–4.1 vol.%).

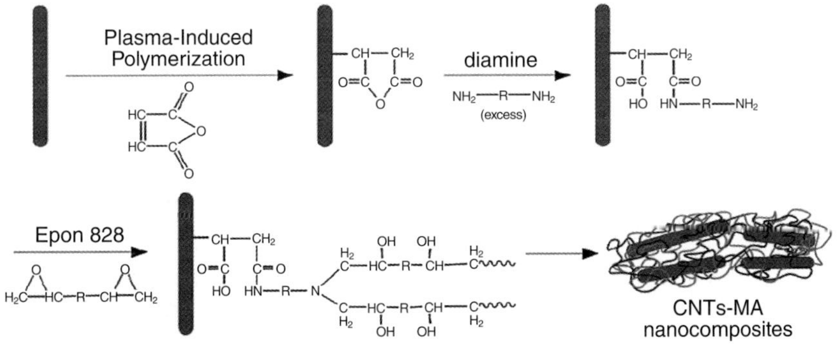

Figure 11.1 Schematic diagram for the preparation of mCNTs and the nanocomposites of mCNTs/epoxy. Reproduced from Ref. [17] with permission from American Chemical Society.

11.2.3
AN System

After the plasma treatment, as mentioned before, large number of radicals were generated on the surface of the CNTs by breaking the C=C sp^2 bonding of the nanotubes, and the acrylicnitrile (AN) monomer with a concentration of 0.1 M dissolved in N,N-dimethylformamide (DMF) was allowed to flow into the reactor. The reaction temperature of grafting polymerization was maintained at 50 °C for 3 h. The mCNTs-grafted AN (known as mCNTs-g-AN) was washed repeatedly with DMF to get rid of excess AN. The mCNTs-g-AN was then dried overnight in a vacuum oven at a temperature of 90 °C to remove traces of the solvent. The amination of mCNTs-g-AN (namely mCNTs-g-mAN) was prepared through the following steps. First, the mCNTs-g-AN was immersed in a hydroxylamine solution (20 wt.%) with the help of a mechanical stirrer at 70 °C for 3 h. Then the mCNTs-g-mAN was rinsed thrice with deionized water. The pristine epoxy resin was prepared by first mixing EPON 828 with diamine curing agent (N,N-bis(2-aminopropyl) polypropyleneglycol) under a temperature of 70 °C for 30 min. This mixture was outgassed in a vacuum oven for 2 h and then subjected to a curing process under a predefined thermal schedule (i.e., 80 °C for 2 h, 120 °C for 2 h, and 140 °C for 4 h). The mCNTs-g-mAN/epoxy nanocomposites were prepared through the following steps. First, the mCNTs-g-mAN were added to the curing agent, and the mixture was shear mixed for 1 h and sonicated for another 15 min. Subsequently, the EPON 828 resins were added and shear mixed with the mCNTs-g-mAN mixture at 80 °C for 30 min. Finally, the resultant mixture was outgassed in a vacuum oven for 2 h and then cast onto an aluminum plate. The curing cycle followed for mCNTs-g-mAN/epoxy nanocomposites was the same as that followed for pristine epoxy. The preparation procedure is illustrated briefly in Figure 11.2.

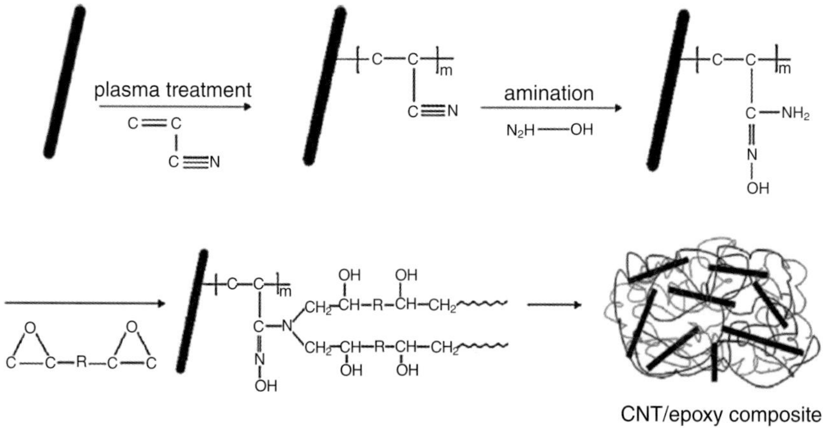

Figure 11.2 Schematic illustration for the preparation of mCNTs-g-mAN/epoxy nanocomposites. Reproduced from Ref. [19] with permission from Wiley.

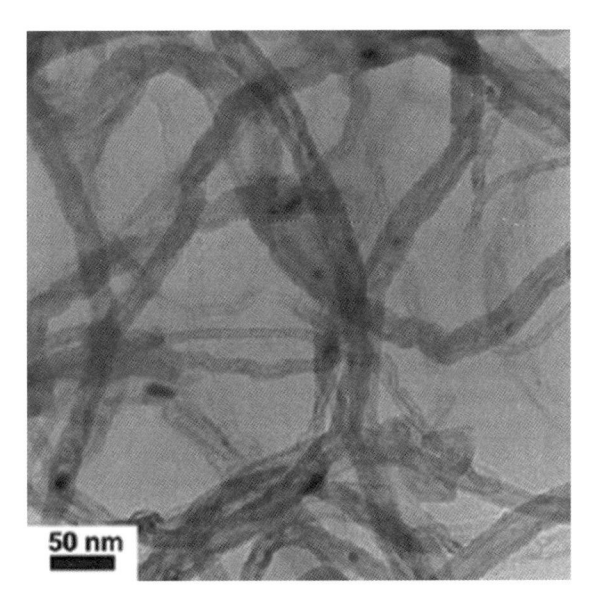

Figure 11.3 TEM image of the functionalized nanotubes (mCNTs). Reproduced from Ref. [17] with permission from American Chemical Society.

11.3
Results

11.3.1
Maleic Anhydride System

11.3.1.1 mCNTs/Epoxy Nanocomposites

CNTs Identification Figure 11.3 depicts the transmission electron microscopy (TEM) image of the functionalized MWCNTs, or the mCNTs after plasma-induced grafting polymerization. The image shows that the mCNTs, instead of showing crowded bundles, showed a less tangled organization, which proves that the grafted MA molecule efficiently promotes the dispersal of the CNTs. Moreover, the mCNTs are still kept in their original length after plasma modification, which is different from the results obtained by chemical acid modification, like HNO_3 and H_2SO_4 treatments [22].

Furthermore, detailed analysis of the X-ray photoelectron spectroscopy (XPS) spectra provides clear evidence that the CNTs have been chemically modified via the novel plasma method. As seen in Figure 11.4, the carbon C 1s peak, observed at 284.3–285.5 eV, is interpreted as the combination of the sp^2 C=C and sp^3 C–C structures of the u-CNTs, that is, the unmodified CNTs. After the plasma modification, mCNTs, the shoulder of the main peak, is composed of two peaks, located at 286.2 and 287.1 eV, which originate from the –C–O and –C=O structures [23, 24], respectively, as depicted in Figure 11.4b. It is believed that the CNTs have been functionalized by MA from these spectra.

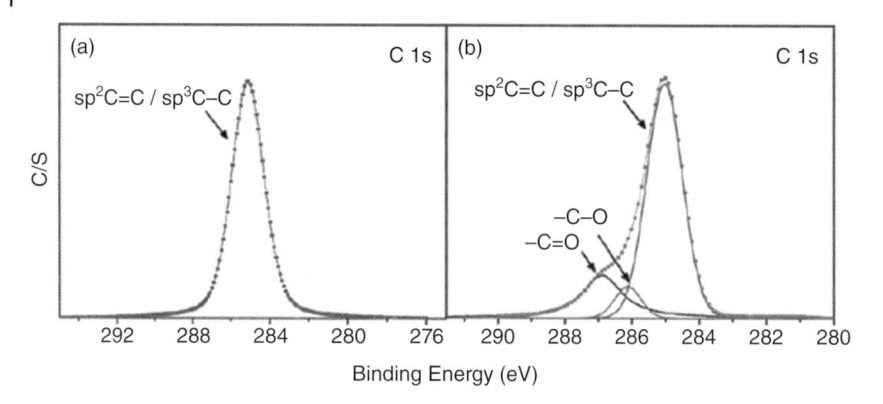

Figure 11.4 XPS core level spectra of C 1s of (a) u-CNTs and (b) mCNTs. Reproduced from Ref. [17] with permission from American Chemical Society.

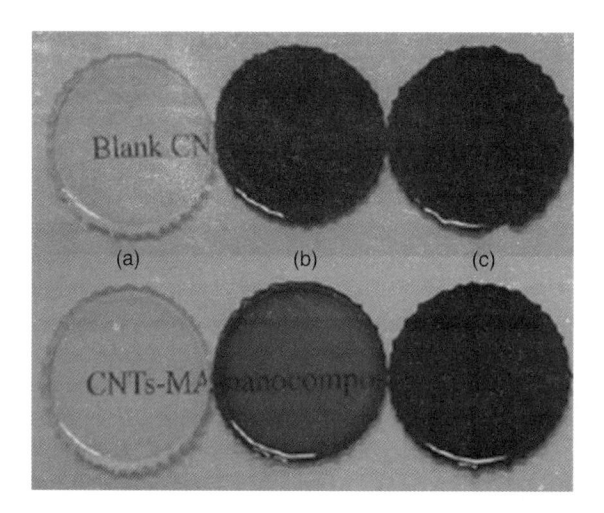

Figure 11.5 Photographs of CNTs/epoxy nanocomposites with various CNTs contents: (a) 0 wt.%, (b) 0.1 wt.%, and (c) 0.3 wt.%. The upper row is the u-CNTs/epoxy nanocomposite, and the lower row is the mCNTs/epoxy nanocomposite. Reproduced from Ref. [17] with permission from American Chemical Society.

Properties of CNTs Nanocomposites In the composite fabrication process, the mCNTs first reacted with the diamine and chemically bonded further with the epoxy via an amidation reaction. The amino groups attached to mCNTs can serve as curing agents, which leads to the formation of a highly cross-linked structure by covalent bonds between the mCNTs and the epoxy matrix. Figure 11.5 presents the snapshots of the CNTs/epoxy nanocomposites with various quantities of CNTs content. Both the u-CNTs/epoxy and the CNTs–MA/epoxy nanocomposites become much darker and opaque with higher CNTs loading. However, the mCNTs/epoxy nanocomposite with 0.1 wt.% is relatively translucent, as compared with the totally opaque u-CNTs/epoxy nanocomposite. The result reveals that the functionalized mCNTs dispersed homogenously on the nanoscale and is more compatible with the epoxy matrix, leading to better optical transparency at low mCNTs contents. In contrast, all the u-CNTs/epoxy nanocomposites appear to be opaque, which could be attributed to the weak chemical bonding between u-CNTs and the epoxy matrix, as well as the fact that they tend to aggregate inside the epoxy matrix. Moreover, as the content of the u-CNTs increases to 1.0 wt.%, the composite showed rougher morphology because of relatively large amount of aggregations of the u-CNTs formed inside the epoxy matrix.

The SEM images of the composite fracture surfaces show the dispersion of the CNTs in the epoxy matrix. The fracture surfaces of the epoxy composites loaded with the unfunctionalized u-CNTs show nonuniform dispersion and the tendency for the nanotubes to entangle as agglomerates. The phenomenon of agglomeration becomes more pronounced as the loading of the u-CNTs increases in Figure 11.6a–e. Besides, most of the u-CNTs show sliding and pulling out

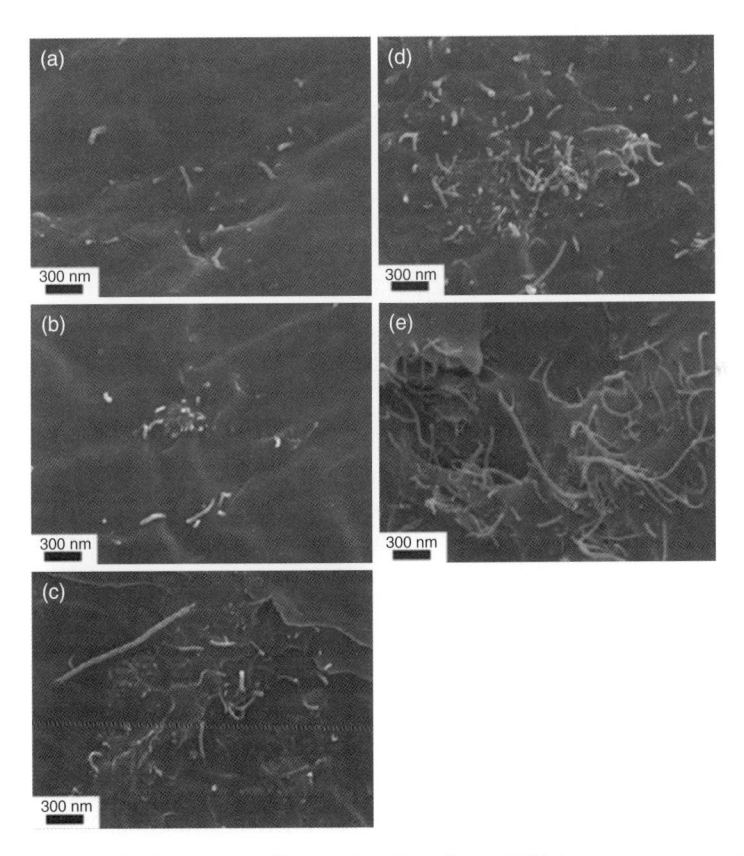

Figure 11.6 SEM images of fractured surface of the u-CNTs/epoxy nanocomposite loaded with different CNTs contents: (a) 0.1 wt.%, (b) 0.3 wt.%, (c) 0.5 wt.%, (d) 0.7 wt.%, and (e) 1.0 wt.%. Reproduced from Ref. [17] with permission from American Chemical Society.

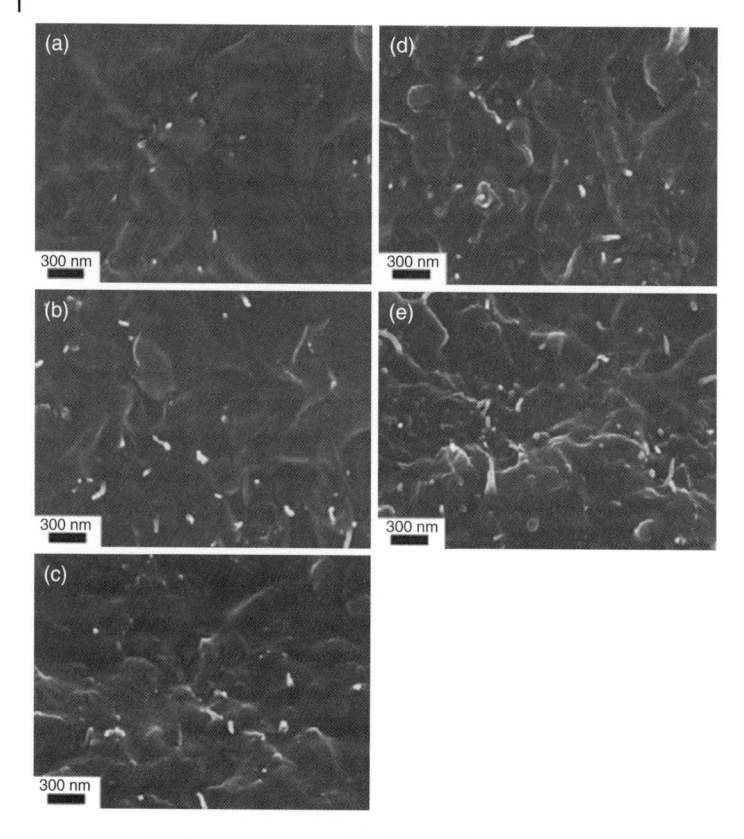

Figure 11.7 SEM images of fractured surface of the mCNTs/epoxy nanocomposites loaded with different CNTs contents: (a) 0.1 wt.%, (b) 0.3 wt.%, (c) 0.5 wt.%, (d) 0.7 wt.%, and (e) 1.0 wt.%. Reproduced from Ref. [17] with permission from American Chemical Society.

at the surface, suggesting a limitation of load transfer. By comparison, the mCNTs system shows good homogeneity and dispersion on the fracture surface. Massive bundles are found to have broken rather than just having pulled out of the surface, demonstrating that strong interfacial bonding exists between the epoxy matrix and the mCNTs. As the loading of the mCNTs increases, good dispersion is still achieved, and most of the mCNTs are embedded and tightly held to the matrix (Figure 11.7a–e).

Figure 11.8 shows the comparative plot of the AC conductivity of the CNTs/epoxy nanocomposite as a function of the CNT content (around 0.1–0.2 wt.%) than the u-CNTs/epoxy system (around 0.5–0.6 wt.%). The reason may be explained as follows: the well-dispersed mCNT inside the epoxy matrix can easily contact each other and, therefore, reduce the amount of CNTs needed to construct a conductive pathway. Moreover, with the increase in CNTs loading, the mCNT/epoxy system can achieve a conductivity of 2.6×10^{-4} S/m with 1.0 wt.% addiction, which is two orders higher than the u-CNTs/epoxy systems (3.5×10^{-6} S/m). A crucial param-

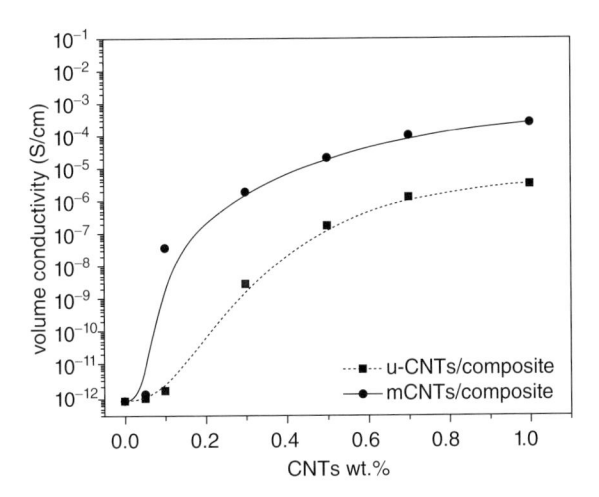

Figure 11.8 Comparative plot of the AC conductivity of the CNTs/epoxy nanocomposite as a function of the CNTs content. Reproduced from Ref. [17] with permission from American Chemical Society.

eter for the production of the conductive epoxy nanocomposites is the control of the degree of the grafting on the tube walls. The degree of grafting should not be too high to avoid significantly disturbing the o-electron system of the tube walls, or even insulating the CNTs from the epoxy matrix. However, it should be sufficient to provide good compatibility between the CNTs and the epoxy matrix. In this study, from XPS results, the percentage of the MA grafted onto the tube walls is 6.4 wt.%, which was controlled according to our designated value to obtain superior mechanical and electrical epoxy nanocomposites. The advantage of the plasma-induced grafting polymerization is that not only the degree of grafting on the CNTs can be controlled by altering plasma-treatment conditions but also that this method can be extended to other polymer systems to obtain various CNTs/polymer nanocomposites.

11.3.1.2 mCNTs/Polyimide Nanocomposites

CNT Identification As described in earlier section, the mCNTs/PI films were prepared in three stages. First, the dianiline compound (ODA) was reacted with the mCNT in order to obtain an mCNTs–ODA solution. This stage is very important in the present work because reaction of the anhydride group on the surface of the mCNTs with the amino group of ODA renders the mCNTs to be dispersible in the ODA solution. This amidation reaction was monitored by FT-IR spectroscopy, as shown in Figure 11.9. The characteristic peaks in the spectrum can be assigned as follows: 3300–3500 cm^{-1} (N–H stretching), 2900–3200 cm^{-1} (COOH, N–H stretching), 1275–1200 cm^{-1} (C–O–C bending), 1383 cm^{-1} (C–N stretching), 1710 cm^{-1} (C=O symmetric stretching), and 1780 cm^{-1} (asymmetric stretching). This IR spectrum thus confirms that the mCNTs reacted with the amine group of

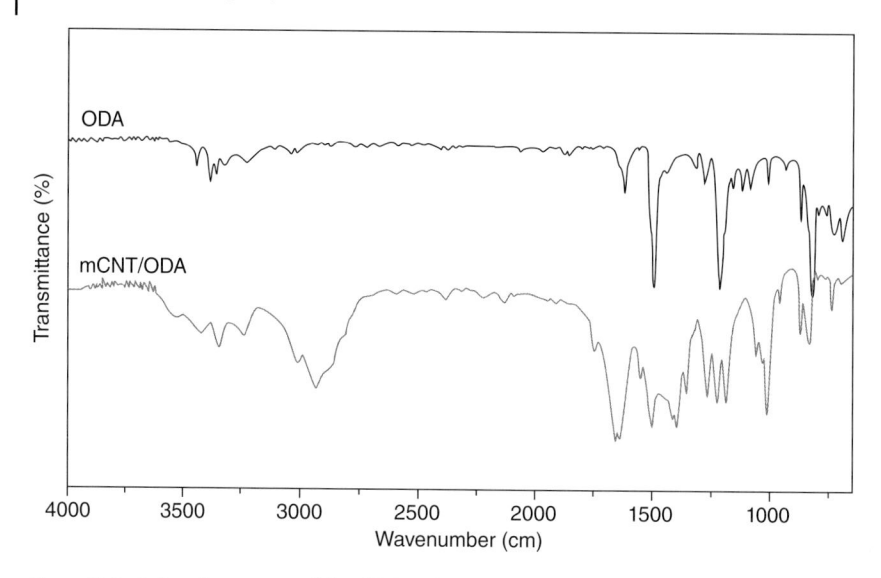

Figure 11.9 Infrared spectrum of the ODA and mCNTs–DA. Reproduced from Ref. [18] with permission from Elsevier.

ODA. In the second stage, the dianhydride compound (PMDA) was added to the mCNTs–ODA solution to prepare the PAA intermediate. Finally, the PAA intermediate was promptly applied to a thoroughly cleaned glass substrate and the cross-linked PI films were prepared by thermal imidization of the corresponding PAA films.

Properties of CNT Nanocomposites The conductivity versus mCNTs concentration curve is presented in Figure 11.10a. It can be seen that the conductivity increased by about five orders of magnitude as the mCNTs content within the PI was increased in the range from 0.0 to 4.1 vol.% (0.0–3.0 wt.%). The conductivity of the pure PMDA/ODA PI is about 1.2×10^{-10} S/cm. A sharp increase is observed as the mCNTs content in the PI is increased in the range between 0.05 and 1 vol.%. The conductivity at 0.5 vol.% is 9.26×10^{-6} S/cm, which is four orders of magnitude higher than the value at 0.05 vol.%. When the loading fraction of mCNTs in the PI exceeds 1 vol.%, the conductivity did not increase further. Therefore, these conductivity data suggest that an infinite network of percolated nanotubes starts to form above 0.5 vol.% [11]. The plot shown in the inset in Figure 11.10 represents a best fit to the experimentally measured conductivity data as a function of $v - v_c$ according to percolation theory [25]

$$\sigma = A(v - v_c)^t$$

where σ is the conductivity of the composite, v is the volume fraction of the CNT in the composite, v_c is the critical volume fraction, and A and t are fitted constants.

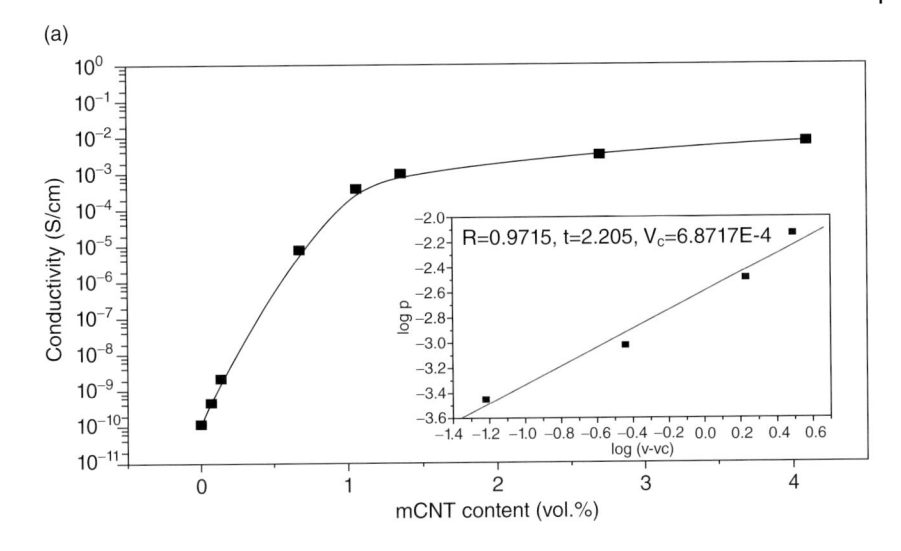

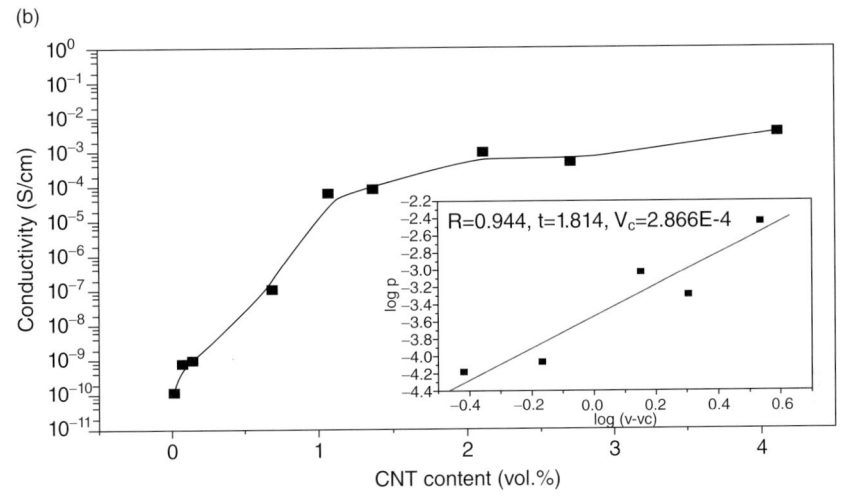

Figure 11.10 Volume conductivity of (a) mCNT/PI films and (b) CNT/PI films. Reproduced from Ref. [18] with permission from Elsevier.

In theory, t is dependent on the dimension of the lattice and the aspect ratio of the filler, with a value in the range of 1.6–2.0 for two- and three-dimensions, while experimental values between 1.3 and 3.1 have been reported [26]. The value of the constant A should approach the conductivity of the CNT [27]. A best fit to the data here resulted in values of $A = 6.872 \times 10^{-4}$ S/cm and $t = 1.814$, where v_c was assumed to be 0.5 vol.%. There were several reports containing the PI composite with CNTs being used as fillers. Zhu et al. [28] prepared the films by modifying the MWCNTs with acid, wherein they pointed out as the content of MWCNTs

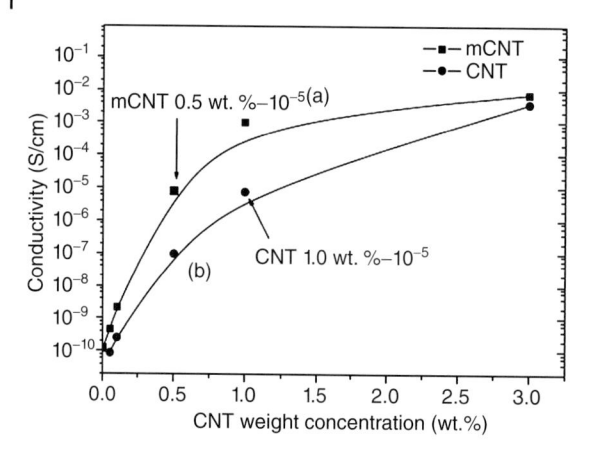

Figure 11.11 Comparison the volume conductivities at different mCNTs and u-CNTs contents in the PI matrix. Reproduced from Ref. [18] with permission from Elsevier.

increased from 7 to 10 wt.%, the conductivity will exhibit a sharp increase. Yuen et al. [29] showed that for amine-MWNT the conductivity clearly increased at 6.98 wt.%. Comparing with those literatures, the mCNTs in the PI matrix have lower percolation threshold. The conductivity versus CNT concentration curve is shown in Figure 11.10b according to percolation theory. The best fit to the data results in $A = 2.866 \times 10^{-4}$ S/cm and $t = 1.814$, where we assumed v_c to be 1.0 vol.%.

Figure 11.11 shows a plot comparing the conductivities of the mCNTs/PI and u-CNTs/PI composites under similar conditions. Dispersal of the mCNT within the PI clearly leads to a lower percolation threshold. This increase in conductivity is believed to be due to the reduced number of filler–filler hops required to cross a given distance by virtue of the good dispersion of mCNTs in the matrix [30].

Figure 11.12 shows TEM micrographs of u-CNTs/PI and mCNTs/PI at 0.5 wt.%. It is apparent that mCNTs showed better dispersion than u-CNTs in the PI matrix. The mCNTs thereby has a more pronounced effect on the percolation threshold than the u-CNTs [28, 31–34]. In the case of the original u-CNTs, they are relatively difficult to disperse in the solution by means of mechanical stirring.

11.3.2
AN System

11.3.2.1 CNT Identification

The CNTs grafted with AN will be denoted as CNTs-g-AN in the following text. The morphology of the u-CNTs and the CNTs-g-AN is shown in Figure 11.13. As shown in Figure 11.13a, average diameter of the u-CNT is about 10–16 nm. Further, in contrast to Figure 11.13b, the polymer layers in Figure 11.13a can be clearly seen and the average diameter of CNTs-g-AN is about 15–20 nm. The images reveal that the thickness of the grafted polymer layer is about 5 nm and there is a fine polymer growth on the surface of CNTs.

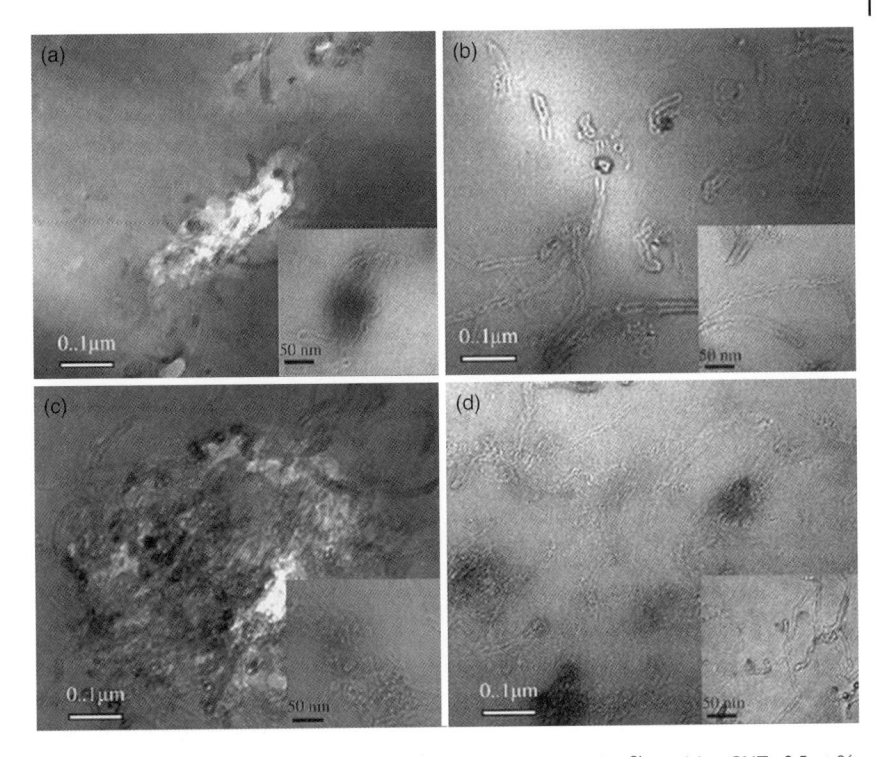

Figure 11.12 TEM images of u-CNTs/PI and mCNTs/PI composite films: (a) u-CNTs-0.5 wt.%, (b) u-CNTs-1.0 wt.%, (c) mCNTs-0.5 wt.% and (d) mCNTs-1.0 wt.%. Reproduced from Ref. [18] with permission from Elsevier.

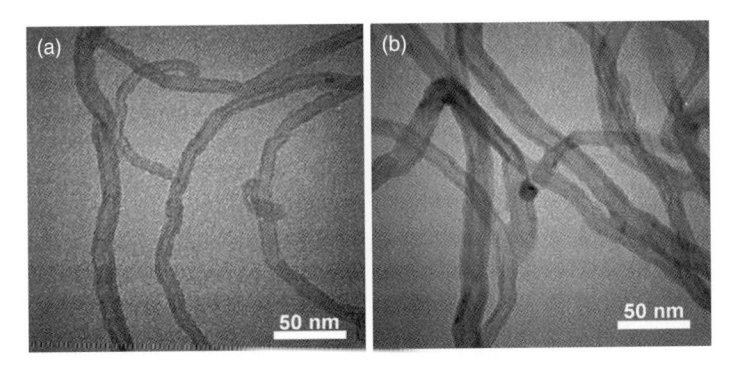

Figure 11.13 Transmission electron microscopy images of (a) u-CNTs and (b) CNTs-g-AN. Reproduced from Ref. [19] with permission from Wiley.

Figure 11.14 Digital photographs of CNTs-g-AN (A) and u-CNT (B) dispersed in DMF for (a) 0h and (b) 1 week. Reproduced from Ref. [19] with permission from Wiley.

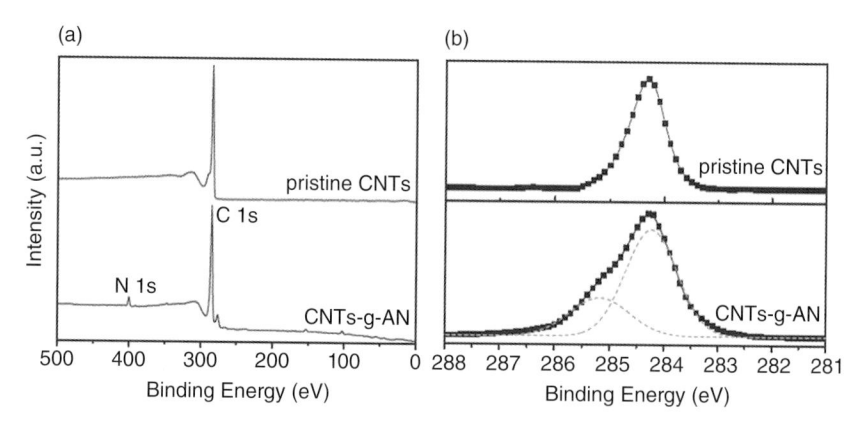

Figure 11.15 X-ray photoelectron spectroscopy survey (a) and deconvoluted (b) spectra of C 1s peak for u-CNT (top) and CNTs-g-AN (bottom). Reproduced from Ref. [19] with permission from Wiley.

Figure 11.14, however, demonstrates the comparison of the varied solubility levels of u-CNTs and the CNTs-g-AN in the DFM solvent. It is evident that all u-CNTs settled down in the dispersion in a week's time. However, in contrast to this, the dispersion of the CNTs-g-AN in the DMF solvent was indefinitely stable, and due to the chemical affinity between the polar-modified groups and the organic solvent, no sedimentation was observed. The presence of CNTs-g-AN was further confirmed by XPS, which provides rich information about the functional groups grafted onto the sidewall of the CNTs.

Figure 11.15 illustrates the comparison of the XPS spectra recorded on the u-CNTs and CNTs-g-AN. Figure 11.15a illustrates a broad scan image obtained (from 0 to 500 eV) from the XPS spectrum of the u-CNT. The carbon (C 1s, C KLL Auger transition) and nitrogen (N 1s, N KLL Auger transition) peaks represent the major constituents of the sample surface. According to the XPS analysis, the N 1s peak appeared in the mCNT-g-AN, which can be attributed to the fact that the polyacrylicnitrile (PAN) was grafted on the CNTs successfully after the plasma-induced grafting polymerization process. Furthermore, a detailed analysis of the XPS spectra (carbon) of the u-CNT and the mCNT-g-AN is shown in Figure 11.15b. As seen in Figure 11.15b, the carbon C 1s peak, observed at 281–292 eV, is interpreted as the combination of the sp^2 C=C and sp^3 C–C structures of the u-CNT. The mCNT-g-AN, also shown in Figure 11.15b, reveals that the C–C groups can be observed clearly at 284.3 eV whereas the C–N groups at 285.5 eV. According to N 1s atomic concentration of the XPS experimental data, the results reveal that the grafting percentage of the PAN onto the CNTs surface is 40.3 wt.%. PAN was grafted on the CNTs successfully after the plasma-induced grafting polymerization process. Furthermore, a detailed analysis of the XPS spectra (carbon) of the u-CNTs and the CNTs-g-AN is shown in Figure 11.15b. As seen in Figure 11.15b, the carbon C 1s peak, observed at 281–292 eV, is interpreted as the combination of the sp^2 C=C and sp^3 C–C structures of the u-CNT. The CNTs-g-AN, also shown in Figure 11.15b, reveals that the C–C groups can be observed clearly at 284.3 eV whereas the C–N groups at 285.5 eV. According to N 1s atomic concentration of the XPS experimental data, the results reveal that the grafting percentage of the PAN onto the CNTs surface is 40.3 wt.%.

11.3.2.2 Properties of CNT Nanocomposites

Figure 11.16 shows the TEM images of the cross-section of the CNTs/epoxy composite. Figure 11.16a represents the dispersion of the u-CNTs in epoxy resin. It was seen earlier that the CNTs agglomerate in the epoxy resin, and Figure 11.16 shows the TEM images of the cross-section of the CNTs/epoxy composite. Figure 11.16a represents the dispersion of the u-CNTs in epoxy resin.

It was seen earlier that the CNTs agglomerating in the epoxy resin and the mCNTs-g-mAN/epoxy composite also increased from 1×10^{-12} to 6.5×10^{-4} S/cm. In particular, when the content of functionalized CNTs is at 0.5 wt.%, the electrical conductivity of CNTs-g-mAN/epoxy composites suddenly increases, as shown in Figure 11.17. Unfortunately, for the u-CNTs, even increasing the CNTs content to

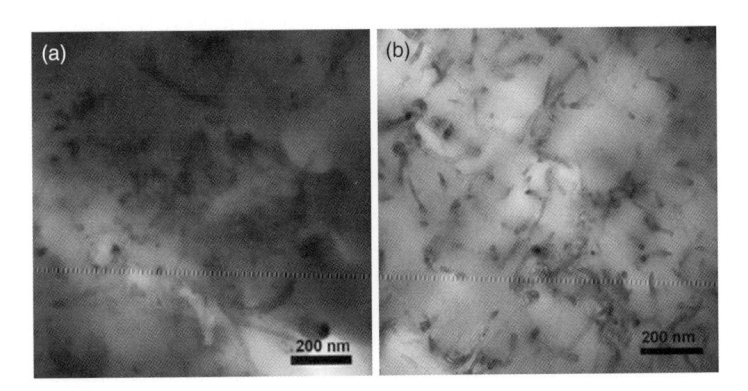

Figure 11.16 Transmission electron microscopy images of the manufactured composite: (a) u-CNT/epoxy; (b) CNTs-g-mAN/epoxy composite (3 wt.%). Reproduced from Ref. [19] with permission from Wiley.

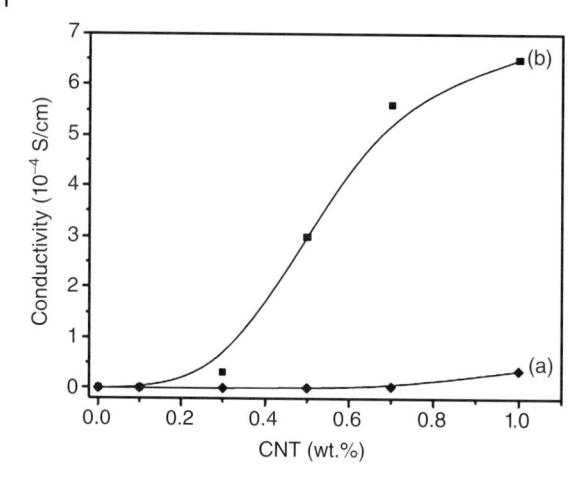

Figure 11.17 Electrical conductivity versus CNTs weight fraction of CNTs/epoxy composite: (a) u-CNTs; (b) CNTs-g-mAN. Reproduced from Ref. [19] with permission from Wiley.

3.0 wt.% cannot improve their electrical conductivity effectively. However, the CNTs-g-mAN/epoxy shows a lower percolation threshold (about 0.3–0.5 wt.%) than the u-CNTs/epoxy (around 0.5–0.7 wt.%). This may be explained with the fact that the well-dispersed CNTs-g-mAN in the polymer matrix can interact with each other easily. Moreover, with the increase in the loading of CNTs, the CNTs-g-mAN/epoxy system can achieve a conductivity of 6.5×10^{-4} S/cm with 1.0 wt.%, which is higher than the u-CNTs/epoxy system (3×10^{-5} S/cm).

11.4
Summary

The MA and AN molecule were successfully grafted onto the CNTs via a plasma-induced method, forming functionalized nanotubes, CNTs–MA and CNTs–AN. The functionalized nanotubes can be incorporated into the epoxy and PI composites through the formation of strong covalent bonding in the course of epoxy curing reactions and, as a result, become an integral structural component of the cross-linked epoxy system and PI system. The results obtained show that the plasma-modified CNTs/polymer nanocomposites, including epoxy and PI, had a good dispersion of nanotubes inside the polymer matrix, as well as an improvement in mechanical properties. Moreover, the plasma-modified CNTs/polymer nanocomposites not only demonstrated superior mechanical properties but also functioned as conductive nanocomposites for the purpose of electrostatic discharge and electromagnetic-radio interference protection materials. Therefore, the method for developing fully integrated nanotube–polymer composites by plasma-induced grafting polymerization can be extended to other polymer systems and provide a variety of hybrid materials.

References

1 Giannelis, E.P. (1996) *Adv. Mater.*, **8**, 29.
2 Agag, T., Koga, T., and Takeichi, T. (2001) *Polymer*, **42**, 3399.
3 Jiang, X.W., Bin, Y.Z., and Matsuo, M. (2005) *Polymer*, **46**, 7418.
4 Ajayan, P.M., Schadler, L.S., Giannaris, C., and Rubio, A. (2000) *Adv. Mater.*, **12**, 750.
5 Yu, A., Hu, H., Bekyarova, E., Itkis, M.E., Gao, J., Zhao, B., and Haddon, R.C. (2006) *Compos. Sci. Technol.*, **66**, 1187.
6 Thostenson, E.T., Ren, Z.F., and Chou, T.W. (2001) *Compos. Sci. Technol.*, **61**, 1899.
7 Cadek, M., Coleman, J.N., Barron, V., Hedicke, K., and Blau, W.J. (2002) *Appl. Phys. Lett.*, **81**, 5123.
8 Koganemaru, A., Bin, Y., Agari, Y., and Matsuo, M. (2004) *Adv. Funct. Mater.*, **14**, 842.
9 Hu, X.G., Wang, T., Wang, L., Guo, S.J., and Dong, S.J. (2007) *Langmuir*, **23**, 6352.
10 Smith, J.G., Connell, J.W., Delozier, D.M., Lillehei, P.T., Watson, K.A., Lin, Y., Zhou, B., and Sun, Y.P. (2004) *Polymer*, **45**, 825.
11 Ounaies, Z., Park, C., Wise, K.E., Siochi, E.J., and Harrison, J.S. (2003) *Compos. Sci. Technol.*, **63**, 1637.
12 Qin, S.H., Qin, D.Q., Ford, W.T., Herrera, J.E., Resasco, D.E., Bachilo, S.M., and Weisman, R.B. (2004) *Macromolecules*, **37**, 3965.
13 Li, H.M., Cheng, F.O., Duft, A.M., and Adronov, A. (2005) *J. Am. Chem. Soc.*, **127**, 14518.
14 Qu, L.W., Lin, Y., Hill, D.E., Zhou, B., Wang, W., Sun, X.F., Kitaygorodskiy, A., Suarez, M., Connell, J.W., Allard, L.F., and Sun, Y.P. (2004) *Macromolecules*, **37**, 6055.
15 You, Y.Z., Hong, C.Y., and Pan, C.Y. (2006) *Macromol. Rapid Commun.*, **27**, 2001.
16 Tseng, C.H., Wang, C.C., and Chen, C.Y. (2006) *Nanotechnology*, **17**, 5602.
17 Tseng, C.H., Wang, C.C., and Chen, C.Y. (2007) *Chem. Mater.*, **19**, 308.
18 Chou, W.-J., Wang, C.-C., and Chen, C.-Y. (2008) *Compos. Sci. Technol.*, **68**, 2208.
19 Chen, I.-H., Wang, C.-C., and Chen, C.-Y. (2010) *Plasma Processes Polym.*, **7**, 59.
20 Tseng, C.H., Wang, C.C., and Chen, C.Y. (2006) *J. Phys. Chem. B*, **110**, 4020.
21 Chou, W.-J., Wang, C.-C., and Chen, C.-Y. (2008) *Polym. Degrad. Stab.*, **93**, 745.
22 Li, Y.-H., Wang, S., Luan, Z., Ding, J., Xu, C., and Wu, D. (2003) *Carbon*, **41**, 1057.
23 Okpalugo, T.I.T., Papakonstantinou, P., Murphy, H., McLaughlin, J., and Brown, N.M.D. (2005) *Carbon*, **43**, 153.
24 Lee, W.H., Kim, S.J., Lee, W.J., Lee, J.G., Haddon, R.C., and Reucroft, P.J. (2001) *Appl. Surf. Sci.*, **181**, 121.
25 Sandler, J.K.W., Kirk, J.E., Kinloch, I.A., Shaffer, M.S.P., and Windle, A.H. (2003) *Polymer*, **44**, 5893.
26 Mark Weber, M.R.K. (1997) *Polym. Compos.*, **18**, 711.
27 Xi, Y., Ishikawa, H., Bin, Y.Z., and Matsuo, M. (2004) *Carbon*, **42**, 1699.
28 Zhu, B.-K., Xie, S.-H., Xu, Z.-K., and Xu, Y.-Y. (2006) *Compos. Sci. Technol.*, **66**, 548.
29 Yuen, S.M., Ma, C.C.M., Lin, Y.Y., and Kuan, H.C. (2007) *Compos. Sci. Technol.*, **67**, 2564.
30 Potschke, P., Fornes, T.D., and Paul, D.R. (2002) *Polymer*, **43**, 3247.
31 Bin, Y.Z., Kitanaka, M., Zhu, D., and Matsuo, M. (2003) *Macromolecules*, **36**, 6213.
32 Imai, Y., Fueki, T., Inoue, T., and Kakimoto, M.A. (1998) *J. Polym. Sci., Part A: Polym. Chem.*, **36**, 1031.
33 Allaoui, A., Bai, S., Cheng, H.M., and Bai, J.B. (2002) *Compos. Sci. Technol.*, **62**, 1993.
34 Bai, J.B., and Allaoui, A. (2003) *Compos. Part A*, **34**, 689.

12
Functionalization of Carbon Nanotubes by Polymers Using Grafting to Methods

Jean-Michel Thomassin, Robert Jérôme, Christine Jérôme, and Christophe Detrembleur

12.1
Introduction

The past two decades have witnessed the steadily increasing attention paid to carbon nanotubes (CNTs) due to great potential in various application fields such as field emission displays [1, 2], EMI shielding materials [3, 4], diodes [5, 6], sensors [7, 8], and hydrogen storage [9, 10]. The emergence of CNTs in these applications is the consequence of unique properties of high aspect ratio, high mechanical strength, and excellent thermal and electrical conductivities. However, one major limitation of CNTs is the lack of solubility in most of the solvents, organic or not, that makes their handling in fabrication processes very difficult. The challenging dispersion of CNTs within most of the polymer matrices accounts for the poor performances of the parent composites.

The pioneering work of Haddon *et al.* [11, 12] about the oxidative treatment of CNTs with the formation of carboxylic acid groups at defect sites of the surface was a decisive step toward surface functionalization. Esters [11] and amides [12] were indeed easily formed at the surface, either or not after the conversion of the carboxylic acid groups into more reactive acyl chlorides [13, 14]. Imparting reactivity to the CNTs surface was essential to promote interactions with surrounding media and to prepare stable fine dispersions within them. Nevertheless, the first step of oxidation by strong acids (HNO_3, H_2SO_4) has the deleterious effect of cutting the CNTs, such that all the unique properties that depend on a high aspect ratio, for example, thermal, electrical, and mechanical properties, are adversely affected, and so are the performances of the materials containing these surface-modified CNTs. Later on, the very rich chemistry of fullerenes [15, 16] was extended to the covalent functionalization of the CNTs sidewalls and substituted for the degrading oxidation treatment. Under appropriate conditions, small molecules were directly attached to the CNTs surface by using highly reactive reagents, that is, carbenes [17–19], nitrenes [20, 21], azomethine ylides [22, 23], Diels–Alder products [24], radicals [25–28], and plasma treatment [29, 30]. In addition to the covalent grafting of functional groups, strategies based on noncovalent interactions of small molecules with the CNTs surface, particularly by π–π interactions,

Surface Modification of Nanotube Fillers, First Edition. Edited by Vikas Mittal.
© 2011 Wiley-VCH Verlag GmbH & Co. KGaA. Published 2011 by Wiley-VCH Verlag GmbH & Co. KGaA.

were also investigated. Aromatic molecules, such as derivatives of pyrene [31–33] and porphyrin [34–36], were extensively used for this purpose. The advantage of this technique is to maintain the surface and thus the intrinsic properties of CNTs unmodified. Reviews have been recently devoted to the methods of covalent and noncovalent modification of CNTs [37, 38].

Because the attachment of small molecules to the CNTs surface was beneficial to the tuning of their properties, the question of the grafting of macromolecules was addressed. Indeed, the large size of grafted polymers is expected to increase the CNTs interaction with the surrounding medium at the benefit of their processability and the production of advanced nanocomposites with unique properties [39]. The large range of reactivity and structure exhibited by the known polymers and their possible post-functionalization contribute advantageously to the tuning of the properties of the CNTs surface.

In this respect, the spectacular progress recently reported in controlled/living polymerization techniques [40–44] accounts for the unprecedented level of diversity and complexity exhibited by synthetic polymers as a result of the fine control of molecular weight, functionality, architecture, and composition. Therefore, the possible grafting of tailored polymers onto CNTs opens a straightforward way to the "on-demand" surface design of CNTs and the merge of their unique properties with those of polymers so leading to materials well suited to novel advanced applications. Most of the covalent and noncovalent methods of grafting small molecules onto CNTs were extended to the surface attachment of polymers. Two main strategies must however be emphasized, that is, the "grafting from" method where the polymer is grown from initiators immobilized onto the CNTs surface [45, 46] and the "grafting to" technique that consists in directly anchoring a preformed polymer [47–54]. To avoid any redundancy with other chapters of this book, the focus will be placed here on the main "grafting to" approaches. Moreover, because of space restriction, all the methods reported in the literature cannot be considered in detail. Representative examples will thus be discussed and illustrated in this chapter.

12.2
Overview of the "Grafting to" Methods

The scientific literature reports on the chemical modification of both single-walled carbon nanotubes (SWNTs) and multiwalled carbon nanotubes (MWNTs) by the aforementioned procedures. The main difference between these two types of functionalized nanotubes has to be found in the analytical techniques used to have them characterized. Reviews about these specific characterization methods are now available [37]. Therefore, no discrimination between SWNTs and MWNTs will be made in this chapter.

The "grafting to" method refers to the attachment of a preformed polymer onto the surface of CNTs. Two techniques have, however, to be distinguished, depending on whether the CNTs have been functionalized before the polymer grafting. So, one of these techniques requires a preliminary step that makes the CNTs

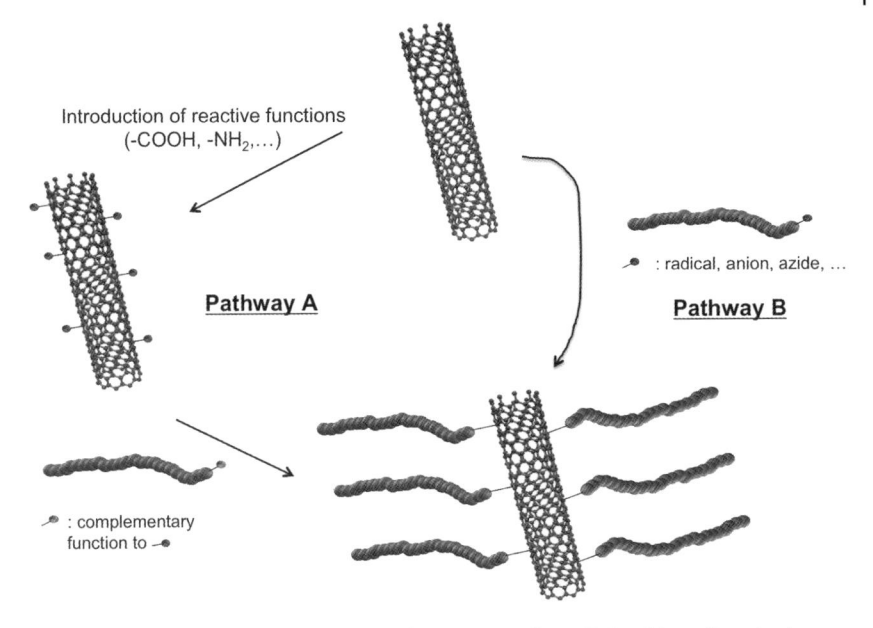

Scheme 12.1 The two-step (pathway A) and one-step (pathway B) "grafting to" methods.

surface reactive toward the functional groups available at the chain-end(s) or along polymer backbones (Scheme 12.1, pathway A). In the second technique, the polymer is directly grafted onto the CNTs by a specific addition reaction onto the sp^2 carbons of the nanotubes (Scheme 12.1, pathway B). Although some authors classify the physical adsorption of polymers onto CNTs by π–π stacking [32, 55–58] as a "grafting to" method, this type of noncovalent functionalization will, however, be the topic of a separate chapter in this book.

12.2.1
The Two-Step Method

The first step consists in activating the CNTs surface for making it reactive toward the polymer to be grafted. A wide range of chemical functions have been considered, which are listed in Table 12.1. The oxidative treatment by a H_2SO_4–HNO_3 mixture (most often under ultrasonic irradiation) remains the most common procedure that leads to hydroxyl- and carboxyl-functionalized CNTs. However, this method is very aggressive, deteriorates the electronic structure of the CNTs by creating numerous defects (sp^3 carbons), and seriously decreases their average length [114]. The carboxylic acid groups may be either converted into more reactive acyl chlorides or activated by, for example, dicyclohexylcarbodiimide (DCC), before being involved in esterification or amidation reactions (Scheme 12.2), with the purpose to have amines [98, 99, 101–103], alkynes [108, 109], pyridyl disulfide (S-S-pyridine) [112], and many other groups available on the surface. Thionyl

Table 12.1 Pairs of mutually reactive groups involved in the two-step "grafting onto" approach.

Reactive groups on the CNTs surface	Reactive groups on the polymer to be grafted	Type of polymer[a]
–COOH	–OH	PEG [59], PVA [60, 61], PIPA [62]
–COOH	–NH$_2$ or –NHR	Chitosan [63], PUR [64], PI [65]
–COOH	-Oxazoline	SAN [66]
–COOH	–Si(OEt)$_3$	PI [67, 68]
–COCl	–OH	PEG [69–74], PLLA [75, 76], PMMA [77], PS [71], Polyester [78], PVA [79], PHEMA [80], PHET [81]
–COCl	–NH$_2$ or –NHR	PEG [82, 83], Chitosan [84, 85], PPEIEI [79, 86, 87], PABS [74, 88]
–COCl	Anion	PS [89, 90], PBD [89]
–OH	–COOH	PCL [91], PLA [92]
–OH	–COCl	PAC [93–96]
–OH	-Anhydride	SMA [97]
–OH	–Si(OEt)$_3$	PI [67, 68]
–NH$_2$	–COOH	PThio [98], PA6 [99]
–NH$_2$	-anhydride	PE-g-MA [100], PP-g-MA [101, 102]
–NH$_2$	-carbonate	PC [103]
–Alkyne	–N$_3$	PS [104, 105], SPS [106], PDMA-PNIPAM [107], P(BrAzPMA) [108, 109]
-Maleic anhydride	–OH or –NH$_2$	PUU [110]
–NCO	–OH	PEG [111]
–S-S-pyridine	–SH	PHPMA [112]
-Anion	–Cl	CPP [113]

a) PEG, poly(ethylene glycol); PVA, poly(vinyl alcohol); PIPA, poly(N-isopropyl acrylamide); PUR, poly(urethane); PI, poly(imide); SAN, poly(styrene-co-acrylonitrile); PLLA, poly(L-lactide); PMMA, poly(methylmethacrylate); PS, poly(styrene), PHEMA, poly(2-hydroxyethyl methacrylate); PHET, poly[3-(2-hydroxyethyl)-2,5-thienylene]; PABS, poly(aminobenzene sulfonic acid); PPEI-EI, poly(propionylethylenimine-co-ethylenimine); PBD, poly(butadiene); PCL, poly(ε-caprolactone); PLA, poly(lactide); PAC, poly(acryloyl chloride); SMA, poly(styrene-co-maleic anhydride); Pthio, poly(thiophene); PA6, polyamide-6; PE-g-MA, poly(ethylene-graft-maleic anhydride); PP-g-MA, poly(propylene-graft-maleic anhydride); PC, poly(carbonate); SPS, sulfonated poly(styrene); PDMA-PNIPAM, poly(N,N-dimethylacrylamide)-poly(N-isopropylacrylamide); P(BrAzPMA), (poly(3-azido-2-(2-bromo-2-methylpropanoyloxy)propyl methacrylate)); PUU, poly(urea-urethane); PHPMA, poly(N-(2-hydroxypropyl)methacrylate); CPP, chlorinated poly(propylene).

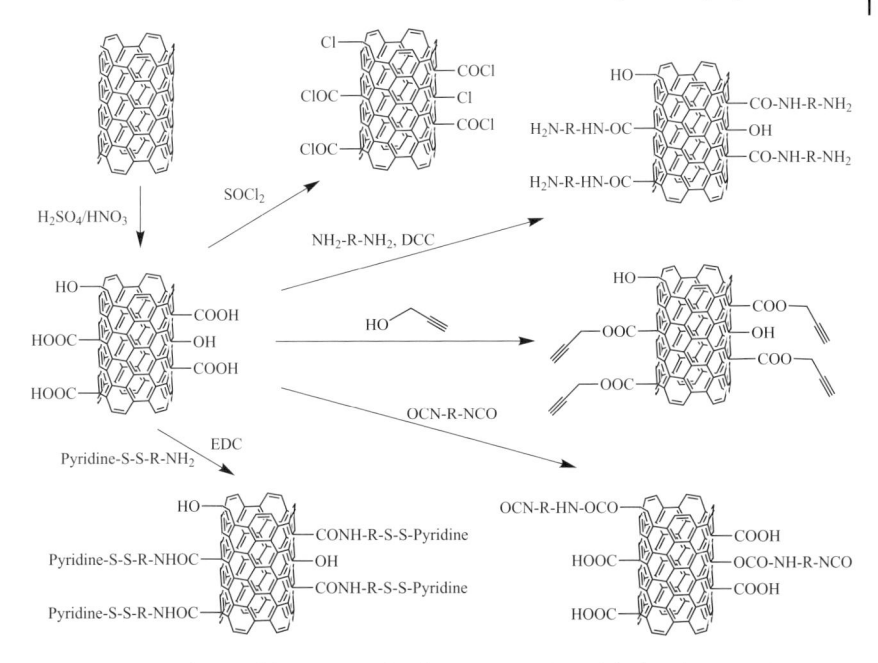

Scheme 12.2 Oxidation of the CNTs surface by HNO$_3$/H$_2$SO$_4$ and further derivatizations.

chloride and oxalyl chloride are commonly used for generating reactive acyl chlorides onto the CNTs surface. The hydroxyl groups at the surface of the oxidized CNTs can be reacted with an excess of diisocyanate (toluene diisocyanate, TDI, for instance), and accordingly, converted into isocyanates [111], reactive toward alcohols and amines.

An alternative pre-functionalization of the CNTs consists of a diazonium salt decomposition chemistry, as reported by Tour et al. [115] and illustrated by the surface attachment of alkyne groups by reaction of native CNTs with p-aminophenyl propargyl ether in the presence of isoamyl nitrite (Scheme 12.3) [104]. The interest paid to alkynes has to be found in the possible anchoring of azide-bearing polymers through the well-established copper(I)-catalyzed Huisgen [2+3] "click" cycloaddition [116, 117], as it will be detailed later.

Radical reaction of maleic anhydride with CNTs is an easy way to prepare anhydride-functionalized CNTs (Scheme 12.4), reactive toward hydroxyl- and amino-containing polymers [110].

The second step of this strategy, thus the polymer grafting onto pre-functionalized CNTs, is discussed hereafter.

12.2.1.1 "Grafting-onto" CNTs Pre-functionalized by Carboxylic Acid Groups (CNTs-COOH)

Esterification and amidation in the presence of coupling reagents, such as DCC, were the first reactions implemented for having polymers grafted onto CNTs-COOH. Among others, poly(hexamethylene carbonate-co-caprolactone)diol [118],

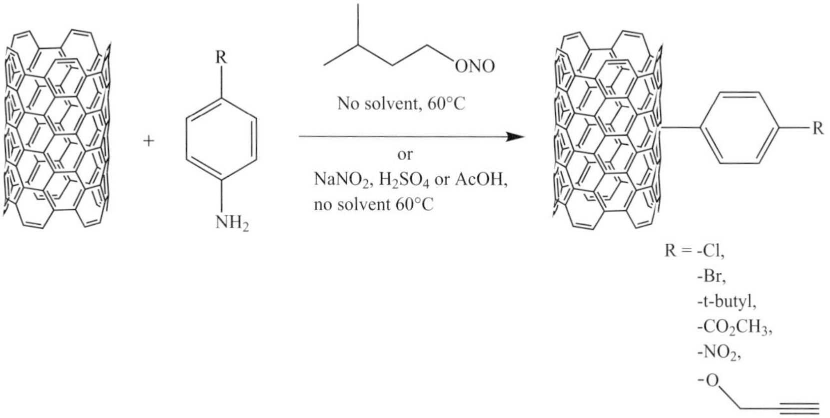

Scheme 12.3 CNTs functionalization by diazonium salt decomposition chemistry.

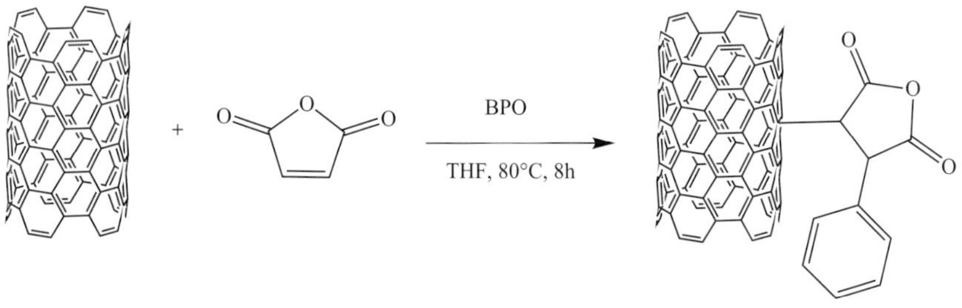

R = -Cl, -Br, -t-butyl, -CO_2CH_3, -NO_2, -O___

Scheme 12.4 Preparation of anhydride-functionalized CNTs.

polycarbonate (PC) [119], polyethylene glycol (PEG) [59], poly(N-isopropyl acrylamide) (PIPA) [62], poly(vinyl alcohol) [60, 61], and chitosan [63] were successfully grafted according to this approach. PEG-grafted CNTs (CNTs-g-PEG) [59] allowed highly selective chemical gas sensors to be built up, whereas the mechanical properties of chitosan were significantly improved by the proper dispersion of 40 wt.% of CNTs-g-chitosan [63]. Polyamidoamine (PAMAM) was also grafted onto CNTs-COOH [120, 121], and the modified CNTs were dispersed in an epoxy resin before curing, so leading to nanocomposites with highly improved mechanical properties.

Carboxylic acids are also known for reactivity toward oxazoline, as illustrated by Shi et al., who converted first the nitrile groups of a styrene-acrylonitrile copolymer (SAN) into oxazolines followed by the copolymer grafting onto CNTs-COOH in a melt-mixing process (Scheme 12.5) [66].

The carboxylic acid and hydroxyl groups available at the oxidized CNTs surface were also used to anchor polymers end-capped by an alkoxysilane group, according to a sol–gel procedure (Scheme 12.6) [67, 68]. Low color and highly flexible polyimide nanocomposite films were accordingly prepared.

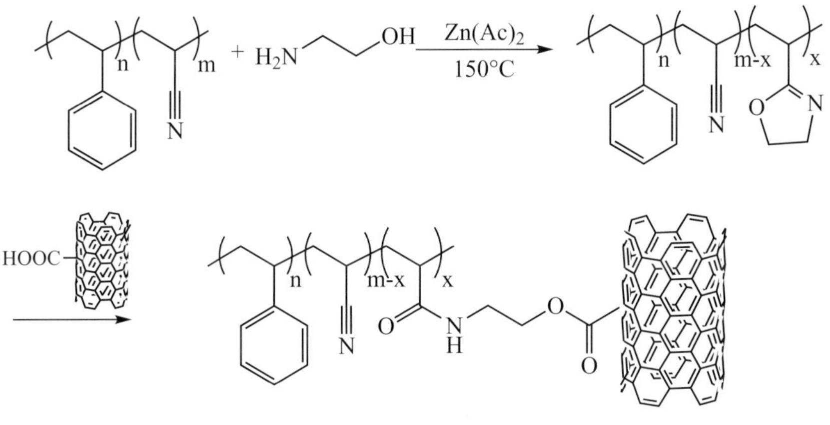

Scheme 12.5 Grafting of oxazoline-bearing polymers onto CNTs-COOH.

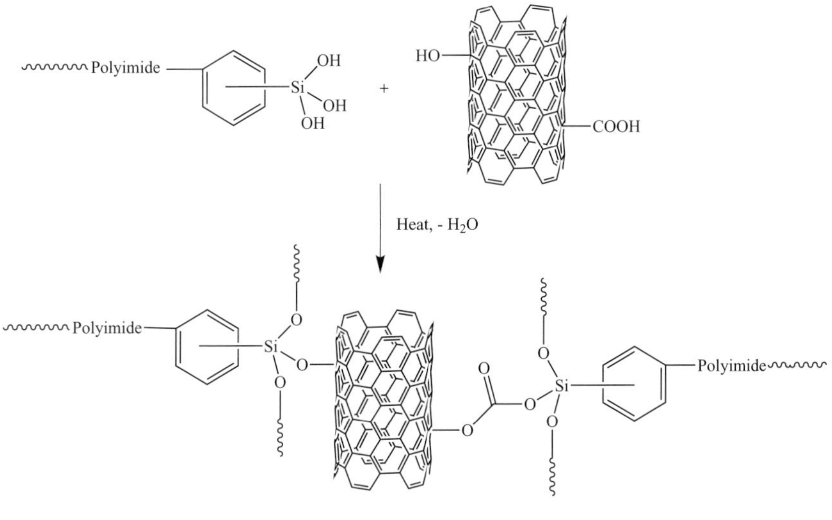

Scheme 12.6 Grafting of polymers end-capped by an alkoxysilane group onto oxidized CNTs.

Recently, Chen *et al.* developed a solid-state grafting method by exposing a polyurethane/CNTs-COOH film to microwave radiations [64]. The ability of CNTs to absorb these radiations accounts for the initiation of the grafting reaction (Scheme 12.7). This approach has the advantage to prevent the nanotubes from being damaged, as they are during conventional physical mixing under high shear [3].

12.2.1.2 Grafting onto CNTs Pre-functionalized by Acyl Chloride Groups (CNTs-C(=O)Cl)

CNTs-C(=O)Cl are by far the preferred modified CNTs for the polymer anchoring. They are indeed more reactive than the carboxylic acid counterparts in esterification and amidation reactions. This is beneficial to the grafting efficiency.

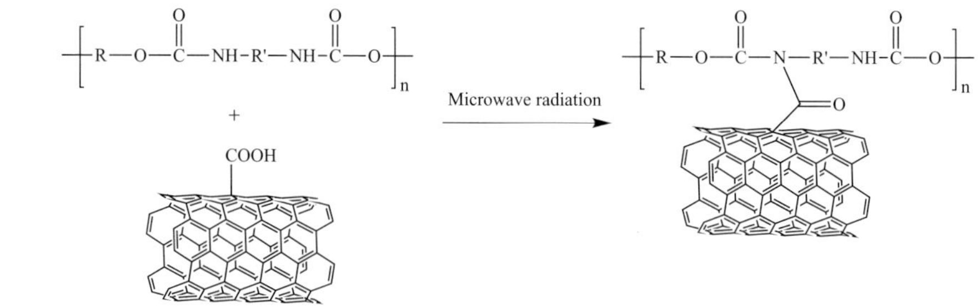

Scheme 12.7 Microwave-assisted solid-state grafting of polyurethane onto CNTs-COOH.

This strategy was mainly used for the grafting of water-soluble polymers, such as PEG [69, 70, 73] and poly(aminobenzene sulfonic acid) (PABS) [74, 88], via the hydroxyl and amino end-groups, respectively. Huang et al. compared the grafting efficiency of α,ω-diamino-PEG onto CNTs-COOH by three different approaches, that is, direct thermal heating, carbodiimide-activated coupling, and acylation amidation [79, 83]. Direct thermal heating was expectedly less efficient than the two other methods that gave similar grafting yields. Poly(L-lactic acid) was also grafted onto CNTs-C(=O)Cl [75, 76]. It must be noted that less polymer was grafted when the molecular weight was increased. Indeed, the access to the reactive sites is restricted by the previously grafted chains. This effect of steric hindrance is the rule whenever "grafting onto" reactions are concerned. Hydroxyl-terminated poly(methyl methacrylate) was also covalently grafted to CNTs-C(=O)Cl in toluene [77]. Polymers with reactive sites distributed along the main backbone, such as chitosan [84, 85], poly(propionylethylenimine-co-ethylenimine) (PPEI-EI), [79, 86, 87] poly(2-hydroxyethyl methacrylate) (PHEMA) [80], poly(vinyl alcohol) [87], and poly[3-(2-hydroxyethyl)-2,5-thienylene] (PHET) [81] were also attached to CNTs by this successful approach. However, deleterious cross-linking by multifunctional polymers had to be restricted by working at high dilution.

Polystyryl or polybutadienyl anions, prepared by anionic polymerization, were also successfully reacted with CNTs-C(=O)–Cl [89, 90]. However, direct reaction of the macromolecular carbanions with pristine CNTs gave similar results.

12.2.1.3 Grafting onto CNTs Pre-functionalized by Hydroxyl Groups (CNTs-OH)

A variety of carboxylic acid-functionalized polymers were grafted onto hydroxyl containing CNTs. For instance, aliphatic polyesters such as poly(ε-caprolactone) (PCL) [91] and polylactide [92], bearing carboxylic acids along the backbone were successfully immobilized onto CNTs-OH by esterification in the melt (Scheme 12.8). To improve the grafting efficiency, the hydroxyl content of oxidized CNTs was increased by converting the surface carboxylic acid groups into acyl chlorides followed by reaction with an excess of 1,6-hexanediol. Compared to pristine CNTs, the modified CNTs improved significantly the thermal stability and the mechanical properties of polymeric matrices (0.5–3 wt.% CNTs).

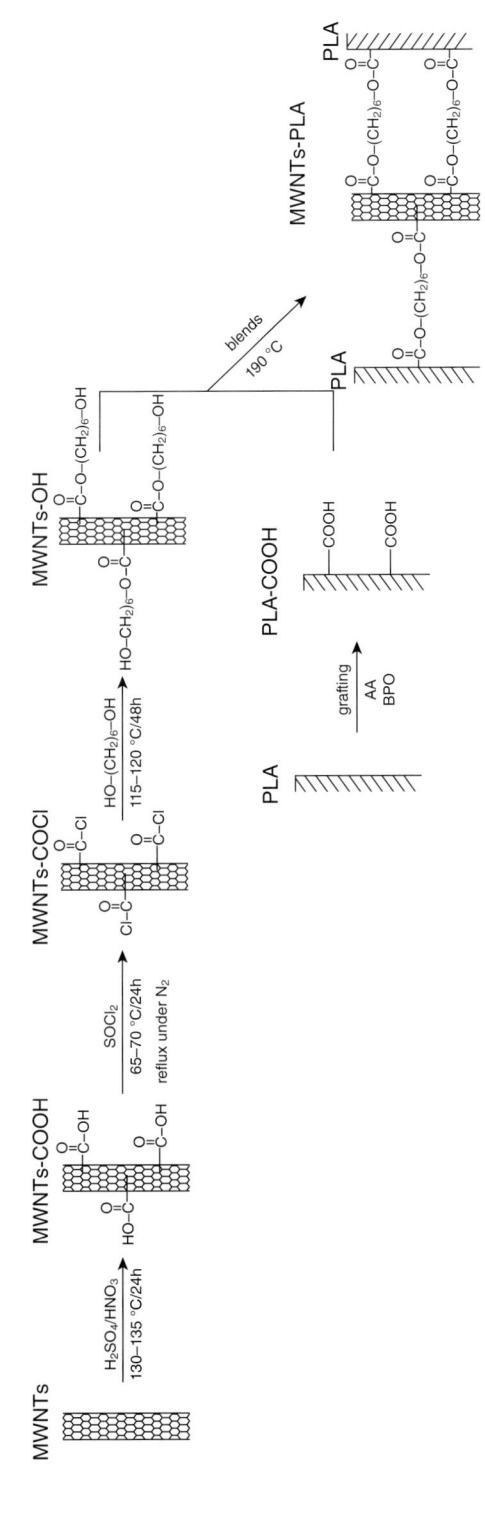

Scheme 12.8 Grafting of carboxylic acid containing polylactide (PLA) onto CNTs-OH in the melt. Figure reproduced from Ref. [92] with permission from Elsevier.

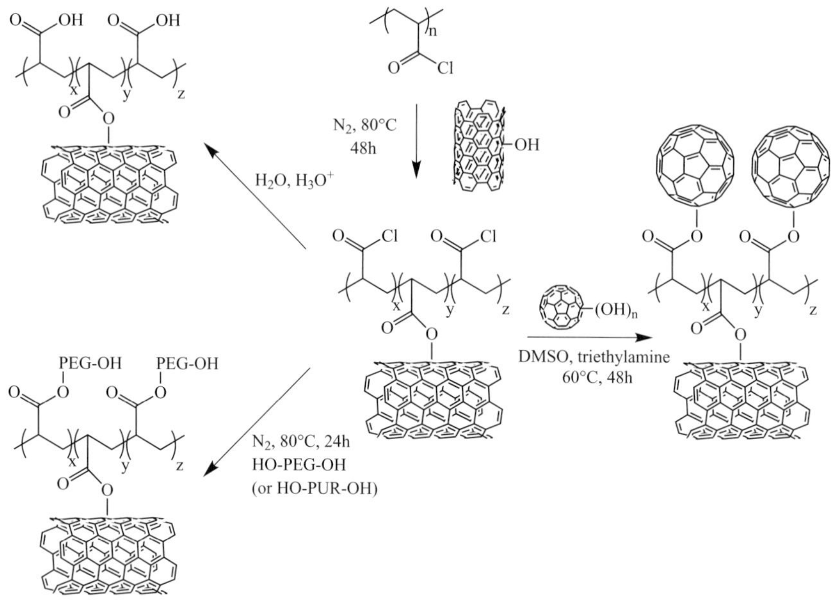

Scheme 12.9 Grafting of poly(acryloyl chloride) onto CNTs and further derivatizations.

Li et al. extensively studied the grafting of poly(acryloyl chloride) (PAC) onto CNTs-OH. The unreacted acyl chloride groups at the CNTs surface were either hydrolyzed with formation of water-soluble CNT-g-polyacrylic acid [94] or used to anchor hydroxyl-containing polymers, such as PEG [93] and polyurethane [95], and fullenerol particles [96] (Scheme 12.9). This strategy has the advantage that the acyl chloride functions are more accessible than chemical groups directly attached to the oxidized CNTs surface, resulting in a very high grafting efficiency (polymer content >80 wt.%).

12.2.1.4 Grafting onto CNTs Pre-functionalized by Primary Amines (CNTs-NH$_2$)

CNTs-NH$_2$ were prepared by amidation of CNTs-COOH with 1,2-diaminoethane in the presence of (N-[(dimethylamino)-1H-1,2,3-triazolo- pyridine-1-ylmethylene]-N-methylmethanaminium hexafluorophosphate N-oxide) [4–6] as a coupling agent [102, 122] (Scheme 12.10). Polyethylene (PE-g-MA) [100] and polypropylene (PP-g-MA) [101, 102] grafted by maleic anhydride were successfully grafted onto CNTs-NH$_2$ in a microextruder at 220 °C (Scheme 12.10). The beneficial impact of these modified CNTs on the mechanical properties of the polyolefins was clearly highlighted. Melt-blending of carboxylic acid end-capped polyamide-6 with CNTs-NH$_2$ was also successfully carried out [99], and the modified CNTs proved to be reinforcing agents for nylon-6.

PC/CNTs–NH$_2$ nanocomposite was also prepared by a solid-state grafting method assisted by microwave irradiation at high temperature (110 °C) [103]. The superiority of this procedure compared to solution reaction without microwave irradiation was clearly established.

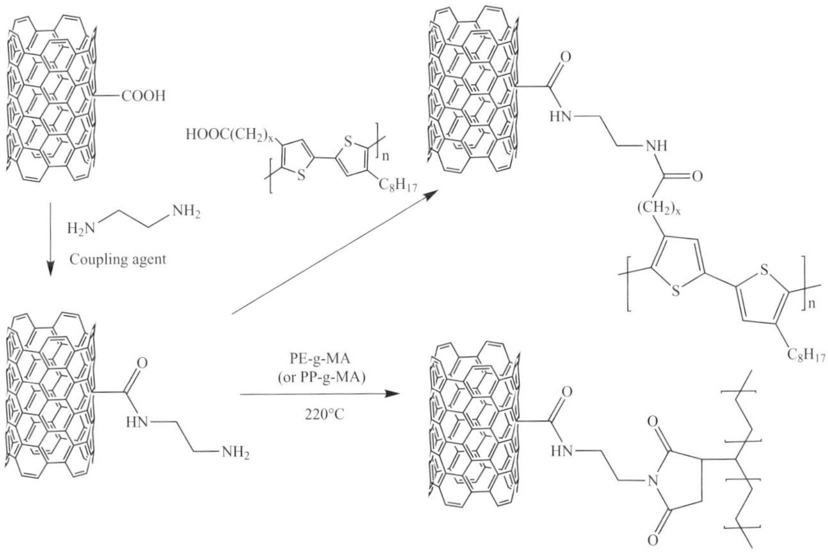

Scheme 12.10 Examples of polymer grafting onto CNTs-NH$_2$.

Finally, CNTs-NH$_2$ were also grafted by carboxylic acid containing polythiophene (Scheme 12.10) [98]. The electrochemical activity of polythiophene was however adversely affected by the grafting reaction, more likely because the π-conjugated system of the polymer was perturbed.

12.2.1.5 Grafting onto CNTs Pre-functionalized by Alkynes (CNTs-alkyne)

Alkyne-functionalized CNTs are building blocks dedicated to the grafting of azide end-capped polymers through the well-known copper(I)-catalyzed Huisgen [2+3] "click" cycloaddition [104]. As an example, the azide end-group of polystyrene (PS-N$_3$) was easily coupled to CNTs-alkyne using Cu(I) as catalyst at 60 °C for 24 h in dimethylformamide (Scheme 12.11). Under the same experimental conditions, no grafting was observed when pristine CNTs were used instead of CNTs-alkyne, which showed that the "click" reaction was responsible for the grafting and not the direct coupling of nitrene onto the CNTs surface. This grafting strategy is not only efficient but also versatile because any polymer or small molecule that bear azido-group(s) can be grafted by this simple process under mild experimental conditions.

Partial sulfonation of the PS-functionalized CNTs made the CNTs pH responsive and soluble in water at pH between 3 and 13 [106]. At high pH, the sulfonate groups at the surface built up an electrostatic repulsion barrier that prevented the CNTs from aggregating.

The versatility of the Huisgen cycloaddition process is also illustrated by grafting onto CNTs-alkyne a polymer bearing both azido-groups (for the click reaction) and activated bromides (initiators for atom transfer radical polymerization, ATRP) along the backbone (poly(3-azido-2-(2-bromo-2-methylpropanoyloxy)propyl

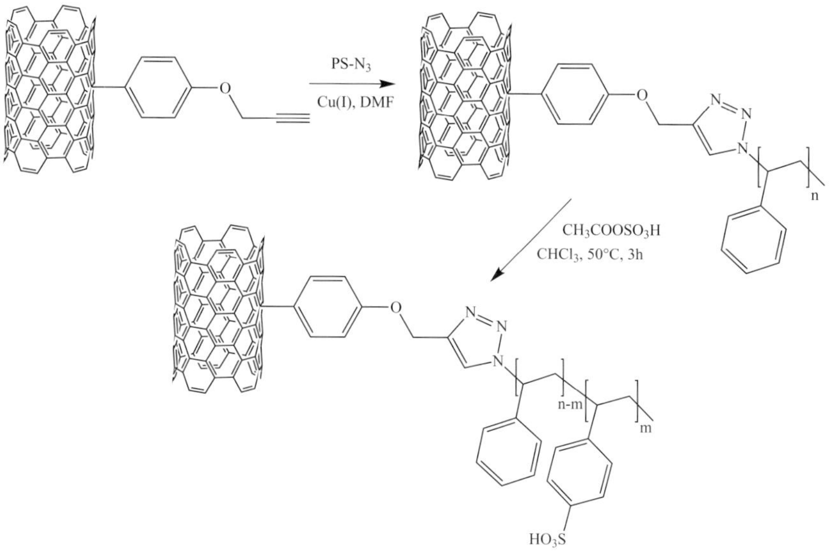

Scheme 12.11 Grafting of polystyrene and sulfonated polystyrene onto CNTs by the copper(I)-catalyzed Huisgen [2+3] click cycloaddition.

methacrylate)) (Scheme 12.12) [108, 109]. The excess of unreacted azido-groups was then used for anchoring alkyne functional polymers by the same reaction, while the activated bromides initiated the polymerization of a vinyl monomer by ATRP. This combination of "grafting onto" and "grafting from" approaches is quite a valuable route to have at least two different types of polymers grafted onto the CNTs surface.

Azide-terminated thermoresponsive poly(N,N-dimethylacrylamide)-b-poly(N-isopropylacrylamide) (N_3-PDMAA-b-PNIPAm) diblock copolymers [107] were also grafted onto CNTs by the Huisgen cycloaddition (Scheme 12.13) in water at 55 °C. Under these conditions, the PNIPAm sequence is hydrophobic, which triggers the self-assembly of the block copolymer into micelles with a hydrophobic PNIPAm core and a hydrophilic PDMAA corona. The azide end-groups of the corona chains were thus exposed at the periphery of the micelles and available at high concentration to reaction with CNTs-alkyne. This elegant coupling between mutually reactive micelles and CNTs is superior to the grafting onto the copolymer coils in solution in terms of grafting density.

Alkynes at the CNTs surface were also reacted with halides bound to polymer chains in the presence of a palladium ($PdCl_2(PPh_3)_2$) catalyst. Mild experimental conditions and high reaction efficiency make this reaction very attractive. It is nicely illustrated by the grafting of PS-bearing pendent benzyl chloride groups (poly(styrene-co-chloro-methylstyrene), PS-co-PCMS) onto CNTs-alkyne at 80 °C for about 6 h [105]. A poly(styrene)-b-poly(styrene-co-chloro-methylstyrene) (PS-b-(PS-co-PCMS)) diblock copolymer was also successfully used as a reactive chain (Scheme 12.14). Importantly, compared to the other "grafting onto" approaches,

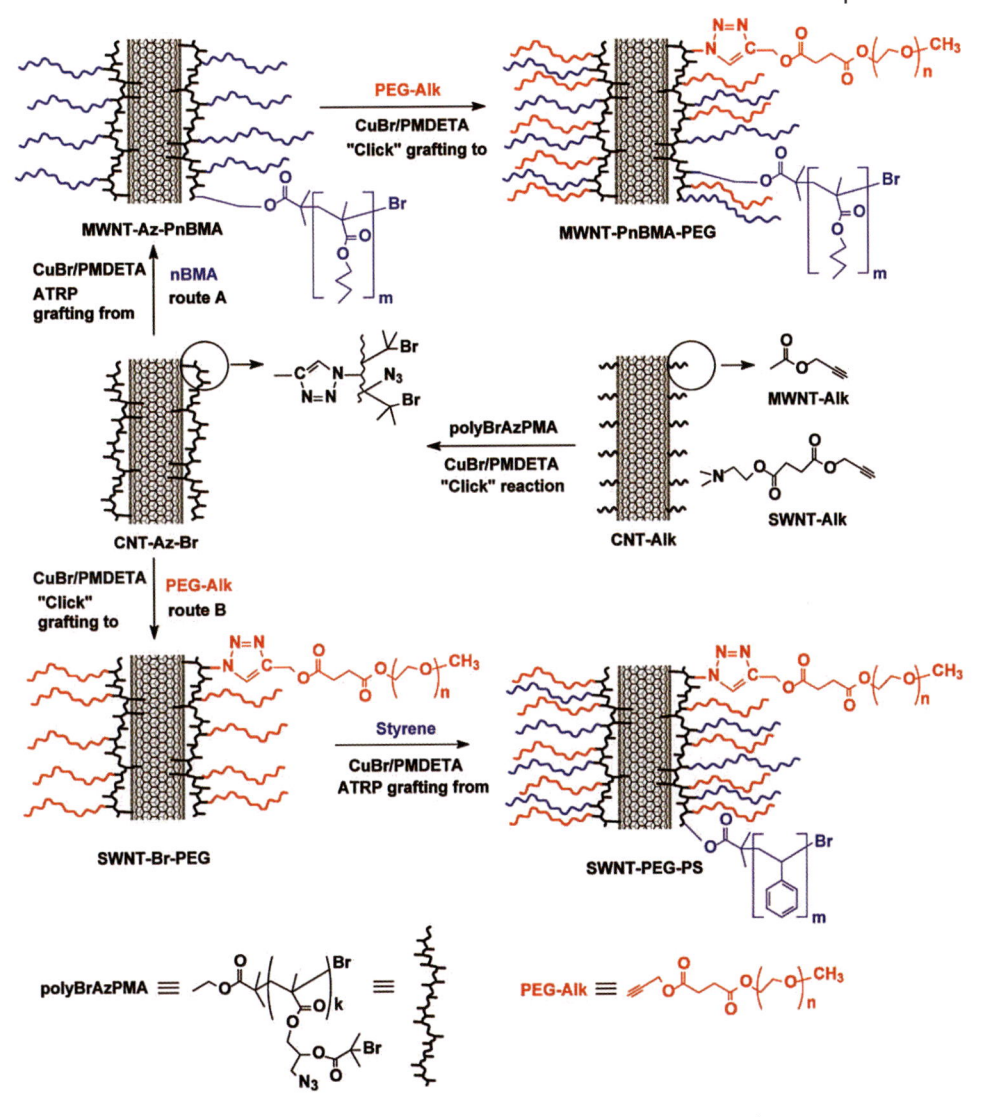

Scheme 12.12 Preparation of multifunctional CNTs by clicking a polymer-bearing azide groups and ATRP initiators. Figure reproduced from Ref. [109] with permission from the American Chemical Society.

the grafting yield is higher, even for polymers with a high molecular weight (~50 000 g/mol). Steric hindrance that is usually responsible for limited grafting yields is at least partly compensated by the multifunctionality of the polymer chains. This method is thus well-suited to the grafting of high molecular weight chains. Moreover, the bonding density and the interfacial layer thickness can be tuned by the molar ratio of the functional monomers in the chains and the

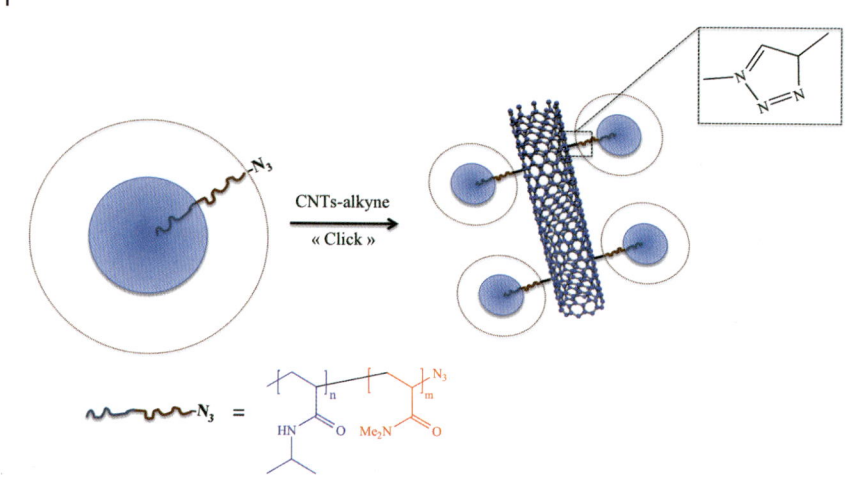

Scheme 12.13 Click coupling of CNTs with decorated azide-functionalized block copolymer micelles.

Scheme 12.14 Palladium ($PdCl_2(PPh_3)_2$)-catalyzed coupling reaction between benzyl chloride substituents of polystyrene and CNTs-alkynes.

molecular weight of the constitutive blocks, respectively. Finally, no significant cross-linking was observed.

12.2.1.6 Other Functionalized CNTs (-NCO, -S-S-Py, -anhydride, -anion)

Hydroxyl-containing polyurethane was grafted onto CNTs pre-functionalized with maleic anhydride in the melt [110]. Hydroxyl end-capped PEG was also grafted onto CNTs-NCO [111].

A thiol coupling reaction was also carried out under mild conditions as illustrated by Scheme 12.15. Disulfide-functionalized CNTs (CNTs-S-S-pyridine) were first prepared by amidation of oxidized CNTs-COOH with an aqueous solution of S-(2-aminoethylthio)-2-thiopyridine hydrochloride. Then, a thiol end-capped poly(N-(2-hydroxypropyl)methacrylate) was conjugated to CNTs-S-S-pyridine under stirring in a phosphate-buffered saline solution at room temperature for 26 h [112].

The addition of butyllithium onto a pristine CNTs (in THF at 10 °C) not only makes them air- and moisture-sensitive but also makes them reactive toward chlorinated polymers, such as chlorinated poly(propylene) (ClPP) (Scheme 12.16) [113]. The surface-grafted nanotubes (CNTs-g-ClPP) [113] are easily dispersed within ClPP, at the benefit of the mechanical properties (threefold increase in the Young's modulus).

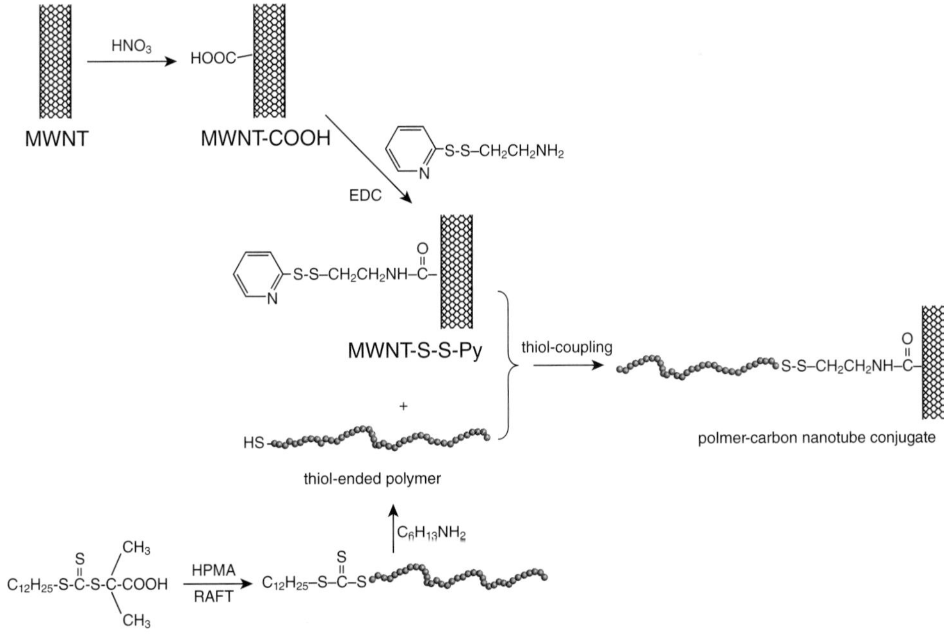

Scheme 12.15 Preparation of thiol-reactive functional CNTs (CNTs-S-S-pyridine) and grafting onto by a thiol-coupling reaction. Figure reproduced from Ref. [112] with permission from Wiley-VCH.

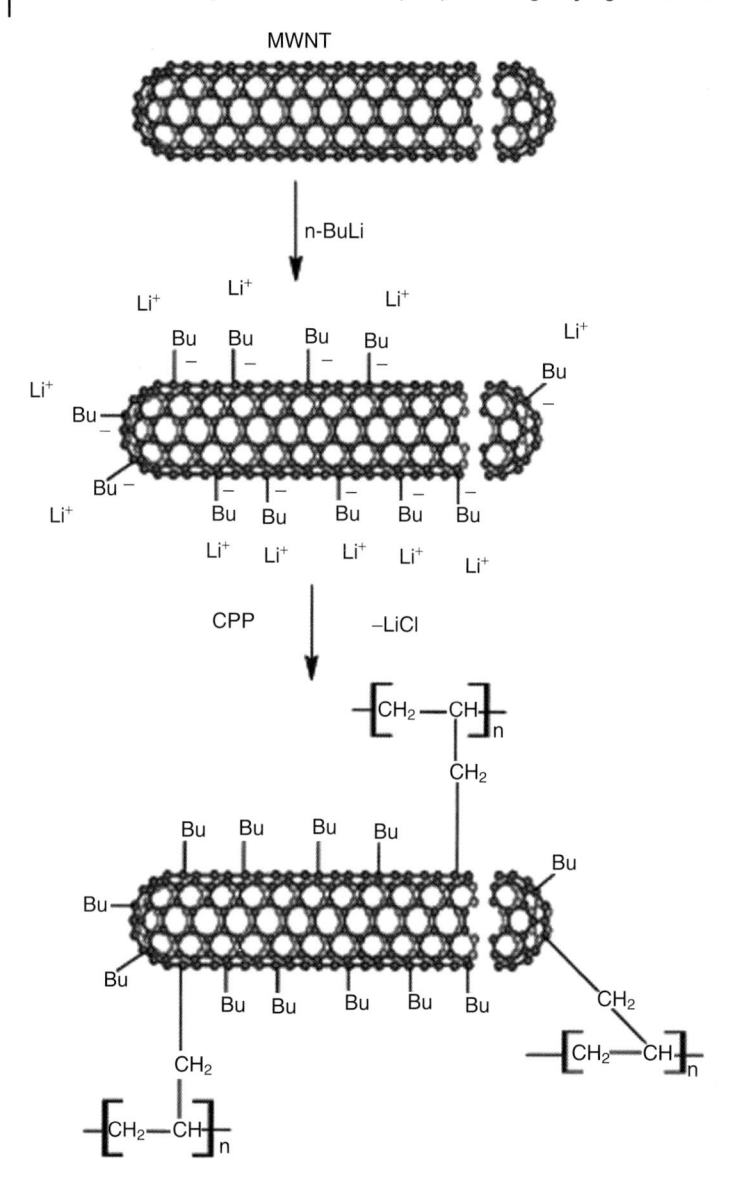

Scheme 12.16 Reaction of CNTs with butyllithium followed by the grafting of chlorinated polypropylene. Figure reproduced from Ref. [113] with permission from the American Chemical Society.

12.2.2
The One-Step Method

Because of the curvature of CNTs, the constitutive sp^2 hybridized carbons are under stress, which is favorable to their conversion into sp^3 hybridized carbons upon the addition of a reactive enough reagent. Therefore, polymers bearing suitably reactive groups can be directly grafted onto native CNTs. This section is devoted to the most powerful one-step methods, which are basically inspired by the very rich addition chemistry on fullerene [15, 16].

12.2.2.1 Radical Grafting

Radical grafting is by far the most extensively used method for the direct coupling of polymers onto CNTs. For the process to be successful, polymer-centered radicals must be generated in the vicinity of the CNTs. Three major routes can be distinguished. The first one, known as the "free radical grafting," consists in initiating the radical polymerization of a vinyl monomer with a conventional initiator in the presence of the CNTs. Growing polymer chains then react with sp^2 carbons at the CNTs surface, which results in covalent bonding to the nanotubes. This technique is *a priori* applicable to any polymer prepared by radical polymerization. Lack of control on the polymer characteristic features (molecular weight, polydispersity, architecture, ...) is a drawback, which is responsible for structural heterogeneity of the polymer-grafted CNTs. The so-called controlled radical grafting method allows bypassing this inconvenience. It consists in first preforming well-defined polymers by controlled radical polymerization (CRP) techniques prior to the radical grafting in a second step. In addition to form polymers with a high level of architectural control, the CRP technique is also chosen to form polymers bearing a radical generating group at their end that promotes the CNTs coupling during the second step. Finally, the third method relies on the use of an external source able to create radicals on preformed polymer chains and on the reaction of these macroradicals with the CNTs.

Free Radical Grafting The first report on this technique is based on the radical polymerization of styrene initiated by benzoyl peroxide or potassium persulfate, in the presence of oxidized CNTs in a water/toluene mixture at 80 °C [123]. During the reaction, CNTs are transferred from the aqueous phase to the organic phase as a result of the grafting of PS onto their originally polar surface. Qin *et al.* demonstrated that this reaction was also effective with pristine CNTs, which allows to avoid the very aggressive oxidation step [124, 125]. Water-soluble CNTs were also prepared by the grafting of sulfonated poly(styrene) (Scheme 12.17) [125].

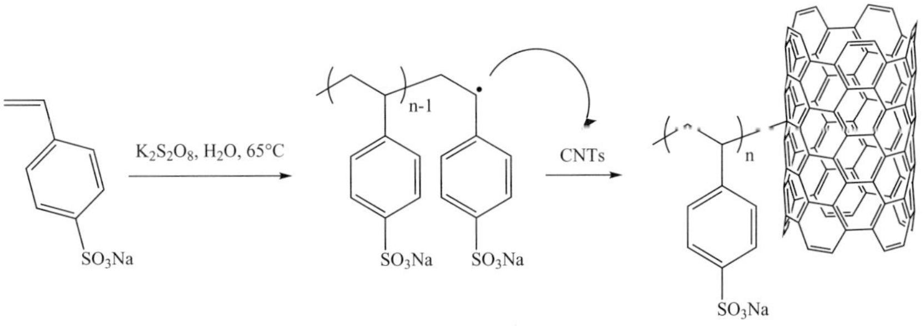

Scheme 12.17 Free radical grafting of growing polystyrene sulfonate onto pristine CNTs.

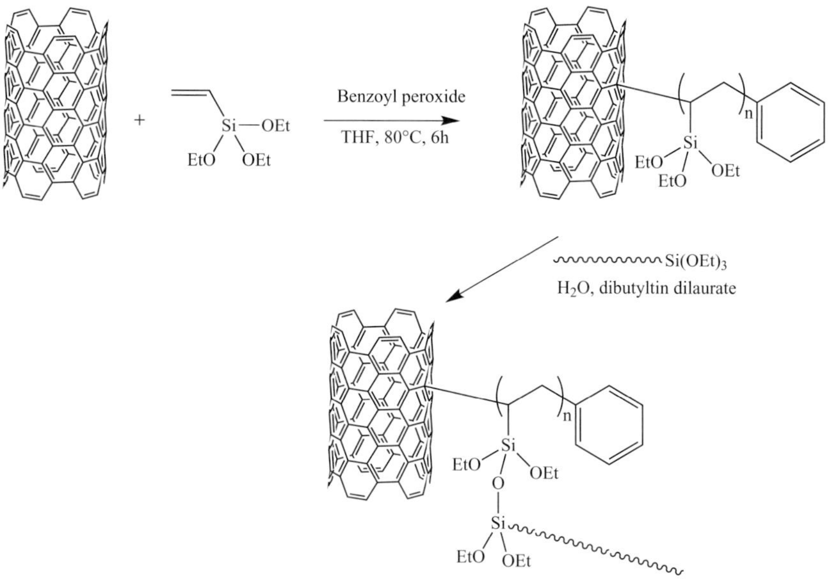

Scheme 12.18 Free radical grafting of vinyltriethoxysilane onto CNTs followed by hydrolytic condensation with the alkoxysilane end-group.

Similarly, poly(4-vinylpyridine) (P4VP) was grafted onto CNTs, and the so-modified nanotubes were used to reinforce polymer films prepared by the layer-by-layer deposition technique of P4VP and poly(acrylic acid) [124].

Free radical polymerization of vinyltriethoxysilane in the presence of CNTs is a way to make the CNTs surface reactive (Scheme 12.18). Indeed, the grafted alkoxysilane groups were successfully co-reacted with the alkoxysilane end-groups of polymers in a hydrolytic condensation reaction. Poly(urea-urethanes) end-capped with aminopropyltriethoxysilane [126] (Scheme 12.18) and low-density poly(ethylene) terminated by trimethoxysilane [127] are examples of polymers grafted by this method.

More recently, poly[(vinylbenzyl)trimethylammonium chloride] (PVBTAn+) was grafted onto CNTs by this one-step process. Negatively charged porphyrins were then deposited on the positively charged CNTs (Scheme 12.19), with the formation of nanohybrids useful in organic photoelectrochemical solar cells [128].

Recently, poly(styrene-co-maleic anhydride) was coupled to CNTs by radical copolymerization of styrene and maleic anhydride initiated by azo-bis-isobutyronitrile (AIBN) in toluene at 80 °C [129]. In a second step, polyamide 6 was grafted onto the modified CNTs. γ-Ray irradiation (^{60}Co) is an alternative method to initiate the radical polymerization of vinyl monomers. For instance, radiolysis of a styrene/CNTs mixture was reported as an effective "grafting onto" method that preserved the structural integrity of the CNTs [130].

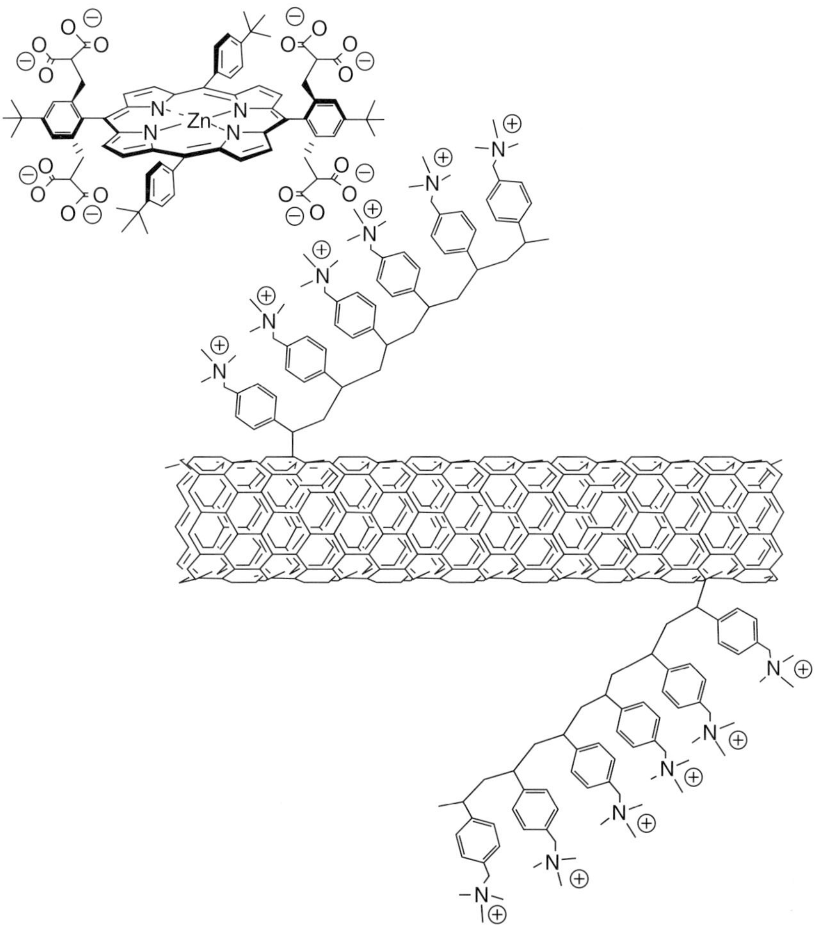

Scheme 12.19 Adsorption of negatively charged zinc porphyrin onto poly[(vinylbenzyl)trimethylammonium chloride] (PVBTAn+) prepared by free radical grafting of (vinylbenzyl)trimethylammonium chloride in the presence of CNTs. Figure reproduced from Ref. [128] with permission from Wiley-VCH.

Controlled Radical Grafting Within the past decades, polymer science has witnessed the discovery and the development of CRP [40–44] techniques that pave the way to the precise synthesis of complex macromolecular architectures with novel properties compared to traditional polymers. The versatility and high efficiency of the CRP techniques allow polymers to be endowed with almost any functionality, structure, and composition with a remarkable precision. This unprecedented structural control is a unique lever to impart on purpose desired properties to CNTs. When they are collected after CRP, the chains are end-capped by a dormant radical species, which can be reactivated on-demand as a carbon-centered radical. In the specific case of nitroxide-mediated radical polymerization

(NMP) [42, 44], the growing radicals are reversibly trapped by a nitroxide, such that an alkoxyamine is the dormant species, which can be thermally reactivated in the presence of CNTs.

Our laboratory was the first to report on the grafting of CNTs by a series of tailored homopolymers and block copolymers [52, 53, 131–133]. For instance, PS, PCL, and poly(ε-caprolactone)-b-poly(styrene) (PCL-b-PS) end-capped by 2,2,6,6-tetramethyliperidinoxy (TEMPO) were successfully grafted onto CNTs in toluene at 130 °C for 24 h (Scheme 12.20) [53]. The molecular weight was in the range from 3000 to 28 500 g/mol.

The grafting ratio (GR: weight ratio of the grafted polymer to the nanotubes) increased with the polymer molecular weight, and the functionalized nanotubes were highly soluble in organic solvents, resulting in dispersions that were stable for weeks, thus without any significant precipitation. Importantly, any polymer prepared by NMP can be grafted by this simple procedure, such that the reactivity/functionality that can be imparted to CNTs is quasi unlimited. A nice illustration of this strategy is the grafting of CNTs by poly(2-vinyl pyridine) end-capped by TEMPO (P2VP-TEMPO). The homogeneous shell of P2VP at the surface of the CNTs makes them dispersible both in organic solvents and in acidic water (Scheme 12.21) [52]. At low pH, the polycationic shell was used to immobilize negatively charged Prussian Blue nanoclusters at the CNTs surface [52, 132]. Similarly, CNTs were selectively deposited onto oppositely charged surfaces, which is a way to integrate CNTs and nano-objects with specific properties into novel nanometer-scale systems [52]. Deposition of aqueous dispersions of CNTs-g-P2VP at low pH on a solid substrate is also an effective method to prepare homogeneous ultrathin, transparent, and conducting films with a tunable CNTs density [133]. These films have potential in touch screens, reflective displays, EMI shielding, and static charge dissipation.

Poly(styrene) and P4VP and their copolymers prepared by NMP were also grafted onto CNTs for the preparation of organic vapor sensors [134]. The extent to which the conductive paths are perturbed by the absorbed molecules, for example, MeOH, THF, and $CHCl_3$, dictates the sensitivity of the detection. Poly(tert-butyl acrylate)-b-poly(styrene) (PtBA-b-PS) of well-defined molecular weights, compositions, and polydispersities were also synthesized by NMP and immobilized at the CNTs surface at 125 °C, which resulted in nanohybrids highly soluble in a variety of organic solvents. Interestingly enough, this solubility can be easily modulated by converting the $tert$-butyl esters of the polyacrylate block into carboxylic acids, so making the grafted block copolymer amphililic [51].

Later on, Adronov et $al.$ studied the effect of the length of grafted polystyrene chains on the grafting density and the solubility of the modified CNTs in THF [49]. The grafting density was observed to decrease and the solubility to go through a maximum upon increasing the molecular weight. Low molecular weight chains (<5000) were unable to stabilize properly CNTs dispersions, which also happened at high molecular weight (>100 000), because of a too low grafting density.

CNTs were also modified by the "grafting onto" of polymers prepared by ATRP, another efficient CRP technique. In contrast to NMP, polymers formed by ATRP

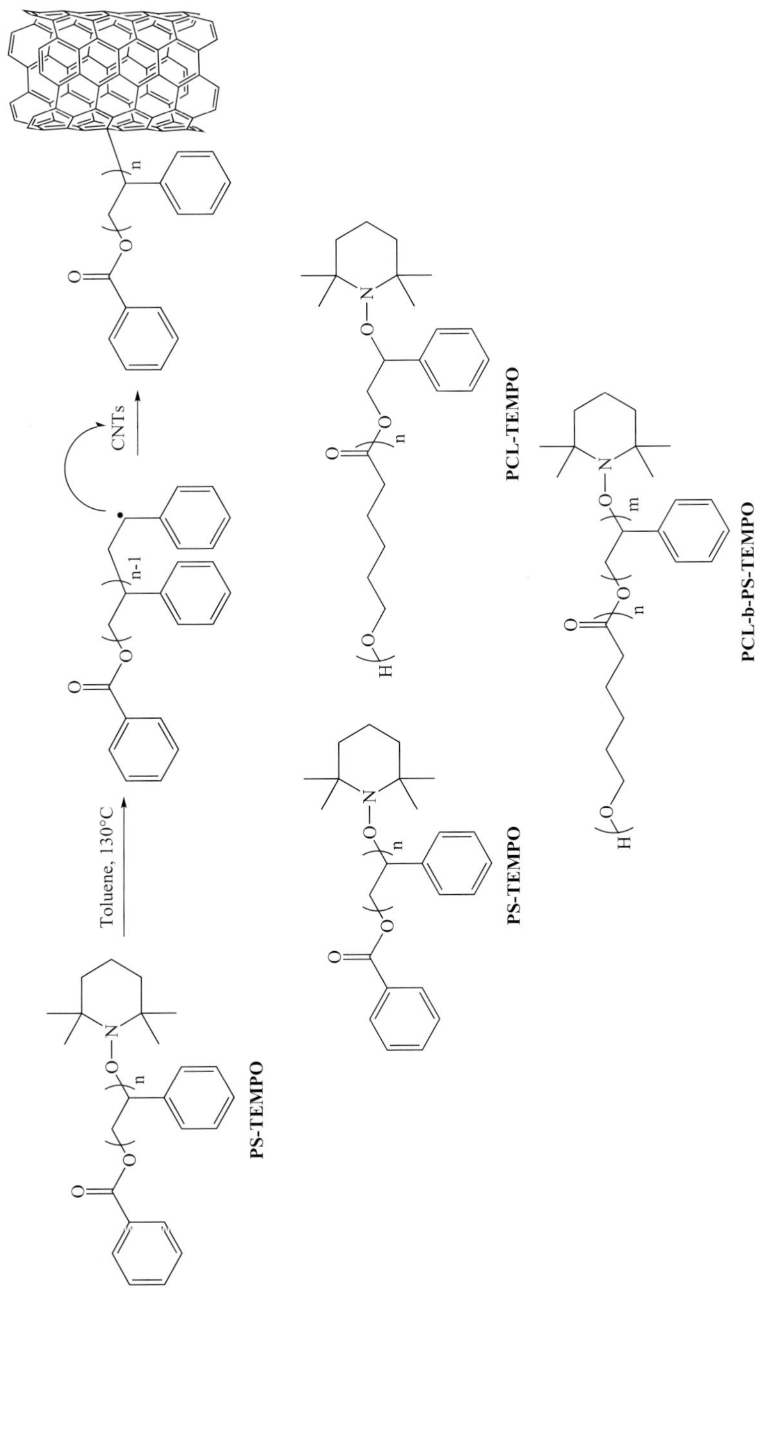

Scheme 12.20 Grafting of alkoxyamine end-capped (co)polymers onto CNTs, and some examples of (co)polymers grafted by this process.

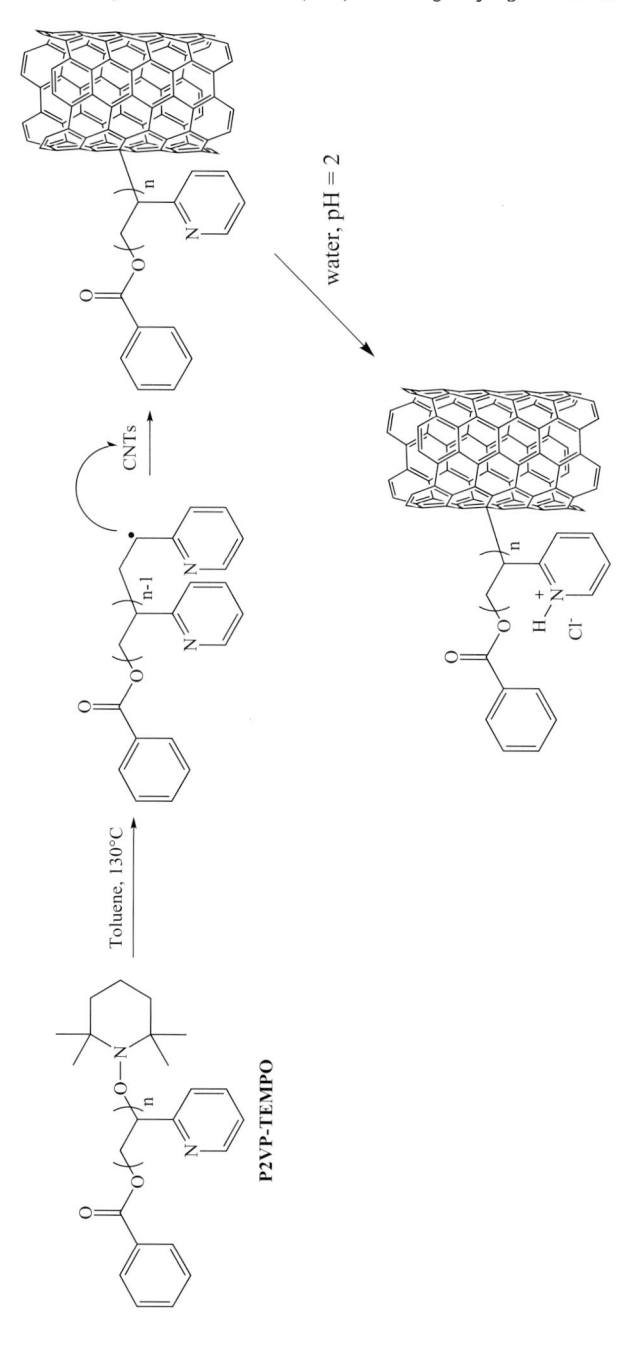

Scheme 12.21 Grafting of alkoxyamine end-capped poly(2-vinyl pyridine) and protonation in an acidic aqueous medium.

are end-capped by a halogen atom (usually, –Cl or –Br), through a stable C-halogen bond. Therefore, they must be reactivated for being grafted onto CNTs. Actually, the halogen is abstracted by a suitable ATRP metal catalyst added to the reaction medium (e.g., CuBr ligated by bipyridine). Well-defined (poly(styrene) [48, 50, 54] and poly(t-butyl acrylate) [135] chains were thus prepared by ATRP and grafted onto CNTs by this simple method (Scheme 12.22).

Organometallic-mediated radical polymerization (OMRP) is an emerging CRP technique [43, 136], based on the momentary deactivation of the growing polymer chains by a complex of a transition metal, for example, titanium [137], vanadium [138], chromium [139], and cobalt [43]. Provided that the C-metal chain-end can be reactivated in the presence of CNTs, the "grafting onto" can occur. This reactivation is possible either thermally or by adding a ligand able to coordinate the transition metal [140, 141]. This concept was recently validated by the so-called cobalt-mediated radical polymerization (CMRP) [142] in which bis-(acetylacetonato)cobalt(II) (Co(acac)$_2$) is the control agent. Poly(vinyl acetate) end-capped by Co(acac)$_2$, (PVAc-Co(acac)$_2$) was prepared by CMRP and added to CNTs in methanol, followed by a low amount of water at 30 °C. The C–Co bond homolytic cleavage was activated by water, thus under mild conditions and the released PVAc macroradicals were rapidly trapped by the CNTs (Scheme 12.23) [142]. After partial hydrolysis of the grafted PVAc, the modified CNTs were used as co-stabilizers in the dispersion polymerization of MMA. PMMA microspheres decorated by CNTs at their surface were accordingly collected. Melt-compression of these microspheres led to nanocomposites with highly improved electrical conductivity compared to PMMA/CNTs nanocomposite prepared by melt-blending. This improvement is the result of a 3D network of connected CNTs formed during melt-compression in relation to the selective localization of CNTs at the surface of the microspheres.

Macroradicals Created by an External Source Radicals can be generated on a polymer backbone by an external source (conventional free radical initiator, sonochemistry, high energy radiations, ...) and used for the "grafting onto" CNTs. For example, a solution of poly(vinylcarbazole) was reacted with CNTs in the presence of AIBN as the radical source [143]. In another example, poly(α-methylstyrene-co-glycidyl methacrylate) (PAMS-co-GMA) and poly(α-methylstyrene-co-styrene) (PAMS-co-St) copolymers were heated in an organic solution of CNTs at 90 °C, so generating radicals along the PAMS blocks that reacted with the nanotubes [144].

Ultrasonication of a poly(vinyl pyrrolidone) (PVP) solution proved efficiency in creating PVP macroradicals that were trapped by CNTs [145]. In addition to polymer grafting, the ultrasonication was responsible for the disentanglement of the CNTs bundles and the shortening of the nanotubes. Moreover, the PVP molecular weight (~600 000 g/mol) decreased rapidly with the sonication time and leveled off at ~25 000 g/mol.

12.2.2.2 Anionic Grafting

The "grafting to" method is also applicable to polymers prepared by living anionic polymerization (Scheme 12.24) [89, 146]. The limitation of this technique has to

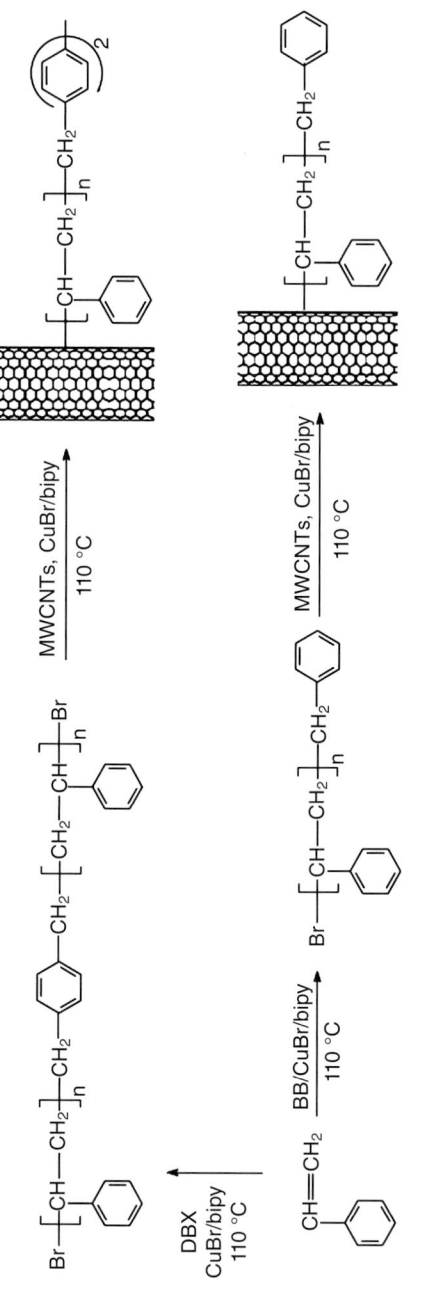

Scheme 12.22 ATRP of styrene and grafting onto CNTs. Figure reproduced from Ref. [54] with permission from Elsevier.

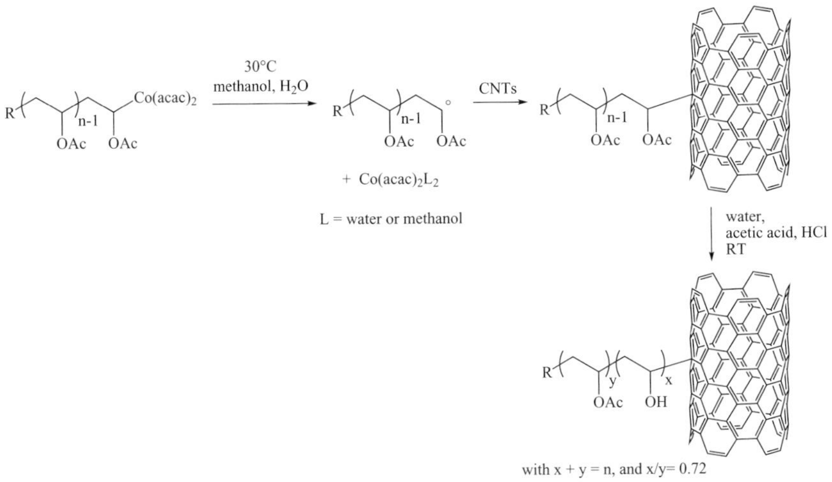

12.2 Overview of the "Grafting to" Methods

Scheme 12.23 Radical grafting of PVAc-Co(acac)$_2$ prepared by CMRP onto CNTs followed by partial hydrolysis of the grafted PVAc chains.

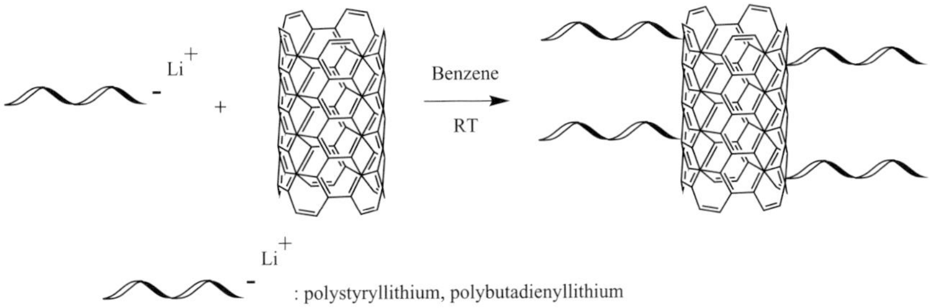

Scheme 12.24 Grafting of living polystyryl chains onto CNTs.

be found in the required experimental conditions, that is, high dilution and dryness of the reaction medium.

12.2.2.3 Azide Grafting

The bromine atom at the ω-chain end of polystyrene prepared by ATRP was converted into an azide by nucleophilic substitution with sodium azide. In solution in dichlorobenzene, this polymer was coupled to CNTs by heating the CNTs/PS–N$_3$ mixture at 130 °C for 60 h [147, 148]. Under these conditions, a [2+1] cycloaddition reaction occurs with the formation of aziridine rings at the polymer/CNTs interface (Scheme 12.25).

This powerful strategy was extended to the grafting of poly(styrene)-b-poly(tert-butyl methacrylate) block copolymer onto CNTs [149]. Upon hydrolysis of the methacrylate block, the CNTs were modified by an amphiphilic copolymer that self-assembled in various solvents.

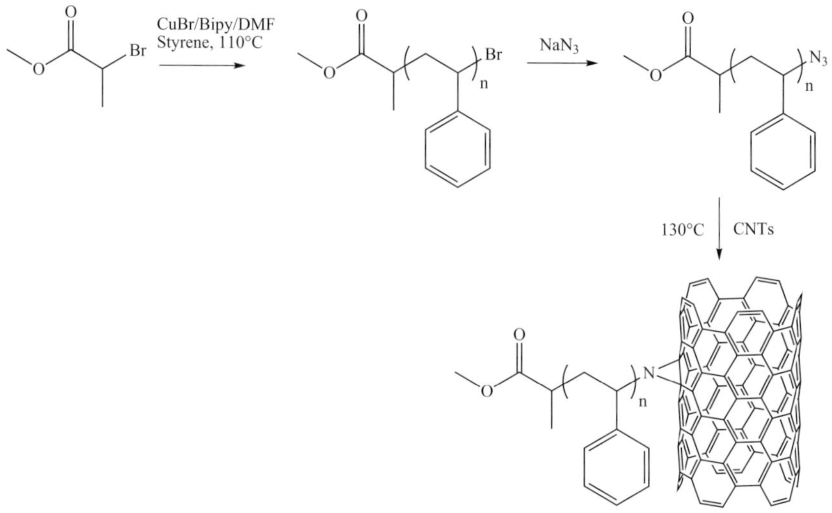

Scheme 12.25 Grafting of PS–N$_3$ onto CNTs.

12.3 Conclusions

The huge research effort devoted to the chemical modification of CNTs by grafted macromolecules is driven by the need to enhance their affinity for a variety of environments, which is a prerequisite for them to play a key role in the emerging technologies. This chapter aimed at reviewing the until now investigated "grafting to" methods used for the surface modification of CNTs by polymers of various functionalities. Two major strategies were emphasized on the basis of the number of needed modification steps (1 vs. 2). The first technique is a two-step one, because the surface of the CNTs must be reactive in a preliminary step, thus before reaction with polymers bearing complementary reactive groups either at one chain-end or along the chain. The second approach is straightforward as a result of the ability of the selected polymer to add directly onto the sp^2 carbons of the CNTs surface. As a rule, the grafting efficiency of the "grafting onto" methods is lower compared to the "grafting from" techniques because of the steric hindrance caused by the first grafted polymer chains that prevent the next ones to reach the reactive sites at the CNTs surface. The major advantage of the "grafting onto" techniques is that the molecular characteristics (molecular weight, polydispersity, functionality, …) of the polymers can be tailored prior to the grafting reaction. Moreover, when using the proper controlled or living polymerization techniques and grafting conditions, polymers of well-defined architecture, functionality, and properties can be easily anchored to the CNTs surface, which makes the on-demand tuning of the chemical properties of CNTs possible.

Acknowledgments

The authors are grateful to the *National Fund for Scientific Research* (F.R.S.-FNRS) and the *Belgian* Federal Science Policy (in the frame of the "Interuniversity Attraction Program" (IAP VI/27) on Supramolecular Chemistry and Supramolecular Catalysis») for financial support.

References

1 Sansom, E.B., Rinderknecht, D., and Gharib, M. (2008) *Nanotechnology*, **19**, 035302/035301–035302/035306.
2 Son, Y.-W., Oh, S., Ihm, J., and Han, S. (2005) *Nanotechnology*, **16**, 125.
3 Thomassin, J.-M., Lou, X., Pagnoulle, C., Saib, A., Bednarz, L., Huynen, I., Jerome, R., and Detrembleur, C. (2007) *J. Phys. Chem. C*, **111**, 11186.
4 Thomassin, J.-M., Pagnoulle, C., Bednarz, L., Huynen, I., Jerome, R., and Detrembleur, C. (2008) *J. Mater. Chem.*, **18**, 792.
5 Kovtyukhova, N.I. and Mallouk, T.E. (2005) *Adv. Mater.*, **17**, 187.
6 Zhang, D., Ryu, K., Liu, X., Polikarpov, E., Ly, J., Tompson, M.E., and Zhou, C. (2006) *Nano Lett.*, **6**, 1880.
7 Kawano, T., Chiamori, H.C., Suter, M., Zhou, Q., Sosnowchik, B.D., and Lin, L. (2007) *Nano Lett.*, **7**, 3686.
8 Tang, X., Bansaruntip, S., Nakayama, N., Yenilmez, E., Chang, Y.-L., and Wang, Q. (2006) *Nano Lett.*, **6**, 1632.
9 Dai, G., Liu, C., Liu, M., Wang, M., and Cheng, H. (2002) *Nano Lett.*, **2**, 503.
10 Nikitin, A., Li, X., Zhang, Z., Ogasawara, H., Dai, H., and Nilsson, A. (2008) *Nano Lett.*, **8**, 162.
11 Hamon, M.A., Hui, H., Bhowmik, P., Itkis, H.M.E., and Haddon, R.C. (2002) *Appl. Phys. A: Mater. Sci. Process*, **74**, 333.
12 Chen, J., Hamon, M.A., Hu, H., Chen, Y., Rao, A.M., Eklund, P.C., and Haddon, R.C. (1998) *Science*, **282**, 95.
13 Hamon, M.A., Chen, J., Hu, H., Chen, Y., Itkis, M.E., Rao, A.M., Eklund, P.C., and Haddon, R.C. (1999) *Adv. Mater.*, **11**, 834.
14 Cao, L., Chen, H., Wang, M., Sun, J., Zhang, X., and Kong, F. (2002) *J. Phys. Chem. B*, **106**, 8971.
15 Diederich, F., and Gomez-Lopez, M. (1999) *Chem. Soc. Rev.*, **28**, 263.
16 Diederich, F. and Thilgen, C. (1996) *Science*, **271**, 317.
17 Lee, W.H., Kim, S.J., Lee, W.J., Lee, J.G., Haddon, R.C., and Reucroft, P.J. (2001) *Appl. Surf. Sci.*, **181**, 121.
18 Kamaras, K., Itkis, M.E., Hu, H., Zhao, B., and Haddon, R.C. (2003) *Science*, **301**, 1501.
19 Holzinger, M., Vostrowsky, O., Hirsch, A., Hennrich, F., Kappes, M., Weiss, R., and Jellen, F. (2001) *Angew. Chem., Int. Ed.*, **40**, 4002.
20 Holzinger, M., Abraham, J., Whelan, P., Graupner, R., Ley, L., Hennrich, F., Kappes, M., and Hirsch, A. (2003) *J. Am. Chem. Soc.*, **125**, 8566.
21 Holzinger, M., Steinmetz, J., Samaille, D., Glerup, M., Paillet, M., Bernier, P., Ley, L., and Graupner, R. (2004) *Carbon*, **42**, 941.
22 Georgakilas, V., Kordatos, K., Prato, M., Guldi, D.M., Holzinger, M., and Hirsch, A. (2002) *J. Am. Chem. Soc.*, **124**, 760.
23 Tagmatarchis, N. and Prato, M. (2004) *J. Mater. Chem.*, **14**, 437.
24 Delgado, J.L., de la Cruz, P., Langa, F., Urbina, A., Casado, J., and Lopez Navarrete, J.T. (2004) *Chem. Commun.*, 1734.
25 Peng, H., Reverdy, P., Khabashesku, V.N., and Margrave, J.L. (2003) *Chem. Commun.*, 362.
26 Peng, H., Alemany, L.B., Margrave, J.L., and Khabashesku, V.N. (2003) *J. Am. Chem. Soc.*, **125**, 15174.
27 Umek, P., Seo, J.W., Hernadi, K., Mrzel, A., Pechy, P., Mihailovic, D.D., and Forro, L. (2003) *Chem. Mater.*, **15**, 4751.

28 Stoffelbach, F., Aqil, A., Jerome, C., Jerome, R., and Detrembleur, C. (2005) *Chem. Commun.*, 4532.

29 Valentini, L., Puglia, D., Armentano, I., and Kenny, J.M. (2005) *Chem. Phys. Lett.*, **403**, 385.

30 Yan, Y.H., Chan-Park, M.B., Zhou, Q., Li, C.M., and Yue, C.Y. (2005) *Appl. Phys. Lett.*, **87**, 213101/213101–213101/213103.

31 Lemek, T., Mazurkiewicz, J., Stobinski, L., Lin, H.M., and Tomasik, P. (2007) *J. Nanosci. Nanotechnol.*, **7**, 3081.

32 McQueen, E.W. and Goldsmith, J.I. (2009) *J. Am. Chem. Soc.*, **131**, 17554.

33 Zu, S.-Z., Sun, X.-X., Liu, Y., and Han, B.-H. (2009) *Chem. Asian J.*, **4**, 1562.

34 Chen, J. and Collier, C.P. (2005) *J. Phys. Chem. B*, **109**, 7605.

35 Cheng, F. and Adronov, A. (2006) *Chem. Eur. J.*, **12**, 5053.

36 Murakami, H., Nomura, T., and Nakashima, N. (2003) *Chem. Phys. Lett.*, **378**, 481.

37 Tasis, D., Tagmatarchis, N., Bianco, A., and Prato, M. (2006) *Chem. Rev.*, **106**, 1105.

38 Hirsch, A. and Vostrowsky, O. (2007) *Functional Organic Materials* (ed. by T.J.J. Mueller and U.H.F. Bunz), Wiley-VCH Verlag GmbH & Co. KGaA, Weinheim, Germany.

39 Spitalsky, Z., Tasis, D., Papagelis, K., and Galiotis, C. (2010) *Prog. Polym. Sci.*, **35**, 357.

40 Matyjaszewski, K. and Xia, J. (2001) *Chem. Rev.*, **101**, 2921.

41 Boyer, C., Bulmus, V., Davis, T.P., Ladmiral, V., Liu, J., and Perrier, S. (2009) *Chem. Rev.*, **109**, 5402.

42 Hawker, C.J., Bosman, A.W., and Harth, E. (2001) *Chem. Rev.*, **101**, 3661.

43 Debuigne, A., Poli, R., Jerome, C., Jerome, R., and Detrembleur, C. (2009) *Prog. Polym. Sci.*, **34**, 211.

44 Sciannamea, V., Jerome, R., and Detrembleur, C. (2008) *Chem. Rev.*, **108**, 1104.

45 Xu, G., Wu, W.-T., Wang, Y., Pang, W., Wang, P., Zhu, Q., and Lu, F. (2006) *Nanotechnology*, **17**, 2458.

46 Zhao, X.-D., Fan, X.-H., Chen, X.-F., Chai, C.-P., and Zhou, Q.-F. (2006) *J. Polym. Sci., Part A: Polym. Chem.*, **44**, 4656.

47 Adronov, A., Homenick, C.M., Liu, Y., and Yao, Z. (2005) *Polym. Prepr.*, **46**, 201.

48 Choi, J.H., Oh, S.B., Chang, J., Kim, I., Ha, C.-S., Kim, B.G., Han, J.H., Joo, S.-W., Kim, G.-H., Kim, H.-K., and Paik, H.-J. (2006) *PMSE Prepr.*, **95**, 492.

49 Homenick, C.M., Sivasubramaniam, U., and Adronov, A. (2008) *Polym. Int.*, **57**, 1007.

50 Jeon, J.-H., Lim, J.-H., and Kim, K.-M. (2009) *Polymer*, **50**, 4488.

51 Liu, Y., Yao, Z., and Adronov, A. (2005) *Macromolecules*, **38**, 1172.

52 Lou, X., Detrembleur, C., Pagnoulle, C., Jerome, R., Bocharova, V., Kiriy, A., and Stamm, M. (2004) *Adv. Mater.*, **16**, 2123.

53 Lou, X., Detrembleur, C., Sciannamea, V., Pagnoulle, C., and Jerome, R. (2004) *Polymer*, **45**, 6097.

54 Wu, H.-X., Tong, R., Qiu, X.-Q., Yang, H.-F., Lin, Y.-H., Cai, R.-F., and Qian, S.-X. (2007) *Carbon*, **45**, 152.

55 Lou, X., Daussin, R., Cuenot, S., Duwez, A.-S., Pagnoulle, C., Detrembleur, C., Bailly, C., and Jerome, R. (2004) *Chem. Mater.*, **16**, 4005.

56 Meuer, S., Braun, L., Schilling, T., and Zentel, R. (2009) *Polymer*, **50**, 154.

57 Nakashima, N., Tomonari, Y., and Murakami, H. (2002) *Chem. Lett.*, 638.

58 Petrov, P., Stassin, F., Pagnoulle, C., and Jerome, R. (2003) *Chem. Commun.*, 2904.

59 Niu, L., Luo, Y., and Li, Z. (2007) *Sens. Actuators B*, **B126**, 361.

60 Paiva, M.C., Zhou, B., Fernando, K.A.S., Lin, Y., Kennedy, J.M., and Sun, Y.P. (2004) *Carbon*, **42**, 2849.

61 Vandervorst, P., Lei, C.H., Lin, Y., Dupont, O., Dalton, A.B., Sun, Y.P., and Keddie, J.L. (2006) *Prog. Org. Coat.*, **57**, 91.

62 Kitano, H., Tachimoto, K., and Anraku, Y. (2007) *J. Colloid Interface Sci.*, **306**, 28.

63 Shieh, Y.-T. and Yang, Y.-F. (2006) *Eur. Polym. J.*, **42**, 3162.

64 Chen, Y., Sai Muthukumar, V., Wang, Y., Li, C., Sivarama Krishnan, S., Siva Sankara Sai, S., Venkataramaniah, K.,

and Mitra, S. (2009) *J. Mater. Chem.*, **19**, 6568.

65 Ge, J.J., Zhang, D., Li, Q., Hou, H., Graham, M.J., Dai, L., Harris, F.W., and Cheng, S.Z.D. (2005) *J. Am. Chem. Soc.*, **127**, 9984.

66 Shi, J.-H., Yang, B.-X., and Goh, S.H. (2009) *Eur. Polym. J.*, **45**, 1002.

67 Smith, J.G., Jr, Connell, J.W., Delozier, D.M., Lillehei, P.T., Watson, K.A., Lin, Y., Zhou, B., and Sun, Y.P. (2004) *Polymer*, **45**, 825.

68 Smith, J.G., Delozier, D.M., Connell, J.W., and Watson, K.A. (2004) *Polymer*, **45**, 6133.

69 Jung, D.-H., Koan Ko, Y., and Jung, H.-T. (2004) *Mater. Sci. Eng. C*, **C24**, 117.

70 Malarkey, E.B., Reyes, R.C., Zhao, B., Haddon, R.C., and Parpura, V. (2008) *Nano Lett.*, **8**, 3538.

71 Baskaran, D., Mays, J.W., and Bratcher, M.S. (2005) *Polymer*, **46**, 5050.

72 Jin, Z., Sun, X., Xu, G., Goh, S.H., and Ji, W. (2000) *Chem. Phys. Lett.*, **318**, 505.

73 Menna, E., Scorrano, G., Maggini, M., Cavallaro, M., Della Negra, F., Battagliarin, M., Bozio, R., Fantinel, F., and Meneghetti, M. (2003) *ARKIVOC*, **12**, 64.

74 Zhao, B., Hu, H., Yu, A., Perea, D., and Haddon, R.C. (2005) *J. Am. Chem. Soc.*, **127**, 8197.

75 Chen, G.-X., Kim, H.-S., Park, B.H., and Yoon, J.-S. (2005) *J. Phys. Chem. B*, **109**, 22237.

76 Kim, H.-S., Park, B.-H., Chen, G., and Yoon, J.S. (2006) *PMSE Prepr.*, **95**, 734.

77 Baskaran, D., Dunlap, J.R., Mays, J.W., and Bratcher, M.S. (2005) *Macromol. Rapid Commun.*, **26**, 481.

78 Wang, X., Liu, H., Jin, Y., and Chen, C. (2006) *J. Phys. Chem. B*, **110**, 10236.

79 Huang, W., Lin, Y., Taylor, S., Gaillard, J., Rao, A.M., and Sun, Y.-P. (2002) *Nano Lett.*, **2**, 231.

80 Ashok Kumar, N., Kim, S.H., Kim, J.T., Lim, K.T., and Jeong, Y.T. (2008) *Surf. Rev. Lett.*, **15**, 689.

81 Philip, B., Xie, J., Chandrasekhar, A., Abraham, J., and Varadan, V.K. (2004) *Smart Mater. Struct.*, **13**, 295.

82 Huang, W., Fernando, S., Allard, L.F., and Sun, Y.-P. (2003) *Nano Lett.*, **3**, 565.

83 Zhou, B., Lin, Y., Li, H., Huang, W., Connell, J.W., Allard, L.F., and Sun, Y.-P. (2003) *J. Phys. Chem. B*, **107**, 13588.

84 Wu, Z., Feng, W., Feng, Y., Liu, Q., Xu, X., Sekino, T., Fujii, A., and Ozaki, M. (2007) *Carbon*, **45**, 1212.

85 Carson, L., Kelly-Brown, C., Stewart, M., Oki, A., Regisford, G., Luo, Z., and Bakhmutov, V.I. (2009) *Mater. Lett.*, **63**, 617.

86 Lin, Y., Rao, A.M., Sadanadan, B., Kenik, E.A., and Sun, Y.-P. (2002) *J. Phys. Chem. B*, **106**, 1294.

87 Riggs, J.E., Guo, Z., Carroll, D.L., and Sun, Y.-P. (2000) *J. Am. Chem. Soc.*, **122**, 5879.

88 Zhao, B., Hu, H., and Haddon, R.C. (2004) *Adv. Funct. Mater.*, **14**, 71.

89 Baskaran, D., Sakellariou, G., Mays, J.W., and Bratcher, M.S. (2007) *J. Nanosci. Nanotechnol.*, **7**, 1560.

90 Huang, H.-M., Liu, I.C., Chang, C.-Y., Tsai, H.-C., Hsu, C.-H., and Tsiang, R.C.-C. (2004) *J. Polym. Sci., Part A: Polym. Chem.*, **42**, 5802.

91 Yeh, J.-T., Yang, M.-C., Wu, C.-J., and Wu, C.-S. (2009) *J. Appl. Polym. Sci.*, **112**, 660.

92 Wu, C.-S. and Liao, H.-T. (2007) *Polymer*, **48**, 4449.

93 Liu, Y.-X., Du, Z.-J., Li, Y., Zhang, C., Li, C.-J., Yang, X.-P., and Li, H.-Q. (2006) *J. Polym. Sci., Part A: Polym. Chem.*, **44**, 6880.

94 Liu, Y.-X., Du, Z.-J., Li, Y., Zhang, C., and Li, H.-Q. (2006) *Chin. J. Chem.*, **24**, 563.

95 Wang, X., Du, Z., Zhang, C., Li, C., Yang, X., and Li, H. (2008) *J. Polym. Sci., Part A: Polym. Chem.*, **46**, 4857.

96 Wei, W., Zhang, C., Du, Z., Liu, Y., Li, C., and Li, H. (2008) *Mater. Lett.*, **62**, 4167.

97 Wang, G., Qu, Z., Liu, L., Shi, Q., and Guo, J. (2007) *Mater. Sci. Eng., A*, **A472**, 136.

98 Pokrop, R., Kulszewicz-Bajer, I., Wielgus, I., Zagorska, M., Albertini, D., Lefrant, S., Louarn, G., and Pron, A. (2009) *Synth. Met.*, **159**, 919.

99 Chen, G.-X., Kim, H.-S., Park, B.H., and Yoon, J.-S. (2006) *Polymer*, **47**, 4760.

100 Yang, B.-X., Shi, J.-H., Li, X., Pramoda, K.P., and Goh, S.H. (2009) *J. Appl. Polym. Sci.*, **113**, 1165.

101 Causin, V., Yang, B.-X., Marega, C., Goh, S.H., and Marigo, A. (2009) *Eur. Polym. J.*, **45**, 2155.

102 Yang, B.-X., Shi, J.-H., Pramoda, K.P., and Goh, S.H. (2008) *Compos. Sci. Technol.*, **68**, 2490.

103 Mormann, W., Lu, Y., Zou, X., and Berger, R. (2008) *Macromol. Chem. Phys.*, **209**, 2113.

104 Li, H., Cheng, F., Duft, A.M., and Adronov, A. (2005) *J. Am. Chem. Soc.*, **127**, 14518.

105 Xie, L., Xu, F., Qiu, F., Lu, H., and Yang, Y. (2007) *Macromolecules*, **40**, 3296.

106 Li, H. and Adronov, A. (2007) *Carbon*, **45**, 984.

107 Liu, J., Nie, Z., Gao, Y., Adronov, A., and Li, H. (2008) *J. Polym. Sci., Part A: Polym. Chem.*, **46**, 7187.

108 Zhang, P. and Henthorn, D.B. (2009) *Langmuir*, **25**, 12308.

109 Zhang, Y., He, H., and Gao, C. (2008) *Macromolecules*, **41**, 9581.

110 Wu, H.-L., Wang, C.-H., Ma, C.-C.M., Chiu, Y.-C., Chiang, M.-T., and Chiang, C.-L. (2007) *Compos. Sci. Technol.*, **67**, 1854.

111 Wang, Y., Xiong, H., Gao, Y., and Li, H. (2008) *J. Mater. Sci.*, **43**, 5609.

112 You, Y.-Z., Hong, C.-Y., and Pan, C.-Y. (2006) *Macromol. Rapid Commun.*, **27**, 2001.

113 Blake, R., Gun'ko, Y.K., Coleman, J., Cadek, M., Fonseca, A., Nagy, J.B., and Blau, W.J. (2004) *J. Am. Chem. Soc.*, **126**, 10226.

114 Liu, J., Rinzler, A.G., Dai, H., Hafner, J.H., Bradley, R.K., Boul, P.J., Lu, A., Iverson, T., Shelimov, K., Huffman, C.B., Rodriguez-Macias, F., Shon, Y.-S., Lee, T.R., Colbert, D.T., and Smalley, R.E. (1998) *Science*, **280**, 1253.

115 Dyke, C.A. and Tour, J.M. (2003) *J. Am. Chem. Soc.*, **125**, 1156.

116 Binder, W.H. and Sachsenhofer, R. (2007) *Macromol. Rapid Commun.*, **28**, 15.

117 Kolb, H.C., Finn, M.G., and Sharpless, K.B. (2001) *Angew. Chem. Int. Ed.*, **40**, 2004.

118 Fernandez d'Arlas, B., Goyanes, S., Rubiolo, G.H., Mondragon, I., Corcuera, M.A., and Eceiza, A. (2009) *J. Nanosci. Nanotechnol.*, **9**, 6064.

119 Mormann, W., Lu, Y., Zou, X., and Berger, R. (2008) *Macromol. Chem. Phys.*, **209**, 2284.

120 Sun, L., Warren, G.L., O'Reilly, J.Y., Everett, W.N., Lee, S.M., Davis, D., Lagoudas, D., and Sue, H.J. (2008) *Carbon*, **46**, 320.

121 Warren, G.L., Sun, L., Hadjiev, V.G., Davis, D., Lagoudas, D., and Sue, H.-J. (2009) *J. Appl. Polym. Sci.*, **112**, 290.

122 Yang, B.-X., Pramoda, K.P., Xu, G.Q., and Goh, S.H. (2007) *Adv. Funct. Mater.*, **17**, 2062.

123 Shaffer, M.S.P. and Koziol, K. (2002) *Chem. Commun.*, 2074.

124 Qin, S., Qin, D., Ford, W.T., Herrera, J.E., and Resasco, D.E. (2004) *Macromolecules*, **37**, 9963.

125 Qin, S., Qin, D., Ford, W.T., Herrera, J.E., Resasco, D.E., Bachilo, S.M., and Weisman, R.B. (2004) *Macromolecules*, **37**, 3965.

126 Wu, H.-L., Yang, Y.-T., Ma, C.-C.M., and Kuan, H.-C. (2005) *J. Polym. Sci., Part A: Polym. Chem.*, **43**, 6084.

127 Kuan, C.-F., Kuan, H.-C., Ma, C.-C.M., Chen, C.-H., and Wu, H.-L. (2007) *Mater. Lett.*, **61**, 2744.

128 Rahman, G.M.A., Troeger, A., Sgobba, V., Guldi, D.M., Jux, N., Tchoul, M.N., Ford, W.T., Mateo-Alonso, A., and Prato, M. (2008) *Chem. Eur. J.*, **14**, 8837.

129 Yan, D. and Yang, G. (2009) *Mater. Lett.*, **63**, 298.

130 Xu, H., Wang, X., Zhang, Y., and Liu, S. (2006) *Chem. Mater.*, **18**, 2929.

131 Bocharova, V., Anton, K., Ulrich, O., Stamm, M., Stoffelbach, F., Jerome, R., and Detrembleur, C. (2006) *PMSE Prepr.*, **95**, 591.

132 Bocharova, V., Gorodyska, G., Kiriy, A., Stamm, M., Simon, P., Moench, I., Elefant, D., Lou, X., Stoffelbach, F., Detrembleur, C., and Jerome, R. (2006) *Prog. Colloid Polym. Sci.*, **132**, 161.

133 Bocharova, V., Kiriy, A., Oertel, U., Stamm, M., Stoffelbach, F., Jerome, R.,

and Detrembleur, C. (2006) *J. Phys. Chem. B*, **110**, 14640.
134 Wang, H.C., Li, Y., and Yang, M.J. (2007) *Sens. Actuators, B*, **B124**, 360.
135 Oh, S.B., Kim, H.L., Chang, J.H., Lee, Y.-W., Han, J.H., An, S.S.A., Joo, S.-W., Kim, H.-K., Choi, I.S., and Paik, H.-J. (2008) *J. Nanosci. Nanotechnol.*, **8**, 4598.
136 Poli, R. (2006) *Angew. Chem. Int. Ed.*, **45**, 5058.
137 Asandei, A.D., and Moran, I.W. (2004) *J. Am. Chem. Soc.*, **126**, 15932.
138 Shaver, M.P., Hanhan, M.E., and Jones, M.R. (2010) *Chem. Commun.*, **46**, 2127.
139 Champouret, Y., Baisch, U., Poli, R., Tang, L., Conway, J.L., and Smith, K.M. (2008) *Angew. Chem. Int. Ed.*, **47**, 6069.
140 Debuigne, A., Champouret, Y., Jerome, R., Poli, R., and Detrembleur, C. (2008) *Chem. Eur. J.*, **14**, 4046.
141 Debuigne, A., Poli, R., Jerome, R., Jerome, C., and Detrembleur, C. (2009) *ACS Symp. Ser.*, **1024**, 131.
142 Thomassin, J.-M., Molenberg, I., Huynen, I., Debuigne, A., Alexandre, M., Jerome, C., and Detrembleur, C. (2010) *Chem. Commun.*, **46**, 3330.
143 Wu, H.-X., Qiu, X.-Q., Cai, R.-F., and Qian, S.-X. (2007) *Appl. Surf. Sci.*, **253**, 5122.
144 Jiang, S., Deng, J., and Yang, W. (2008) *Macromol. Rapid Commun.*, **29**, 1521.
145 Li, Q., Ma, Y., Mao, C., and Wu, C. (2009) *Ultrason. Sonochem.*, **16**, 752.
146 Mountrichas, G., Pispas, S., and Tagmatarchis, N. (2008) *Mater. Sci. Eng. B*, **152**, 40.
147 Qin, S., Qin, D., Ford, W.T., Resasco, D.E., and Herrera, J.E. (2004) *Macromolecules*, **37**, 752.
148 Wang, G., Dong, Y., Liu, L., and Zhao, C. (2007) *J. Appl. Polym. Sci.*, **105**, 1385.
149 Wang, G. and Liu, Y. (2009) *Macromol. Chem. Phys.*, **210**, 2070.

13
Organic Functionalization of Nanotubes by Dipolar Cycloaddition
Vassilios Georgakilas and Dimitrios Gournis

13.1
Introduction

Carbon nanotube (CNT) could be considered as a carbon nanomaterial or an organic macromolecule depending on which point of view is preferred between their physical properties or their chemical reactivity. The structure and the chemical composition of CNTs limit the possible organic reactions that could be applied on them.

Dipolar cycloaddition is a widely used organic reaction that has been applied successfully in carbon nanostructures, starting from the fullerenes [1–5] and following for the derivatization of single- (SWCNT) and multi- (MWCNT) walled CNTs. The double carbon–carbon bonds on the sidewalls of SWNT or the external surface of MWNT can act as dipolarophile and reacting with dipoles form five-membered rings which lie perpendicular to the CNT surface. Usually such five-membered rings – pyrrolidine, pyrazoline, etc. – could bear several organic functional groups offering to the nanotubes interesting properties such as dispersibility in organic solvents, compatibility with polymers, optical properties, etc.

The reactivity of the double C=C bonds as dipolarophile is low in CNTs, in contrast with C_{60} fullerene. This is probably due to the low dispersibility of CNT in organic solvents, their extended aromatic system, and the fact that the C–C bonds are not as curved as in the case of C_{60}. The obstacle of the low reactivity is partially surpassed using extended reaction time, excess of reactants, and usually the number of the added groups per 60 carbon atoms do not exceed one, although in C_{60} five added groups per 60 carbon atoms are easily achieved. However, in many cases this ratio is quite enough to make the CNT dispersible in organic solvents.

In this chapter, we will focus on the most recent developments on the use of 1,3-dipolar cycloaddition reaction for the sidewall functionalization of CNTs (SWNT or MWNT). The dipole, in the 1,3-dipolar cycloaddition, can be an organic heterocyclic compounds such as nitrile oxide, aziridine, nitrone, or a reactive intermediate that occurred during the reaction by the proper precursors such as azomethine and pyridinium ylides, nitrile imines, etc.

Surface Modification of Nanotube Fillers, First Edition. Edited by Vikas Mittal.
© 2011 Wiley-VCH Verlag GmbH & Co. KGaA. Published 2011 by Wiley-VCH Verlag GmbH & Co. KGaA.

13.2
The Case of Azomethine Ylide

The addition of azomethine ylide using 1,3-dipolar cycloaddition was one the first organic reactions that have been successfully applied on CNTs [6, 7]. The dipole here is a reactive intermediate that is formed by the thermal condensation of an aldehyde and an α-amino acid. The reaction of azomethine ylide with the double C=C bonds on the CNT surface results in the formation of pyrrolidine rings. Using several different techniques, it has been estimated that one pyrrolidine moiety is introduced per 100 carbon atoms in the CNT network (Figure 13.1). The triethylene glycol groups that are introduced in the pyrrolidine ring using proper α-amino acid make the modified CNT highly dispersed in organic solvents, whereas amorphous carbon, metallic nanoparticles, and other impurities are not part of the reaction and thus can remove as insoluble material (Figure 13.2).

Figure 13.1 The formation of azomethine ylide and the cycloaddition on CNT [6].

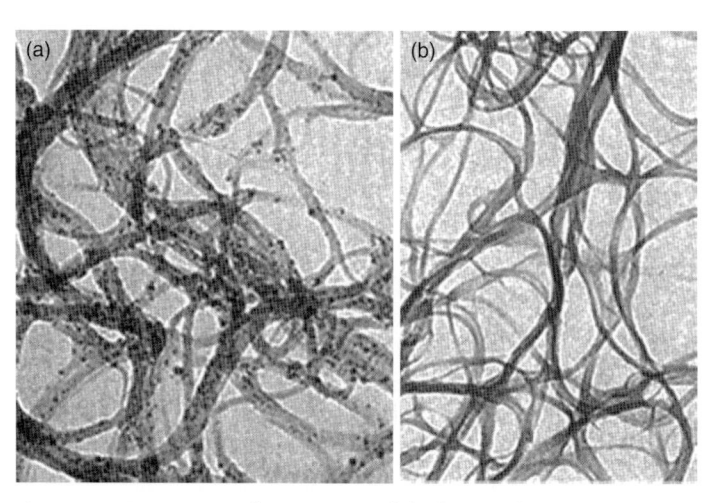

Figure 13.2 TEM images of (a) pristine and (b) functionalized SWNTs. Reproduced with permission from Ref. [7].

Using analogous modified α-amino acid, the final CNT derivatives could bear through the pyrrolidine rings a great variety of organic functional groups (Figure 13.3). Utilizing paraformaldehyde and ferrocene-modified glycine, soluble f-CNT decorated with ferrocene units (Fc) were obtained. The photoexcitation of these derivatives leads to an electron transfer from the ferrocene units to CNTs with slow charge recombination. Such system could be useful in solar energy conversion devices [8, 9].

Alternatively, the most simple α-amino acid N-methyl glycine – could be used in combination with aldehydes such as substituted benzaldehydes, in the place of paraformaldehyde. Using, for example, 3,4-dihydroxy benzaldehydes and N-methyl glycine, the functionalized CNTs (SWCNT-f-OH or MWNT-f-OH) are highly loaded with phenol groups (Figure 13.4a) [10].

The hydroxylated CNTs are easily dispersible in ethanol, DMF (Figure 13.4b), and other polar organic solvents. The functionalization changes the hybridization

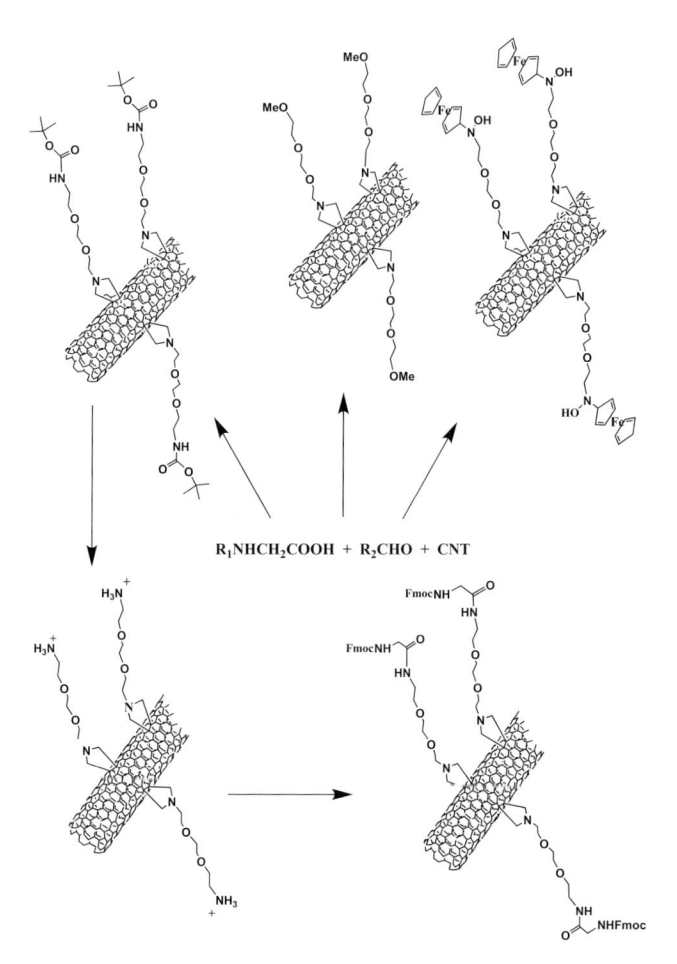

Figure 13.3 Derivatization of CNT through 1,3-dipolar cycloaddition with several organic functional groups. [8]–Reproduced with permission from The Royal Society of Chemistry.

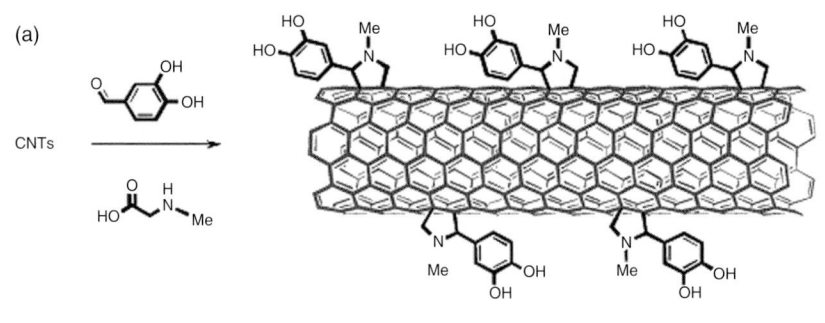

Figure 13.4 (a) Schematic representation of 1,3-dipolar cycloaddition to CNTs using 3,4-dihydroxybenzaldehyde. (b) Photograph of a solution of SWCNT-f-OH in DMF. (c) Concentrated and dilute solutions of hydroxylated MWNT in ethanol. Reproduced with permission from Ref. [10].

of the reacted carbon atoms from sp^2 to sp^3 and as observed in Raman spectroscopy the I_G/I_D ratio was reduced significantly from 28 in the pristine SWNT to 10 for the SWCNT-f-OH.

The TEM and AFM images of SWCNT-f-OH reveal the appearance of thin bundles of tubes throughout the scanned areas. Comparison of these images with those for the raw material illustrated that debundling took place, providing evidence for the success of the functionalization (Figure 13.5).

The thermogravimetric analysis for pristine SWCNTs and SWCNT-f-OH exhibited a weight loss of 28% at 800 °C, which roughly corresponds to one functional group every 79 carbon atoms (Figure 13.6).

The fully hydroxylated CNTs apart from their dispersibility in polar organic solvents are highly compatible with polymers such as polyacrylonitrile, poly(ethylene vinyl acetate), or clays. Thus, transparent films or gels can be formed by simple mixing in such a polymer or in clay, respectively (Figure 13.7).

The phenol groups can be further substituted offering a great variety of organic derivatives. Perfluoro-octadecyl-silyl chains can be attached on CNT through a Si–O–C bridge with the phenolic groups (Figure 13.8a) [11]. The presence of perfluoroalkyl groups is extended to almost all the surface of the nanotube forming a thin layer which is easily observed by TEM and transforms the hydrophilic CNTs

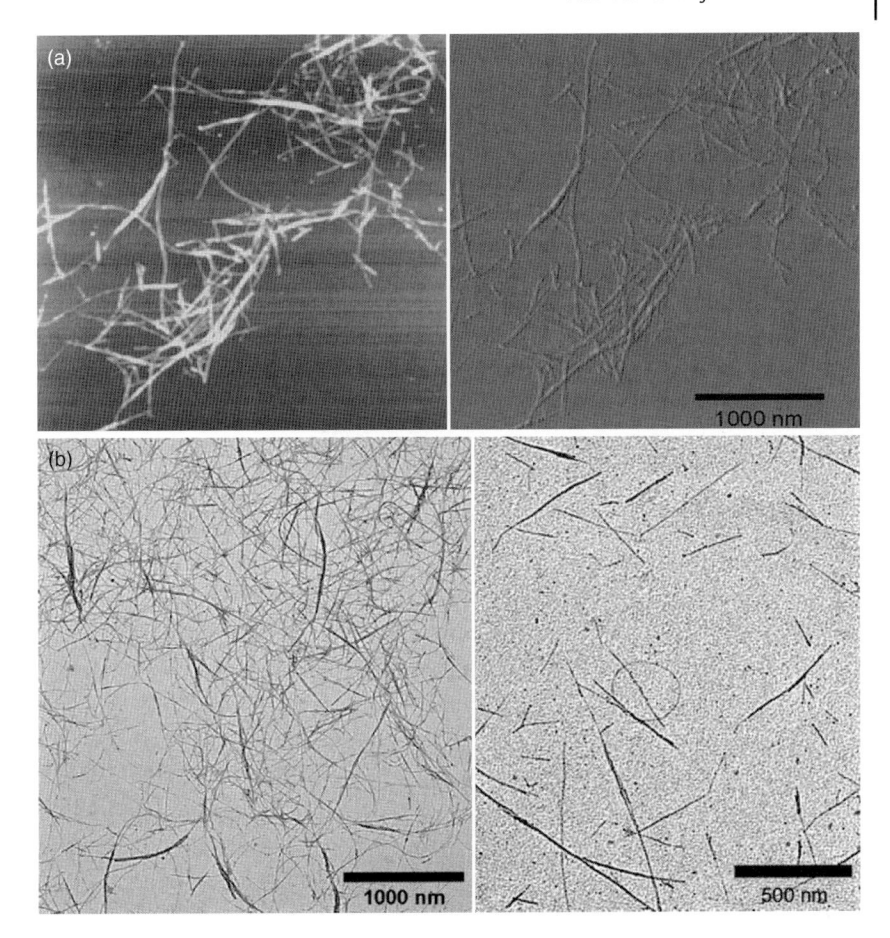

Figure 13.5 (a) AFM and (b) TEM images of SWCNT-f-OH. Reproduced with permission from Ref. [10].

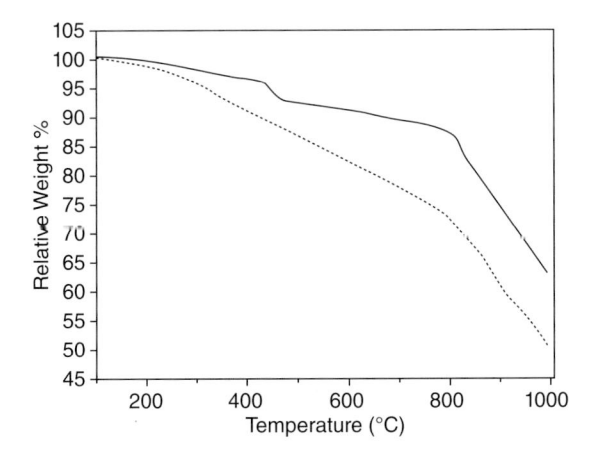

Figure 13.6 TGA curves of (solid line) pristine SWCNTs and (dashed line) SWCNT-f-OH. Reproduced with permission from Ref. [10].

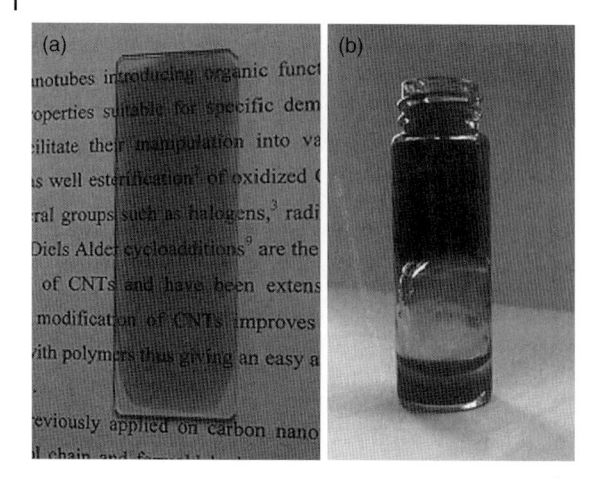

Figure 13.7 (a) Photograph of optically transparent polymer-composite films obtained by homogeneous dispersion of MWCNT-f-OH in PAN. (b) Laponite/MWCNT-f-OH composite gel. Reproduced with permission from Ref. [10].

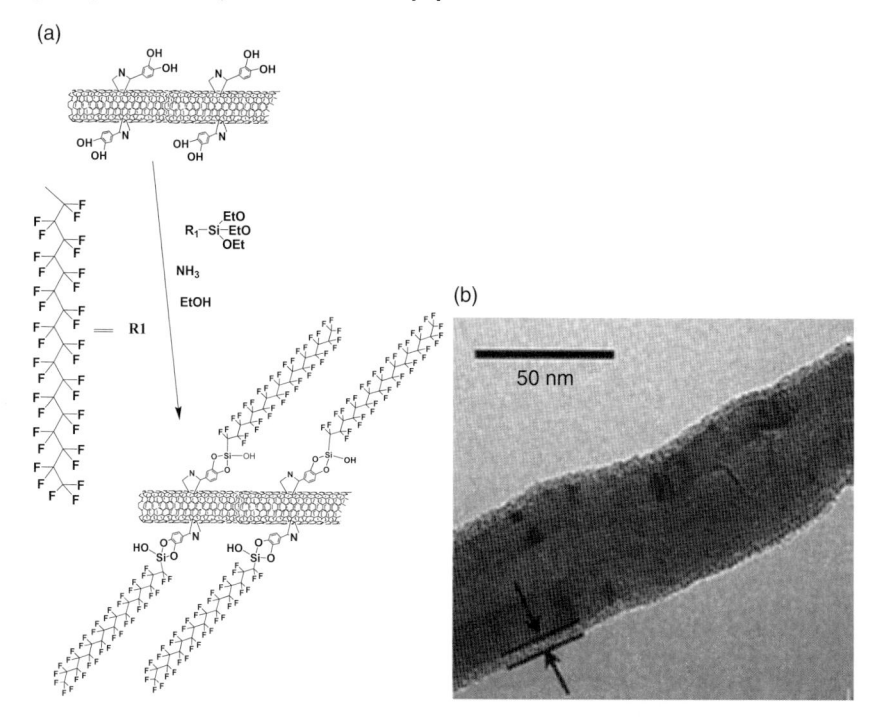

Figure 13.8 (a) Synthesis of the perfluorinated CNT. (b) The TEM image depicts a surface-coated CNT. Reproduced with permission from Ref. [11].

to superhydrophobic (Figure 13.8b). The contact angle of a water droplet on a film of neat superhydrophobic CNTs is about 170° (Figure 13.9b) [11].

The attachment of the perfluoroalkyl groups on the surface of the nanotubes was revealed by FT-IR spectroscopy (Figure 13.9a). The intense band at 1212 cm^{-1}

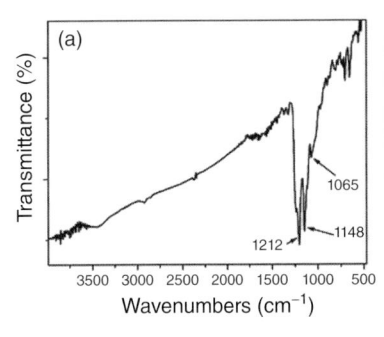

Figure 13.9 (a) FTIR spectrum of the perfluorinated CNTs. (b) Water droplet (diameter ~2 mm) over a superhydrophobic glass after its coating with perfluorinated MWNTs. Reproduced with permission from Ref. [11].

is characteristic of the stretching vibration of C–F bonds, whereas the bands at 1148 and 1065 cm^{-1} of the formation of Si–O–C bonds.

Using a similar method Adronov, Goward, and coworkers [12, 13] employed 4-hydroxylphenyl glycine and octanal to create phenol-bearing pyrrolidine rings, followed by a reaction of the phenols with 2-bromoisobutyryl bromide. The resulting α-bromo esters serve as initiators for the grafting of polymers such as poly(methylmethacrylate) or poly(*tert*-butyl acrylate) (Figure 13.10). AFM analysis of the products indicated large globular areas connected by small nanotube bundles or individual nanotubes [12, 13].

Dendrimers are spherical macromolecular nanostructures with highly controlled size and shape, with high importance in drug delivery. They are engineered to carry small molecules such as drugs in their internal cavity or attached to their surface. After the 1,3-dipolar cycloaddition, amino-terminated functional groups on the pyrrolidine rings serve as links for the attachment of dendrimers through amide bonds as presented in the Figure 13.11 [14]. Morphologically, the attached dendrimer generates a wrapping of 8 nm thick around the CNTs walls where the estimated theoretical length of the dendrimer molecule is around 6.4 nm.

1,3-Dipolar cycloaddition of azomethine ylide has been used by Prato and coworkers [15] to introduce on the nanotube surface two different functional groups. As described in the referred article, the aim of this work is to introduce a fluorescent probe, fluorescein isothiocyanate (FITC) and an anticancer agent, methotrexate (MTX), around the CNT sidewalls (Figure 13.12). The subsequent presence of the drug and the fluorescent probe on the carrier could help to control of the drug-delivery procedure [15].

Wang et al. [16] have used microwave as an alternative of heating to reduce the time and the solvent of the 1,3-dipolar cycloaddition reaction. Apart from the combination of an α-amino acid and aldehyde, aziridines are well-known precursors to azomethine ylides. Prato *et al.* [17] have used microwave as an alternative of heating to reduce the time and the solvent of the 1,3-dipolar cycloaddition reaction (Figure 13.13).

The application of pyrrolidine ring decorated CNT can be further exploited, employing conjugated polymers, to areas such as organic light-emitting diodes for

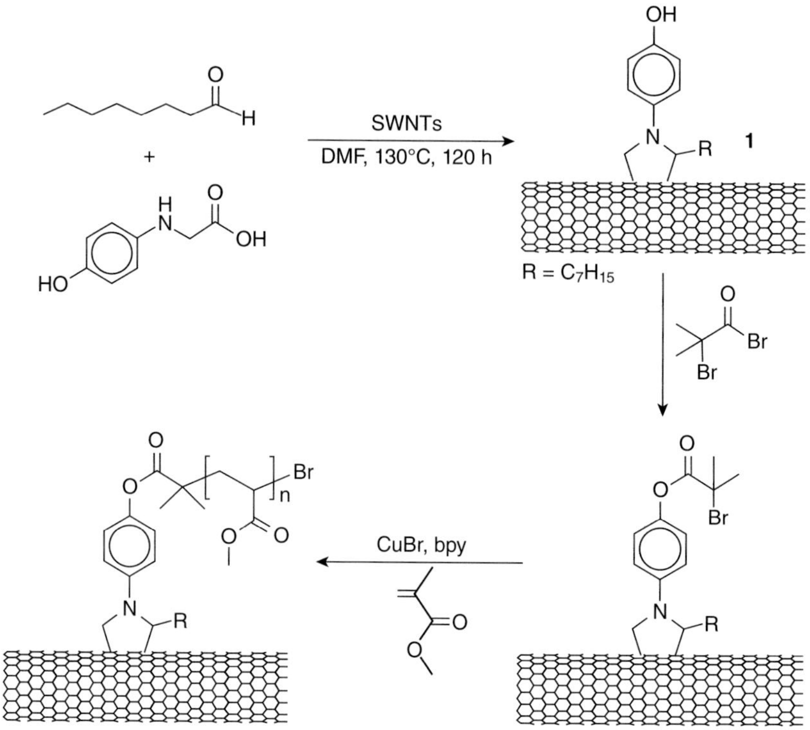

Figure 13.10 Grafting of polymer on phenol-substituted CNTs. Reproduced with permission from Ref. [12].

flat panel displays, photovoltaic cells for solar energy conversion, thin-film transistors, and chemical sensors. Bo et al. [18] have applied the 1,3-dipolar cycloaddition of azomethine using 7-bromo-9,9-dioctylfluorene-2-carbaldehyde and L-serine on MWNT. The growth of conjugated polyfluorene chains on the surfaces of MWCNTs was achieved by the polycondensation of the functionalized CNT and 2-(2-bromo-9,9-dioctylfluoren-7-yl)-4,4,5,5-tetramethyl-1,3,2-dioxaborolane (Figure 13.14). The fluorescence of polyfluorenes was completely quenched by the MWCNTs, indicating a fast photo-induced electron transfer from polyfluorenes to MWCNTs (Figure 13.15) [18].

13.3
The Case of Pyridinium Ylides

A similar to the pyrrolidine ring is formed by the 1,3-dipolar cycloaddition of pyridinium ylides to the double C=C bonds of the CNT surface [19]. The pyridinium ylide is generated from N-(ethoxycarbonylmethyl)-pyridinium bromide, commonly named as Kröhnke salt in a basic solution. The reaction is terminated in

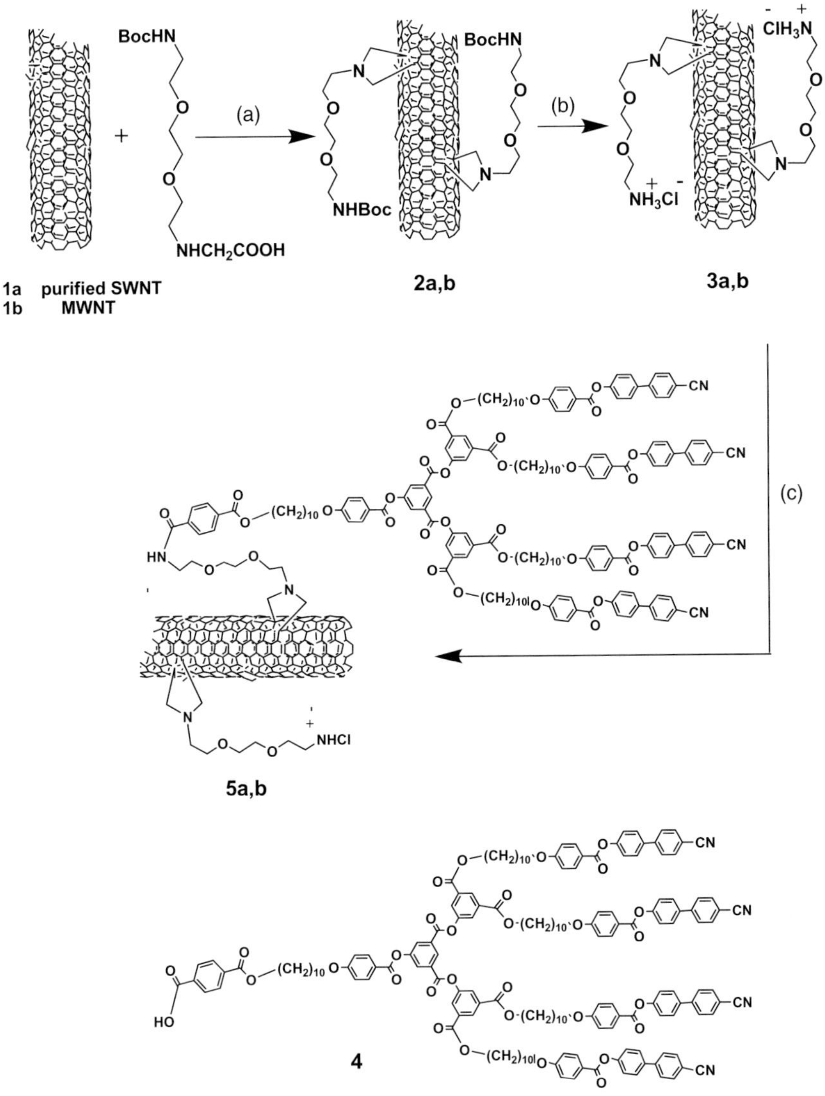

Figure 13.11 The attachment of dedrimer molecules on amine-terminated functionalized CNT through amide bond. Reproduced with permission from Ref. [14].

two steps: the addition of pyridinium ylide on the double C–C bond directly on the surface of CNT forming a five-membered ring and a further addition of the same intermediate on the aromatic system of the added groups forming finally an indolizine group (Figure 13.16).

This procedure has been carried out using common heating or microwaves to reduce the reaction time. The efficiency of the derivatization of CNT was among else identified by the FTIR spectra of the products where characteristic bands of

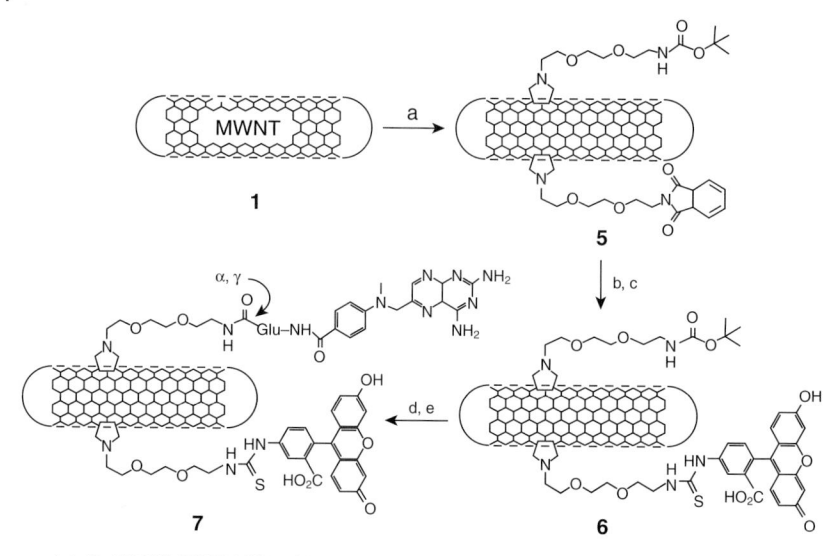

(a) R–NHCH$_2$COOH/(CH$_2$O)n in DMF, 130 uC;
(b) hydrazine in EtOH;
(c) Fitc in DMF;
(d) HCl 4M in dioxane;
(e) MTX, Bop/DIEA in DMF.
R = Boc–NH(CH$_2$CH$_2$O)$_2$–CH$_2$CH$_2$–
and Pht–N(CH$_2$CH$_2$O)$_2$–CH$_2$CH$_2$–.

Figure 13.12 Introduction of fluorescein isothiocyanate and methotrexate on amine-terminated functionalized CNT through amide bonds. [15] – Reproduced with permission from The Royal Society of Chemistry.

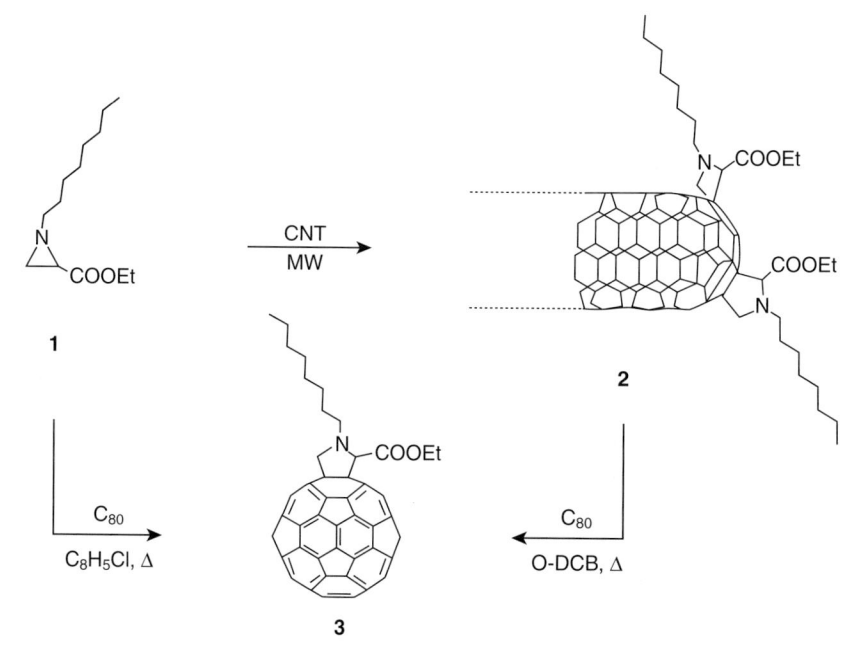

Figure 13.13 Reaction of aziridine with CNT under microwave and C$_{60}$ by heating. Reproduced with permission from Ref. [17].

Figure 13.14 Synthesis of the PF-functionalized MWCNTs. Reproduced with permission from Ref. [18].

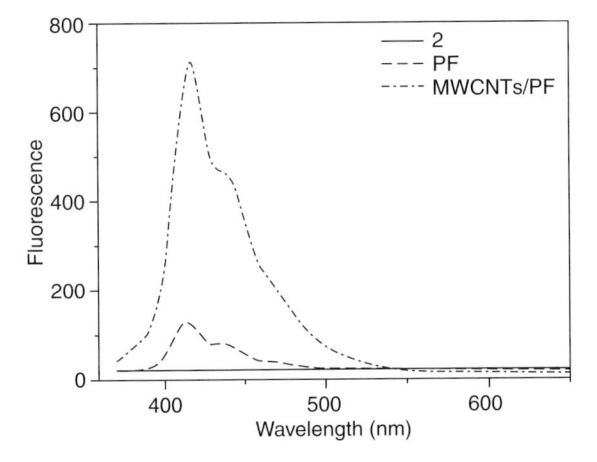

Figure 13.15 Photoluminescent spectra of polyfluorene, the PF-functionalized MWCNTs (2), and the acid-treated MWCNTs/polyfluorene (MWCNTs/PF) dissolved or dispersed in THF. Reproduced with permission from Ref. [18].

ester group ($\nu(CO)$) at 1730 cm^{-1} and indolizine group at 1632 and 1541 cm^{-1} were recorded. The degree of functionalization derived from the thermogravimetric analysis ranges between 1 functional group per 133 and 96 carbon atoms depending on the presence of heating or microwaves in the procedure, respectively [19]. The Raman spectra of the final product lead to the observation that the 1,3-dipolar cycloaddition of the pyridinium ylides was preferable for metallic and large-diameter semiconducting SWNTs. The functionalized CNTs upon excitation at 335 nm emit blue light at 416 nm similarly to the emission of indolizine.

The presence of esters on the indolizine offers the opportunity for further modification of the derivatized CNT through the known organic chemistry of ester or carboxylic acids, which occurred after the hydrolysis of ester. As an example, the reaction of N,N-dimethylethylenediamine with the ester group leads to the appearance of the positively charged $-N^+(Me_3)$ end groups on the indolizine (Figure 13.17a). Positively charged CNT can then interact with negatively charged components such as citrate-stabilized gold nanoparticles forming a CNT/gold nanoparticles composite material (Figure 13.17b) [19].

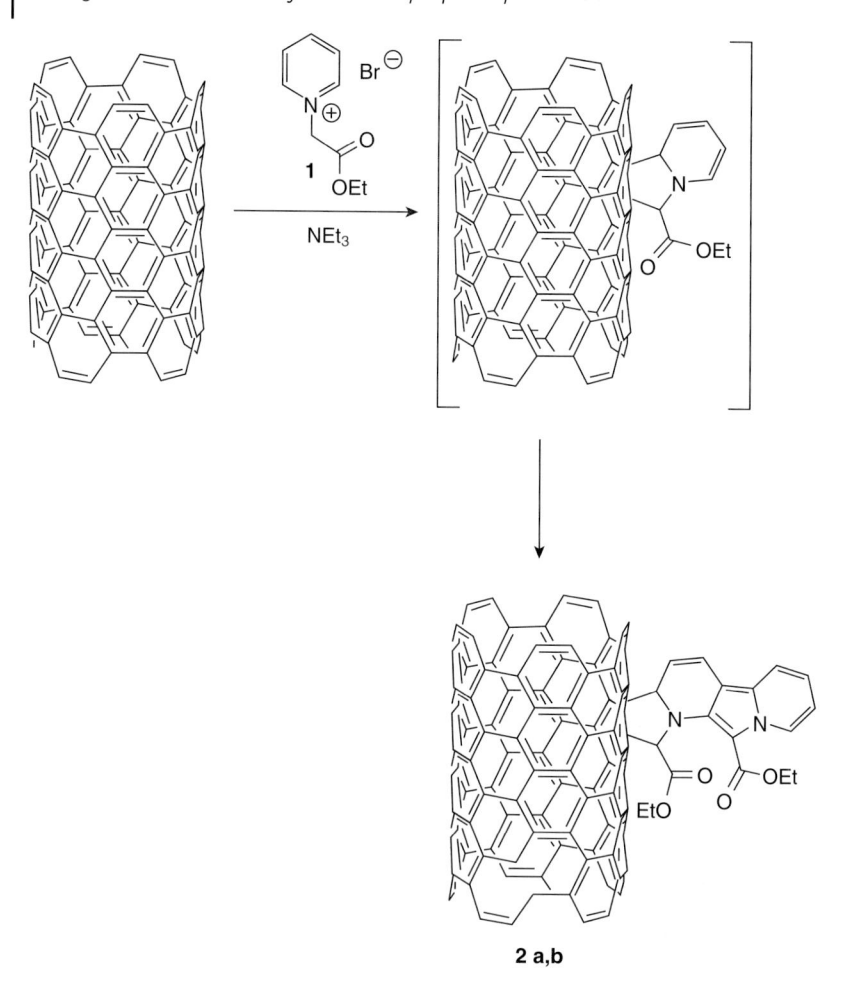

Figure 13.16 Schematic representation of the formation of indolizine groups on SWNT through 1,3-dipolar cycloaddition of a pyridinium ylide. Reproduced with permission from Ref. [19].

13.4
The Case of Nitrile Oxide

Apart from azomethine ylide, nitrile oxide has been also used with 1,3-dipolar cycloaddition on CNTs. Here the reactive intermediate – hydroxamic chloride – is prepared by treating a benzaldehyde oxime with N-chlorosuccinimide. By the 1,3-cycloaddition of hydroxamic chloride on CNT in the presence of triethylamine, a five-membered ring is formed on the external surface of CNT. Apart from CNT, this reaction can be formed on C_{60} or carbon-encapsulated iron oxide nanostructures (Figure 13.18) [20].

13.4 *The Case of Nitrile Oxide* | 301

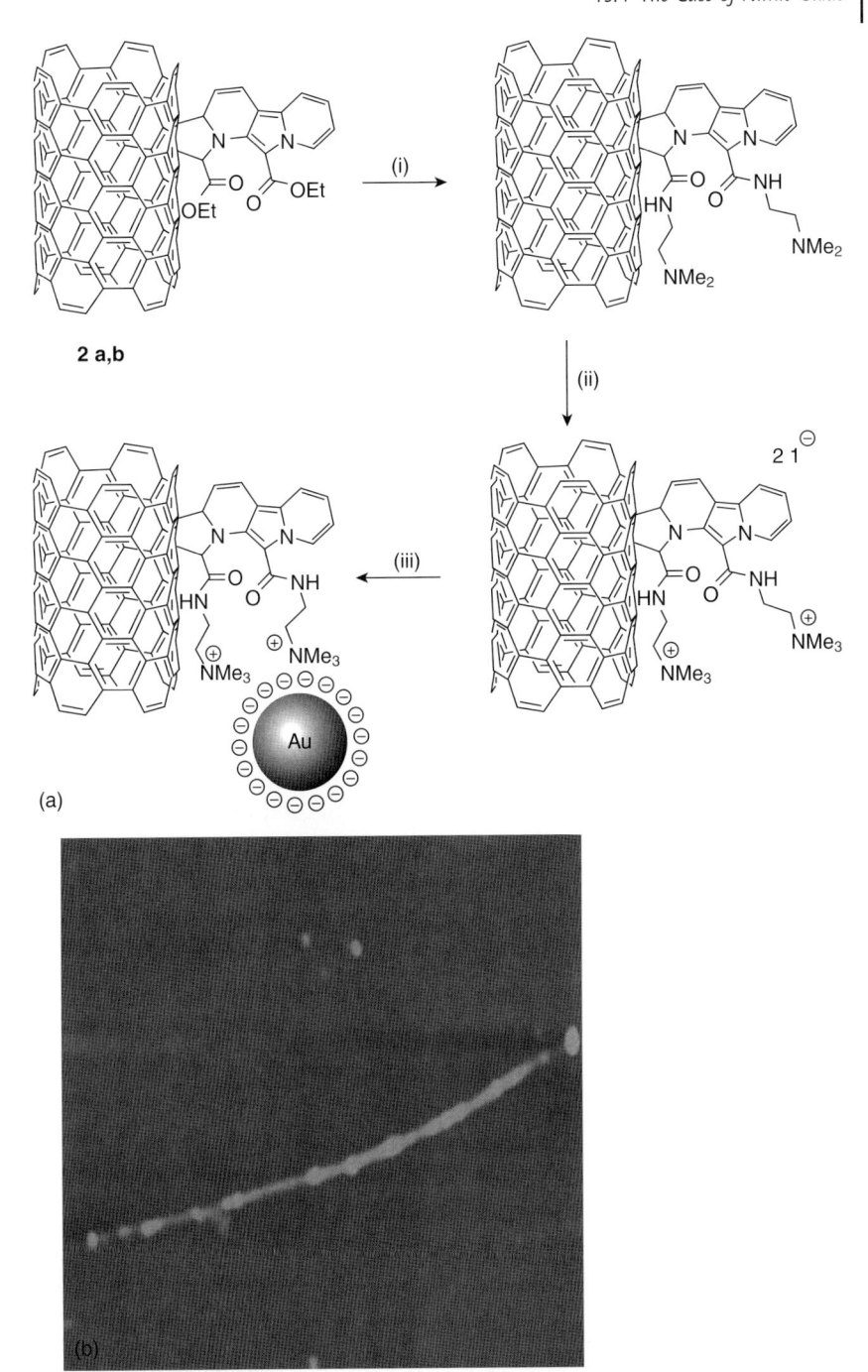

Figure 13.17 Schematic representation of the electrostatic interaction of gold colloids with indolizine functionalized SWNTs **(2a,b)**. A typical AFM image of an indolizine- functionalized SWNT **(2a,b)** with positively charged tertiary amines after exposure to citrate-stabilized Au colloids (4–6 nm). Reproduced with permission from Ref. [19].

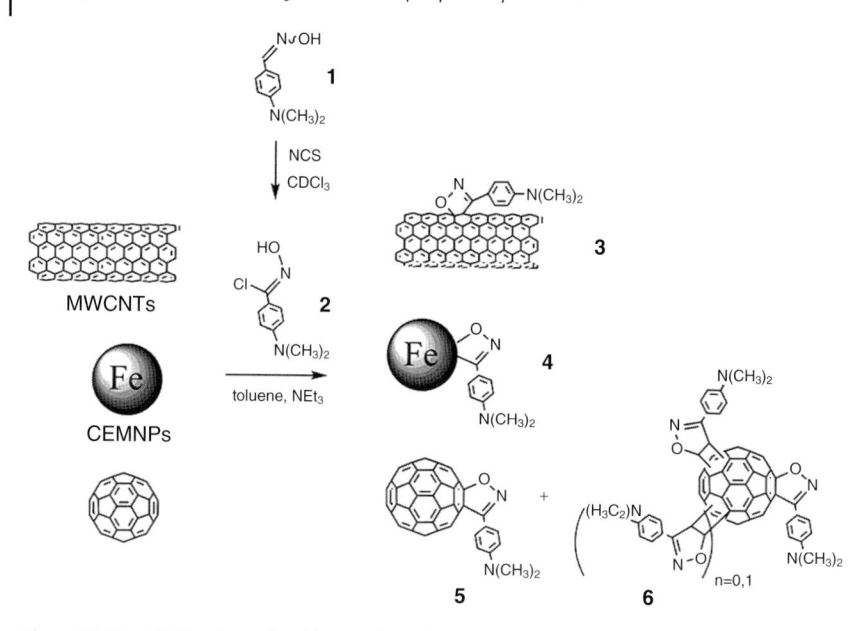

Figure 13.18 1,3-Dipolar cycloaddition of nitrile oxide to CNTs, C_{60}, and carbon-encapsulated iron oxide nanostructures. Reproduced with permission from Ref. [20].

A similar case has been presented by Echegoyen et al. [21] where SWNTs were functionalized at the end tips with pentyl esters to increase their dispersibility and on the walls by pyridyl isoxazoline groups using nitrile oxide reactants. The pyridyl groups capture zinc porfyrin (ZnPor) forming a complex (Figure 13.19) [21].

For the evaluation of this complex, ZnPor/Py–SWNT between ZnPor and the pyridine-functionalized carbon nanotubes (Py–SWNT), ^{1}H NMR, and UV-vis spectroscopy have been used. The UV-vis spectra of ZnPor in CH_2Cl_2 are characterized by three absorption bands, which are clearly red-shifted after the addition of Py–SWNT and the complexation (Figure 13.20). Moreover, the protons corresponding to the pyridyl group in the ^{1}H NMR spectra are shifted to higher field after the complexation with ZnPor.

13.5
The Case of Nitrone

Nitrone is also a reactive heterocyclic organic molecule for 1,3-cycloaddition. Cyclic nitrones are useful intermediates in organic synthesis and very reactive with 1,3-cycloaddition even with weak dipolarophiles. The result of the addition of cyclic nitrones on the double bonds of CNTs through the 1,3-dipolar cycloaddition is the formation of five-membered pyrrolidine N-oxide rings (Figure 13.21). The as-produced functionalized CNTs appear a remarkable high solubility in DMF (10 mg/ml) [22].

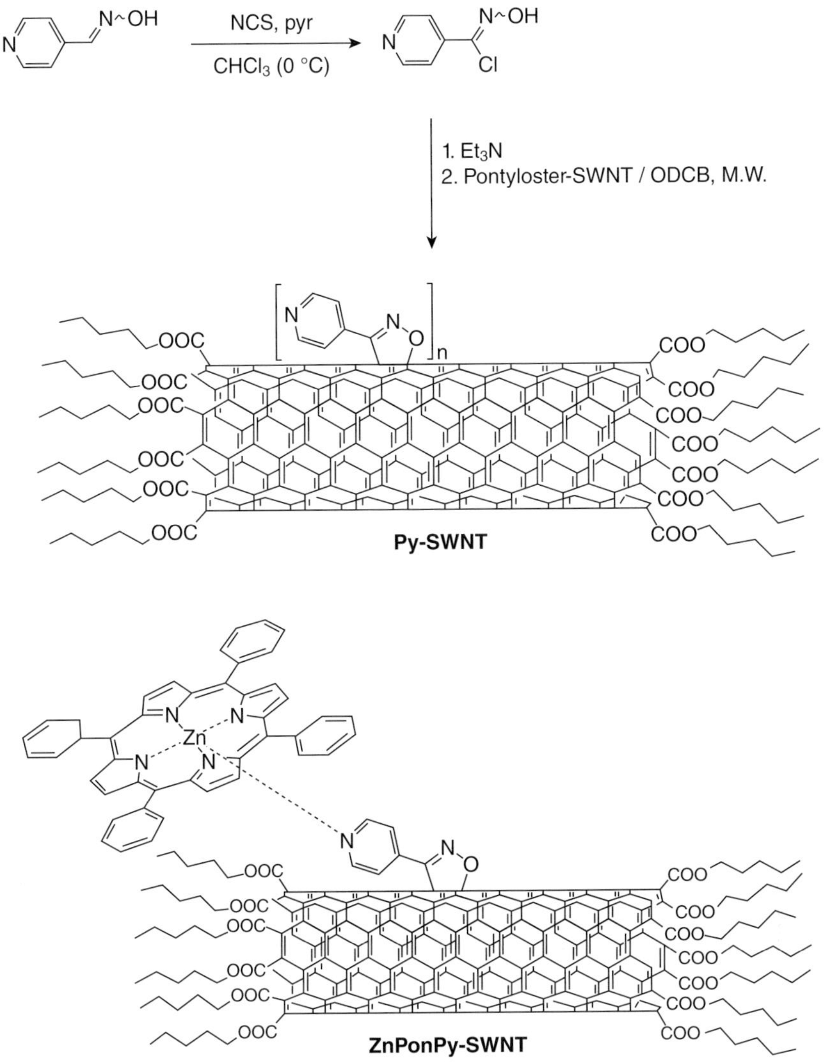

Figure 13.19 Synthetic route for the preparation of Py–SWNT. Chemical structure of ZnPor/Py–SWNT. Reproduced with permission from Ref. [21].

The characterization of the CNTs derivatives in this chapter [22] includes a combination of TGA profiles with mass spectrometry analysis of the volatile fragments that produced during the thermal decomposition. As observed by the TG profiles of pristine and functionalized MWCNTs presented in Figure 13.21a, a weight loss of 4.57% is recorded for the f-MWCNTs in comparison with the pristine MWCNTs, in a temperature range from 96 °C to 575 °C. The volatile fragment identified by mass spectrometry to be emitted in this temperature range was isobutene and thus the observed weight loss is fairly attributed to the removal of

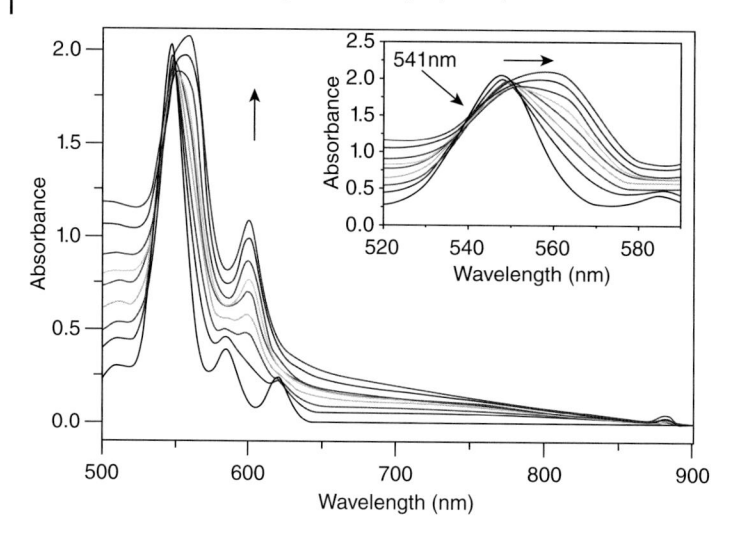

Figure 13.20 Absorption spectra of ZnPor in solution in CH$_2$Cl$_2$ and the red shift upon the addition of increasing amounts of Py–SWNT. The arrows indicate the direction of Py–SWNT concentration increases. The inset shows an enlargement of the visible bands and the red shift of the band maximum. Reproduced with permission from Ref. [21].

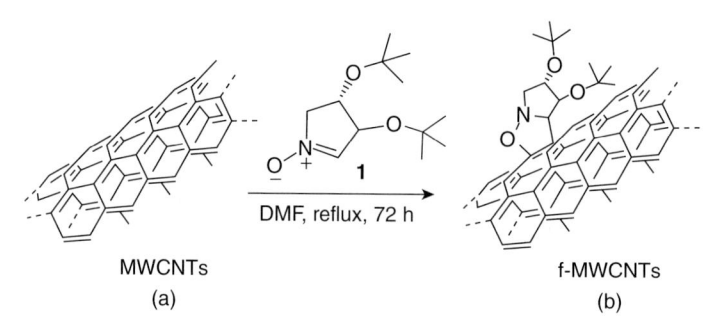

Figure 13.21 Nitrone 1,3-dipolar cycloaddition onto MWCNTs. [22] – Reproduced with permission from The Royal Society of Chemistry.

the added nitrones from the f-MWCNTs. Based on these remarks, it is concluded that the grafted nitrones in f-MWCNTs represent 9.4% of the total weight, which roughly corresponds to 1 di-*tert*-butoxypyrrolidine-*N*-oxide group every 184 carbon atoms. The organic functionalization of the MWCNTs results in the decrease in their bundle diameter (debundling), which influences positively their dispersibility in organic solvents. In Figure 13.22B, characteristic TEM and AFM images of pristine MWCNTs and f-MWCNTs are presented. In the AFM image (d), short and thin bundles of f-MWCNTs spread in the total area are observed, in contrast with a dense aggregate of bundles of pristine MWCNTs presented in AFM image (c).

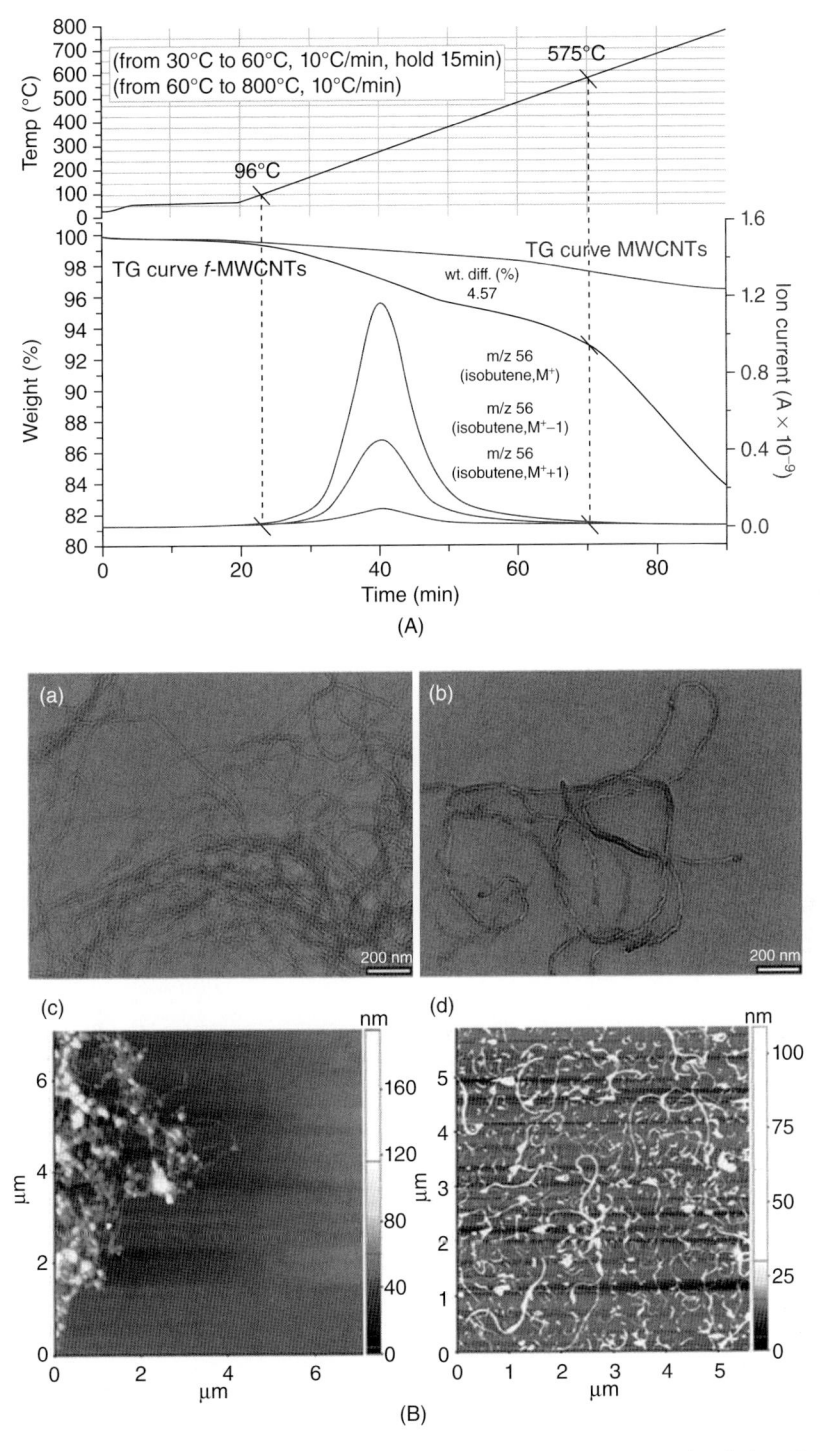

Figure 13.22 (A) TGA profiles of pristine MWCNTs and f-MWCNTs associated with the MS analysis of volatiles. (B) TEM (top) and AFM (bottom) images of pristine (a, c) and functionalized (b, d) MWCNTs. [22] – Reproduced with permission from The Royal Society of Chemistry.

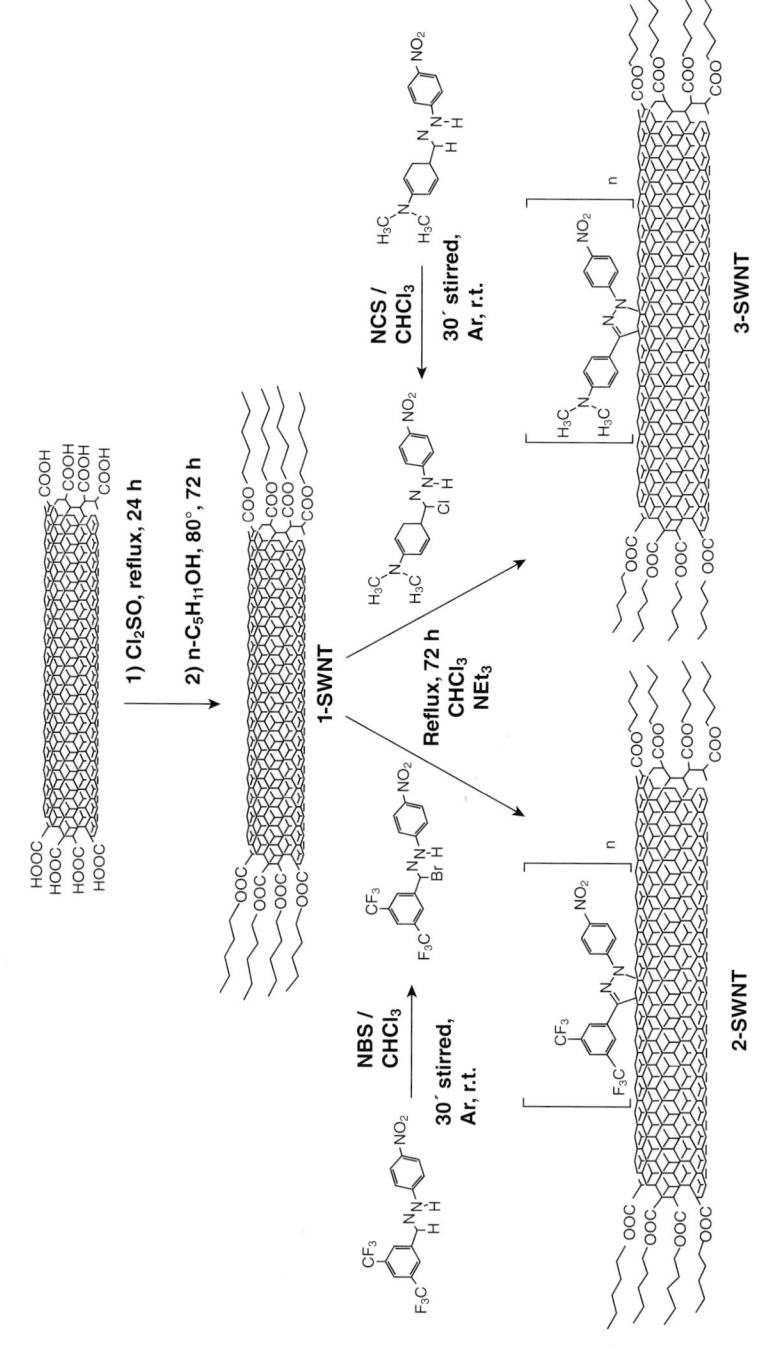

Figure 13.23 Synthetic route for the preparation of soluble, photoactive 2-SWNT and 3-SWNT. Reproduced with permission from Ref. [23].

13.6
The Case of Nitrile Imines

Nitrile imines are also reactive intermediates that can attack double C=C bonds. They are formed from substituted α-bromo hydrazones in basic environment. The 1,3-dipolar cycloaddition of nitrile imine on the carbon double bonds of the CNT surface results in the formation of pyrazoline rings. A characteristic paradigm is described by Alvaro et al. [23] as presented in Figure 13.23. The authors have selected the starting hydrazones here in that way to have in the final functionalized nanotubes either an electron poor (3,5-bis(trifluoromethyl)phenyl) or an electron-rich (4-(N,N-dimethylamino)phenyl) group on the pyrazoline rings [23].

13.7
Conclusions

Dipolar cycloaddition has been approved a powerful tool as regards the derivatization of carbon nanostructures including fullerenes, nanohorns, and CNTs. The flexibility as regards the added functional groups offers the opportunity to develop a great variety of organic derivatives especially in CNTs. The addition of the several reactive intermediates that have been used in 1,3-dipolar cycloaddition with C=C bonds on CNTs surface results in the appearance of one five-membered heterocyclic ring every 100–200 carbon atoms showing that the reactivity of this reaction is not strongly depended from the reactive intermediate. The heterocyclic rings could bear organic functional groups which can remarkably alter the physical and the chemical properties of the functionalized CNTs.

References

1 Prato, M. and Maggini, M. (1998) *Acc. Chem. Res.*, **31**, 519.
2 Maggini, M., Scorrano, G., and Prato, M. (1993) *J. Am. Chem. Soc.*, **115**, 9798.
3 Wilson, S.R. and Lu, Q. (1995) *J. Org. Chem.*, **60**, 6496.
4 Hirsch, A. (1993) *Angew. Chem. Int. Ed.*, **32**, 1138.
5 Taylor, R. and Walton, D.R.M. (1993) *Nature*, **363**, 685.
6 Georgakilas, V., Kordatos, K., Prato, M., Guldi, D.M., Holzinger, M., and Hirsch, A. (2002) *J. Am. Chem. Soc.*, **124**, 760.
7 Georgakilas, V., Voulgaris, D., Vazquez, E., Prato, M., Guldi, D.M., Kukovecz, A., and Kuzmany, H. (2002) *J. Am. Chem. Soc.*, **124**, 14318.
8 Tagmatarchis, N. and Prato, M. (2004) *J. Mater. Chem.*, **14**, 437.
9 Guldi, D.M., Marcaccio, M., Paolucci, D., Paolucci, F., Tagmatarchis, N., Tasis, D., Vazquez, E., and Prato, M. (2003) *Angew. Chem. Int. Ed.*, **42**, 4206.
10 Georgakilas, V., Bourlinos, A., Gournis, D., Tsoufis, T., Trapalis, C., Mateo-Alonso, A., and Prato, M. (2008) *J. Am. Chem. Soc.*, **130**, 8733.
11 Georgakilas, V., Bourlinos, A.B., Zboril, R., and Trapalis, C. (2008) *Chem. Mater.*, **20**, 2884.
12 Cahill, L.S., Yao, Z., Adronov, A., Penner, J., Moonoosawmy, K.R., Kruse, P., and Goward, G.R. (2004) *J. Phys. Chem. B*, **108**, 11412.

13 Yao, Z.L., Braidy, N., Botton, G.A., and Adronov, A. (2003) *J. Am. Chem. Soc.*, **125**, 16015.

14 Garcia, A., Herrero, M.A., Frein, S., Deschenaux, R., Munoz, R., Bustero, I., Toma, F., and Prato, M. (2008) *phys. status solidi A*, **205**, 1402.

15 Pastorin, G., Wu, W., Wieckowski, S., Briand, J.P., Kostarelos, K., Prato, M., and Bianco, A. (2006) *Chem. Commun.*, 1182.

16 Wang, Y.B., Iqbal, Z., and Mitra, S. (2005) *Carbon*, **43**, 1015.

17 Brunetti, F.G., Herrero, M.A., Munoz, J.D.M., Giordani, S., Diaz-Ortiz, A., Filippone, S., Ruaro, G., Meneghetti, M., Prato, M., and Vazquez, E. (2007) *J. Am. Chem. Soc.*, **129**, 14580.

18 Xu, G.D., Zhu, B., Han, Y., and Bo, Z.S. (2007) *Polymer*, **48**, 7510.

19 Bayazit, M.K. and Coleman, K.S. (2009) *J. Am. Chem. Soc.*, **131**, 10670.

20 Poplawska, M., Zukowska, G.Z., Cudzilo, S., and Bystrzejewski, M. (2010) *Carbon*, **48**, 1318.

21 Alvaro, M., Atienzar, P., la Cruz, P., Delgado, J.L., Troiani, V., Garcia, H., Langa, F., Palkar, A., and Echegoyen, L. (2006) *J. Am. Chem. Soc.*, **128**, 6626.

22 Ghini, G., Luconi, L., Rossin, A., Bianchini, C., Giambastiani, G., Cicchi, S., Lascialfari, L., Brandi, A., and Giannasi, A. (2010) *Chem. Commun.*, **46**, 252.

23 Alvaro, M., Atienzar, P., de la Cruz, P., Delgado, J.L., Garcia, H., and Langa, F. (2004) *J. Phys. Chem. B*, **108**, 12691.

Index

a

adsorption on nanotube surfaces   142
– encapsulation of organic molecules   105
– of aromatic amino acids   104
– of charged surfactants   91
– of chloroform   103
– of electrolytes   91
– of metalloporphyrin complexes   104, 275
– of nanoparticles on charged CNT-g-polyelectrolytes   189, 275
– of nitrogen   148
– of polymers by π-π stacking   259
– of triglycerides   150
aminated acrylicnitrile (mAN)   242
amine functionalization of CNTs
– application of   143
– limitations of   154f
– strategies for   139ff, 143
– via 1,3-dipolar cycloaddition of azomethine ylide   296
3-aminopropyltriethanoxysilane (APTES)   7
3-aminopropyltrimethoxysilane (APTS)   13
application of functionalized CNTs   37, 40, 45ff, 67, 123, 257
– application of CNT/polymer composites   81f, 294
– as catalyst in organic reactions   150f
– as CO sensor   151
– as composite electrolytes in PEM fuel cells   154
– as DNA detector   151
– as fluorescent probe   296
– as molecular transporters for biomolecules   126f, 139
– as nanoelectronic devices   135
– as vapor sensor   164f
– in drug delivery   234f, 296
– in molecular electronics   152f
– in semiconductor chemisty   153
– manipulation of solubility   149f
– of amine-functionalized CNTs   149ff
– of CNT-QD in the detection of proteins   131
– of MWCNT-g-PEI   233ff
– of nanoparticles-conjugated CNTs composites   129f, 135
– of pyrrolidine ring decorated CNTs in microelectronics   297
– of sulfonated CNTs   136
– superhydrophobic coating for glass   295
APTES, see 3-aminopropyltriethanoxysilane
APTS, see 3-aminopropyltrimethoxysilane
atom transfer radical polymerization (ATRP)   15f, 30, 35, 45, 53, 69, 83, 168, 179
– CNT-based macroinitiators for   180f, 197
– CNT-initiated ATRP of (meth)acrylates   183ff
– CNT-initiated ATRP of acrylamides   195f
– CNT-initiated ATRP of styrenes   194ff
– controllability of CNT-initiated   197
– sequential   199
– vinyl monomers for   183
ATPR, see atom transfer radical polymerization
azide-alkyne click chemistry   35f, 203
azomethine ylide
– from α-amino acid and aldehyde   291f
– from aziridines   297

b

biocompatibility of functionalized CNTs   153
biosensor   45ff
biotin   3, 39
2,2′-bipyridyl (bpy)   180
boron nitride nanotube (BNNT)   207f

Surface Modification of Nanotube Fillers, First Edition. Edited by Vikas Mittal.
© 2011 Wiley-VCH Verlag GmbH & Co. KGaA. Published 2011 by Wiley-VCH Verlag GmbH & Co. KGaA.

c

cadmium selenide nanoparticle
– mercaptoacetic acid coating of 114f
cadmium telluride nanoparticle 44
carbon nanotube (CNT) 25
– amidation of 143
– amine functionalization of 138ff
– as charged polymer-functionalized nanosubstrate 189ff
– bearing substituted pyrrolidine rings
– bioapplications of 45ff
– buckling of 96f
– defect site functionalization of 140f
– dispersants for 68
– donor acceptor assemblies based on 43f
– filling with biomolecules 127
– fluorination of 148
– free-radical-functionalized 106f
– functionalization strategies for 136ff
– hyperbranched polymer-grafted 204
– interaction with aromatic amino acids 104f
– interaction with dispersive conjugated polymers 68
– interaction with pyrene-containing block copolymer 52ff, 57
– interaction with pyrene-containing random copolymer 57
– LbL modification of 28ff
– linear block-copolymer-grafted 199f
– linear homopolymer-grafted 183
– mechanism of CNT-block copolymer interaction 72ff
– modification by π-π stacking 52ff
– modification of the electronic band gap 148f
– nylon-functionalized 163
– oxidized 31
– phophonic acid-modified 153f
– pH-responsive polymer-functionalized 188, 267
– polycation-grafted 43, 189
– polyelectrolyte-grafted 188
– polymer-grafted 31, 179ff
– processability of 51, 67
– properties of 67
– QD-conjugated 114ff, 153
– tip functionalization of 140f
cationic ring opening polymerization
– active chain-end (ACE) mechanism 218ff, 221
– active monomer (AM) mechanism 216ff, 221
– for preparation of hyperbranched polyether-*grafted* CNTs 229f
– in the presence of chain transfer reagents in one pot 216f
– of 1,3-dioxepane/ethyleneglycol 218
– of aziridine for preparation of functional polyamine 224ff
– of glycidol
– propagation mechanisms 217
chain transfer reagent 216ff
characterization of amino-functionalized CNTs 142f
characterization of functionalized nanotubes 9, 13, 35f, 118
– degree of functionalization 147f
– IR spectra of functionalized CNTs 143ff
– of CNTs-g-AN 250f
– of mCNTs/epoxy nanocomposites 243ff
– of mCNTs/polyimide nanocomposite films 247
– of SWCNT-*f*-OH 293
– using AFM measurements 146
– via Raman spectrospcopy 145f
– via thermogravimetric analysis (TGA) 147
– via X-ray photoelectron spectroscopy (XPS) 146f
chemical vapor deposition 1, 7
chitosan 19
clickable polymer 35
CNT, *see* carbon nanotube
CNT-COOH 83ff
CNT-based core-shell nanostructure 185, 187
CNT-g-maleic anhydride (mCNT) 240
colloid stabilization principle 2
complex formation by π-π interactions 91
compression loading of CNTs 96
conductivity of functionalized nanotubes 33f, 246, 248
controlled radical polymerization (CPR) 179, 273, 275f
copper catalyst 180
copper(I)-catalyzed azide/alkyne click reaction 35, 267f
covalent functionalization of nanotubes 12ff, 67
– attachment of macromolecules 52ff
– "grafting from" approach, *see also* surface-initiated polymerization 53, 91, 106, 179
– "grafting to" approach 53, 57, 91, 106, 179

– layer-by-layer (LbL) deposition of
   polymeric thin films 25ff
– "random grafting" 179
– with biodegradable polymers 160ff, 192ff
– with magnetite nanoparticles 14, 119ff
– with PAA and PSS 30
covalent sidewall functionalization of CNTs
– anhydride functionalization 262
– attachment of amino acids 291f
– attachment of dendrimer molecules 296
– attachment of phenol groups 292f
– by diazonium salt decomposition 262
– coupling of CNTs-COOH with
   polyamines 261ff
– coupling of CNTs-COOH with polyols
   261ff
– coupling with flourescence marker 297
– direct surface functionalization 137
– electrochemical 137
– "gemini grafting" strategy 203
– multifunctionalization 269
– surface oxidation by HNO_3/H_2SO_4 261
– via [4+2] cycloaddition 175
– via 1,3-dipolar cycloaddition of
   azomethine ylide 290
– via fluorination 100, 137
– via surface-initiated ROP 160ff
– via the introduction of anions 102f
– via the introduction of free radicals 102
– with amide groups 100f
– with amine groups 100f, 136ff, 140
– with biodegradable polymers via ATRP
   192ff
– with carboxylic groups 100f
– with gold nanoparticles 123f, 299
– with nitro groups 101
– with poly(amidoamine) dendrimers 126f
– with polymers via RAFT 208
– with quantum dots 114ff
– with transition metal complexes 101f
cover effect 239
CPNT, see electrically conductive polymer
   nanotube
CPR, see controlled radical polymerizsation
cross-linked polymer bilayer 9
cytotoxicity of PEI-g-CNTs 233f

d
debundling 279, 304
deformation of nanotubes 95
– bending deformation of CNTs 98f
– deformation modes 96f
– tensile deformation of CNTs 96
– torsional deformation of CNTs 97

defunctionalization of CNTs 155
degrafting 59
dendron-nanotube complex 59
density functional theory (DFT) 92, 104
deposition of conductive LbL films 26f
deposition of enzymes 38f
deposition of hyperbranched polymers 31, 160ff
deaggregation of CNTs 68
dipolar cycloaddition 289ff
– dipolarophiles for 289
– formation of indolizine-group decorated
   SWCNTs 299f
– formation of pyrazoline-ring decorated
   CNTs 305f
– formation of pyrrolidine-ring decorated
   CNTs 290ff
– microwave-assisted 297f, 299
– of azomethine ylides to CNTs 291ff
– of cyclic nitrones to CNTs 302f
– of nitrile imines to CNTs 307
– of nitrile oxide to CNTs 300
– of pyridinium ylide to CNTs 298ff
dispersant 68
dispersion of nanotubes
– dispersion of CNTs 71ff, 91
– using conjugated block copolymer
   dispersants 68, 83ff
– using linear-dendritic polymer hybrids
   59
– using P3HT-*b*-PS 71ff
donor acceptor assembly 43
doping of nanotubes 105
double-hydrophylic block copolymer 60ff

e
electrical conductivity of CNTs 239, 246, 248f
electrically conductive polymer nanotube
   (CPNT) 51ff
electromagnetic interference (EMI)
   shielding 82
electron transfer from enzymes to CNTs
   44
electronic band structure of CNTs 148f
electronic tunneling resistance of CNTs 34
endocytosis 139, 152
energy dispersive X-ray (EDX) analysis 72

f
ferritin 9, 39
ferrocene 291
field-flow fractionation device 5
finite size effect 51

fluorescein isothiocyanate (FITC)   234f, 296f
fluorescence quenching   75
free radical grafting   273f
fullerene   25, 106, 289
functionalization of CNTs through mechanical deformation   95ff

g

gas sensor   43, 151
– based on nitro-functionalized CNTs   101
gel permeation chromatography (GCP)   70
gold nanoparticle   9f, 113, 123ff, 299
– methyl mercaptoacetate modification of   125f
grafting ratio   276

h

helical wrapping   68, 75
high shear mixing   91, 263
high-temperature evaporation   1
HiPco SWNT   2f
honeycomb structure   81
Huisgen [2+3] click cycloaddition   267f
hybridization change   98, 100
hydrolyzed-poly(styrene-*alt*-maleic anhydride) (*h*-PSMA)   9
hyperbranched polymer   160ff, 204f, 224f
– degree of branching (DB) of   230f

i

immobilization of nanoparticles on nanotube surfaces   13
immobilization of porphyrins   44
immobilization of pyrene derivatives   52
indium trioxide nanotube
– as gas sensor   43
– synthesis of   41f
inorganic nanoparticle for stabilization of dispersed nanotubes   3
intercavity filling   136
ionic self-assembled multilayer (ISAM)   25f
irreversible adsorption on nanotubes   11
ISAM, *see* ionic self-assembled multilayer
isokinetic thermostatting
– Berendsen thermostat scheme   93f
– Evans-Hoover scheme   93f
– Gaussian feedback scheme   93
– using a hybrid quantum mechanics/molecular mechanics approach   95

l

laser ablation   1, 25
layer-by-layer (LbL) modification of CNTs   28ff, 190
– deposition of biological molecules   38f
– for template development   41
– on vertical aligned (VA) MWNTs   37f
– using click chemistry (LbL-CC)   35ff
LbL modification, *see* layer-by-layer modification of CNTs
LbL-CC process, *see* layer-by-layer modification of CNTs
local-density approximation (LDA)   92

m

macroradical   279
magnetic nanohybrid material   47
magnetite nanoparticle (MNP)   13f, 119f
– aminosilane coating of   121
– polyamidoamine dendrimer-modified   127
mAN, *see* aminated acrylicnitrile
mCNT, *see* CNT-*g*-maleic anhydride
mCNTs-*g*-acrylicnitrile   242
mechanical exfoliation   67
membrane   154
metallic nanotube   149, 299
metal-oxide nanotube   41
metananotube   136
methotrexate (MTX)   296
microwave-generated N_2/NH_3 plasma   139
MNP, *see* magnetite nanoparticle
MNP/CNT nanohybrid   119ff
modeling of nanotube functionalization
– using *ab-initio* quantum mechanics   92
– using molecular dynamics   92f
moisture/vapor barrier material for electronic devices   82
molecular dynamics simulation   92, 95, 139
multi-walled carbon nanotube (MWCNT)
– binary functionalized Janus-MWCNT   168
– dendrimer-modified   126ff
– dispersibility of surface-modified   54
– fluorescent   163
– formation of Au-MWNT conjugates   125f
– functionalization with chitosan   19
– functionalization with PAA-PSS   29
– functionalization with polyetherimide   18
– functionalization with pyrene-functionalized block copolymers   55ff
– glycopolymer-functionalized   193
– growth of polyfluorene chains on   297f
– hyperbranched polymer-grafted   205f
– inclusion complex bound on   165f
– magnetic   45
– methylimidazolium-tethered   168

- P3HT-*b*-PAA dispersed   83ff
- P3HT-*b*-PPEGA dispersed   85f
- P3HT-*b*-PS dispersed   71ff
- perfluorinated   295
- poly(epoxychlorpropane)-grafted   168
- polyfluorene-functionalized   298f
- structure and morphology of polymerfunctionalized   186ff, 201f
- through multisite ionic interactions with block copolymer   63
- vertical aligned (VA)   37
- water-soluble PEG-*g*-MWCNT   32

MWCNT, *see* multi-walled carbon nanotube
MWCNT-COOH   31, 63, 83
MWCNT-*g*-(PCL)   164
MWCNT-*g*-(PEI)   226f, 232f
MWCNT-*g*-(PEO-*b*-PCL)   169f
MWCNT-*g*-PDEAEMA   188
MWCNT-*g*-PLLA   173f
MWCNT-*g*-PMMA via *in situ* ATRP   185f
MWCNT-*g*-PNIPAAm   195ff, 198
MWCNT-*g*-PS   194f
MWCNT-NH$_2$ 115
MWCNT-PAA   31
MWCNT-PSS   31

n

nanoparticle (NP)   113
- loading by polyelectrolyte-grafted CNTs   189

nano-sized filler effect   51
nanotube encapsulation   12
nanotube nanocomposite, synthesis of   13
nanotube surface
- binding of proteins on   2f
- coated with luminescent polymers   12
- coated with watersoluble polyvinylalcohol   12
- DNA immobilization on   20
- hydrophobic interactions with surfactants   3
- norbornene polymerization on   7, 53

nanotube-bound anion   103
nanotube-bound radical   102
nitroxide-mediated radical polymerization (NMRP)   179, 207
NMRP, *see* nitroxide-mediated radical polymerization
noncovalent functionalization of nanotubes
- by multiple ionic interactions with double-hydrophilic block copolymers   60ff
- by π-π stacking with block polymers   52ff

noncovalent modification of CNTs   103ff, 142

noncovalently functionalized nanotube   142
- swelling behaviour of   8
- synthesis of   2ff

nonhelical adsorption   68
NP, *see* nanoparticle
NPs-conjugated CNTs composite   113f

o

organometallic-mediated radical polymerizaiton (OMRP)   279

p

P3HT, *see* poly(3-hexylthiophene) block copolymer
PAA, *see* polyacrylic acid
PAH, *see* poly(allylaminehydrochloride)
PAN, *see* polyacrylnitril
PANI nanotube   60ff
- doping of   62
- stabilization of aqueous dispersions of   60ff

PANI, *see* polyaniline
PCL, *see* poly(ε-caprolactone)
PDDA, *see* poly(diallyldimethylammonium chloride)
PDEAEMA, *see* poly(2-(N,N-diethylaminoethyl) methacrylate
PDMAEMA, *see* poly(2-(N,N-dimethylaminoethyl) methacrylate
PEG, *see* poly(ethylene glycol)
PEI, *see* poly(ethyleneimine)
N,N,N,N,N-pentamethyldiethylenetriamine (PMDETA)   35, 180
percolation threshold   78f
physical adsorption, *see* noncovalent functionalization of nanotubes
π-π interaction
- spectroscopic analysis of   72ff

π-π stacking   2
plasma treatment of CNTs   240
- electrode reactor for   240
- morphology of plasma-modified mCNTs   243ff
- properties of CNTs-*g*-mAN/epoxy nanocomposites   253f
- properties of mCNTs nanocomposites   244, 248

PLLA, *see* poly(L-lactide)
PMDETA, *see* N,N,N,N,N-pentamethyldiethylenetriamine
PMMA, *see* poly(methyl methacrylate)
PNIPAAm, *see* poly(*N*-isopropylacrylamide)

polarized infrared internal reflectance spectroscopy 9
poly(allylaminehydrochloride) PAH 32f
poly(AzEMA), see poly(2-azidoethyl methacrylate)
poly(2-azidoethyl methacrylate) (poly(AzEMA) 35
poly(ε-caproloactone) (PCL) 160
poly(diallyldimethylammonium chloride) (PDDA) 29, 32, 45
poly(2-(N,N-diethylaminoethyl) methacrylate (PDEAEMA) 187
poly(2-(N,N-dimethylaminoethyl) methacrylate (PDMAEMA) 31, 45, 53
poly(p-dioxanone) (PPDX) 160
poly(ethylene glycol) (PEG) 3
poly(ethyleneimine) (PEI) 216
poly(3-hexylthiophene) (P3HT) block copolymer, synthesis of 69
poly(N-isopropylamide) (PNIPAAm) 195
poly(L-lactide) (PLLA) 160, 172
poly(L-lysine) 172
poly(methyl methacrylate) (PMMA) 53, 79
poly(PgMA), see poly(propargyl methacrylate)
poly(propargyl methacrylate) (poly(PgMA) 35
poly(vinyl pyrrolidone) (PVP) 4f, 39
poly(vinylimidazole) (PVI) 37
poly-[(vinylbenzyl)trimethylammonium] chloride (PVBTA^{n+}) 43f
polyacrylic acid (PAA) 9, 30, 266
polyacrylnitrile (PAN) 252
polyamidoamine (PAMAM) 127, 129, 262
polyaniline (PANI) 60
polyelectrolyte deposition 25ff
polyethylene 106
polymer adsorption 9
polymer doping 51, 62
polymer grafting 1, 105f
– anionic 279, 281
– azide 281
– click coupling of CNTs with block copolymer micelles 270
– click grafting 35f, 203
– combination of "grafting onto" and "grafting from" approaches 268
– combinations of surface-functionalized CNTs and polymers for 260
– free radical grafting 273
– "gemini grafting" 202f
– "grafting from" approach, see also surface-initiated polymerization 53, 91, 106, 179, 215, 258f

– "grafting to" approach 53, 57, 91, 106, 179, 215, 258f
– microwave-assisted solid-state grafting 264, 266
– of alkoxyamine end-capped (co)polymers onto CNTs 277f
– of chlorinated polymers onto CNTs 271
– of diblock copolymers on MWCNTs 169f
– of linear glycopolymers from MWCNTs 193
– of oligo-N-vinyl carbazol 109
– of poly(acryloyl chloride) onto CNTs 266
– of polyethylene 106ff
– of polypeptides on MWCNTs 171f
– of propylene oxide 108f
– on boron nitride nanotubes by ATRP 207
– on CNTs by ARTP 180, 279
– on CNTs by NMRP 179, 207
– on CNTs by plasma deposition 240ff
– on CNTs by RAFT 208
– on phenol-substituted CNTs 295
– one-step methods 272ff
– onto CNTs-alkyne 267, 270
– onto CNTs-anhydride 271
– onto CNTs-COCl 263f
– onto CNTs-COOH 262ff
– onto CNTs-NCO 271
– onto CNTs-NH2 266f
– onto CNTs-OH 264f
– onto CNTs-S-S-pyridine 271
– two-step "grafting onto" approach 259ff
polymer matrix for nanotubes 79
polymer micelle for nanotube encapsulation 11f
polymer wrapping method 4ff
polymer-wrapped nanotube 5f, 30, 296
polypropylene oxide 108
polystyrene sulfonate sodium salt (PSS) 29, 32
polystyrene-g-(glycidyl methacrylate-co-styrene) (PS-g-(GMA-co-St)) 7
PPDX, see poly(p-dioxanone)
protein binding on nanotube surfaces
– nonspecific 2
– specific 3
protein detection by functionalized CNTs 131, 151
PS-g-(GMA-co-St), see polystyrene-g-(glycidyl methacrylate-co-styrene)
h-PSMA, see hydrolyzed-poly(styrene-alt-maleic anhydride)

PSS, see polystyrene sulfonate sodium salt
purification of nanotubes 2
PVBTA^{n+}, see poly-[(vinylbenzyl) trimethylammonium] chloride
PVI, see poly(vinylimidazole)
PVP, see poly(vinylpyrrolidone)
pyrene-based block copolymer 52ff
1-pyrenebutanoic acid, irreversible adsorption of 11

q
QD, see quantum dot
quantum confinement effect 51
quantum dot (QD) 113f
– mercaptoacetic-acid-capped cadmium selenide 114f

r
radial breathing mode (RBM) 76, 145
RAFT polymerization, see reversible addition-fragmentation chain transfer polymerization
reversible addition-fragmentation chain transfer (RAFT) polymerization 35, 55, 179
ring opening polymerization (ROP) 160, 216ff
rod-coil copolymer 80
ROP, see ring-opening polymerization

s
SCVP, see self-condensing vinyl polymerization
SDS, see sodium dodecyl sulfate
self-assembly of block copolymers into micelles 268
self-assembly of P3HT-b-PS 80
self-assembly of polyelectrolyte nanoshells 28f
self-condensing vinyl polymerization (SCVP) 205
semiconducting nanotube 149, 299
separation of semiconducting and metallic CNTs 138f, 154
sidewall functionalization, see also covalent sidewall functionalization of CNTs and surface functionalization 52ff
silica coating of nanotubes 7
silver nanoparticle 190
single-walled carbon nanotube (SWCNT) 4
– adsorption of chloroform on 103f
– adsorption of metalloporphyrin complexes on 104
– air-stable amphoteric doping of 105
– armchair 93f, 105
– caprolactam-functionalized 164
– electrostatic LbL deposition of enzymes on 38f
– encapsulation of organic molecules by 105
– fluorinated 20
– formation of SWCNT-f-OH via cycloaddition of azomethine ylides 291f
– functionalization of 4f, 19, 57, 99ff, 160
– interaction with a linear-dendritic hybrid polymer 59
– P3HT-b-PS dispersed 71ff
– photoactive 306
– PVP-SWCNT system 5
– shortened sidewall-functionalized 57
– zigzag 93f, 102
sodium dodecyl sulfate (SDS) 4, 60ff
solubility of CPNTs 51, 58f
solubilization of CNTs 138, 149f, 252
stabilization media for nanotubes 2f
stabilization of CNT dispersions 53
starched nanotube 12
Stone-Wales defect 101
streptavidin 2f, 9, 39, 139
stress-strain relationship 93f
styrene-acrylonitrile copolymer (SAN) 262
superhydrophobic film 81f, 294
surface functionalization , see also covalent sidewall functionalization of CNTs 2
– by anionic polymerization 19, 279
– by controlled living polymerization 15
– by free radical melt grafting of polymers 7, 106
– by in situ polymerization 13, 19
– by long-chain polymers 12
– by noncovalent adsorption 99, 103ff, 142, 148, 189, 259, 279
– by oxidation 137
– by π-stacking 9
– by polymer multilayers 8
– covalent chemical modification of sidewalls 99ff
– covalently binding of amines 14
– epoxy functionalization 18
– immobilization of biomolecules 9, 13
– microwave-induced acid functionalization 20
– noncovalent, of CNTs 103ff
– noninvasive 2
– of acid-treated nanotubes 8
– of CNTs with oligonucleotide/antibody 131

- of graphite 10
- reactive groups on CNTs for polymer grafting 260
- using an inorganic silica layer 6
- using aromatic polyimides 12
- using inorganic nanoparticles 2f
- using polmer wrapping 4ff

surface modification, *see* surface functionalization
surface-force measurement 26
surface-initiated polymerization 53, 91, 106, 179
- anionic 174f, 279

surface-initiated ring-opening polymerization 160f
- anionic 163, 170
- cationic 167
- precursour initiator for 161

SWCNT, *see* single-walled carbon nanotube
SWCNT-CONH$_2$ 101
SWCNT-COOH 101
SWCNT-*g*-PS 195
SWCNT-NH$_2$ 101
SWCNT/PS nanocomposite thin film 78
SWNT field-effect transistor 105
composite material, nanotube-based 13, 78, 81f, 113f, 119ff, 129, 135, 154,191, 241ff, 253f, 294
synthesis
- covalently binding of CNTs with gold nanoparticles 123f, 299
- covalently binding of MWNTH-COOH with MNP-NH$_2$ 119ff
- covalently binding of MWNTH-NH$_2$ with CdSe-COOH 114ff, 117
- of 1D molecular nanocomposites 119ff
- of 3D molecular nanocomposites 121
- of biodegradable CNTs hybrids 165ff
- of CNT-based macroinitiators 181f
- of CNT-COOH 83
- of colloidal PANI nanoparticle 60
- of functional polyamine via cationic ROP 224ff
- of hyberbranched polyether-*grafted* CNTs 229f
- of Janus-type structures 169
- of mCNTs/epoxy nanocomposites 241
- of mCNTs/polyimide nanocomposites 241f
- of mCNTs-g-mAN/epoxy nanocomposites 242
- of metal-oxide nanotubes 41f
- of methylimidazolium-tethered MWCNTs 168
- of multifunctional CNTs 269
- of multihydroxl dendritic macromolecules on MWCNTs 121
- of multilayered protein/nanoparticle CNT conjugates 40
- of MWCNT-polymer/metal hybrid nanocomposites 191f
- of nanotubes 1, 25
- of noncovalent protein-nanotube conjugates 122
- of P3HT block copolymer 70
- of P3HT-*b*-poly(poly(ethylene glycol) methyl ether acrylate) (P3HT-*b*-PPEGA) 71
- of PEI-*g*-MWCNTs 226f
- of perfluorinated CNT 294
- of photoactive SWCNTs 306
- of P3HT-*b*-PAA 71
- of P3HT-*b*-PMMA 70f
- of P3HT-*b*-PS 70
- of poly(amidoamine) dendrimers on CNTs 126f
- of poly(epoxychloropropane)-grafted MWCNTs 168
- of PS-*b*-PAA-grafted MWCNTs 203
- of semiconducting SWCNTs 299

t

template, CNT-based 41f
TGA weight loss curve 200, 202
thermostat atom 93
thin film deposition 35
triazole linkage 35
triton-X surfactant 3

u

ultrasonication 4f, 11, 68, 91, 241, 279

v

van der Waals attraction 38, 67f, 91, 103, 136
vapor sensor 164f, 276
volatile organic compound (VOC) sensor 82

z

zirconium dioxide nanoparticle 2